The Invisible Crash

Also by James Dines

How the Average Investor Can Use Technical Analysis for Stock Profits

JAMES DINES

The Invisible Crash

What It Is,
Why It Happened,
How to Protect Yourself Against It

RANDOM HOUSE
New York

Library of Congress Cataloging in Publication Data
Dines, James.
The invisible crash; what it is, why it happened,
how to protect yourself against it.
1. Inflation (Finance)—United States.   367p.   c1975
2. Finance, Personal. 3. Gold. I. Title.
HG538.D54   332.4'1   75-875
ISBN 0-394-49881-X
Manufactured in the United States of America

98765432

First Edition

4/12/76 —$10.95

To You Uncomfortables Out There
Also Ahead of Your Time

Go placidly amid the noise & haste & remember what peace there may be in silence. As far as possible without surrender be on good terms with all persons. Speak your truth quietly & clearly, and listen to others, even the dull & ignorant; they too have their story. Avoid loud & agressive persons, they are vexations to the spirit. If you compare yourself with others, you may become vain & bitter; for always there will be greater & lesser persons than yourself. Enjoy your achievements as well as your plans. Keep interested in your own career, however humble; it is a real possession in the changing fortunes of time. Exercise caution in your business affairs, for the world is full of trickery. But let this not blind you to what virtue there is; many persons strive for high ideals, and everywhere life is full of heroism. Be yourself. Especially, do not feign affection. Neither be cynical about love; for in the face of all aridity & disenchantment it is perennial as the grass. Take kindly the counsel of the years, gracefully surrendering the things of youth. Nurture strength of spirit to shield you in sudden misfortune. But do not distress yourself with imaginings. Many fears are born of fatigue & loneliness. Beyond a wholesome discipline, be gentle with yourself. You are a child of the universe, no less than the trees & the stars; you have a right to be here. And whether or not is clear to you, no doubt the universe is unfolding as it should. Therefore be at peace with God, whatever you conceive Him to be, and whatever your labors & aspirations, in the noisy confusion of life keep peace with your soul. With all its sham, drudgery & broken dreams, it is still a beautiful world. Be careful. Strive to be happy.

Found in Old Saint Paul's Church, Baltimore, dated 1692.

Contents

Preface

I should have written this book much sooner. The idea for it first came to me in 1970 when I appeared on *The Joe Franklin Show* with Harry Browne, author of *You Can Profit from a Monetary Crisis* and *How You Can Profit from the Coming Devaluation*. In my weekly stock market publication, *The Dines Letter*, I have been writing and warning about a coming gold crisis since 1961. You will see my current predictions in the "Odyssey" chapter, precisely where I was right and wrong, and what I now see ahead. This book will also tell you what you can do to protect yourself during the coming crises.

My life was the preparation for this unusual book. I have told it like it is, and it will stand or fall on its own merits. Above all, my aim was to present a simple book for the voting layman, even at the risk of oversimplification. This book is written especially for the generation who forgot gold, but also for those who, for one reason or another, denied its value and in today's world can no longer do so. It is written for those who feel uncomfortable with textbook math or economics.

I have given up in exasperation trying to awaken Wall Street or the Washington Economic Establishment. That is why I want to by-pass them and the other currency manipulators and concentrate instead on the next generation. It is important to study the current period because history could soon repeat itself, and the old mistakes will be there to make once again. Hopefully, this book will help the world by exposing the problems. If enough of the right people grasp its thrust and channel their energies properly, the twenty-first century will indeed be a better time for those who will live in it. If not, there will be yet another boom, followed by a market crash and depression.

This book is about how an obscure young security analyst blew the whistle on the U.S. Treasury itself, whose "gold bubble" I predict will turn out to be the scandal of the century. No official pronouncement swayed me from insisting that the emperor wore no clothes. The suppression of gold is a particularly sad and negative story. My persistence and determination, when I knew I was right, buoyed me through difficult times as I warned my readers of what was happening. This book is a lesson in how courage of conviction can enable one to withstand even one's government. Whenever I doubted, I remembered the function of gold. Governments cannot create gold at will in order to spend more money than they actually possess. This restraint throws a permanent body block at their efforts to inflate and buy votes. Gold is the dagger aimed right at the heart of inflation. So note well which politicians denigrate gold.

America will awake one anguished morning to find itself close to bankruptcy. It will have happened slowly; yet its end will be swift. An "invisible crash" will have arrived on stealthy feet. Those who did not heed the signs of the crash's advent will discover that the function of a depression is to return property to its rightful owners. Few are describing what is now happening as a "crash." Not only will it be described as one, but I predict it will be referred to as "the Crash of '69." Not only was this crash invisible, but the function of gold was as well. You will see that the dangers of inflation, among other things, began to register in the American consciousness far too late—in the summer of 1974.

The real culprits are the currency manipulators and the Washington Economic Establishment, with their stubborn refusal to accept the business cycle. The business cycle can be conquered, but the price for that is your freedom. Is it worth it? Attempts to conquer the business cycle succeed only in bringing too much government, and thereby a loss of individual freedom.

Clearly, something is wrong with the United States. Double-digit inflation became the norm in 1974. No democracy has ever survived an inflation of more than 20% for very long. There is a malaise in the land; a crisis of confidence in political integrity—a suspicion of business, a new militancy of labor, and attacks on our most cherished institutions such as religion, the military, and marriage. All-time high interest rates in 1974 strongly suggested that the U.S. economy was recklessly out of control. In what kind of country do utilities ask customers *not* to use their service? You get what you invite.

You will read about how the Italian government nearly went bankrupt. It was a scene from Ayn Rand's *Atlas Shrugged*, in which grocery prices shocked housewives, higher interest rates devastated Wall Street, and inflation capriciously redistributed income, overturned social priorities and misallocated resources. The socialist system we now have is at a dead end. Governments have simply run out of tricks. Presidents Nixon and Ford called for Americans to save more, leaving unsaid "so that you can be embezzled by inflation more easily." Average people have simply run out of money to save. They are being ruined by inflation progressing faster than their incomes can rise. Political parties have begun to fall all over the world.

> *If five years ago an ordinary informed American citizen had been told that a serious international monetary problem was in the making, he would not have had the slightest idea what was being talked about. Indeed, five years ago, the United States government, on the rare occasions that the subject was mentioned, indignantly denied that there was anything wrong. The idea of world monetary reform was taboo. Today monetary reform is one of the top items on the international agenda.*
>
> The New York Times
> *July 18, 1974*

In the "Odyssey" chapter, you will see how bitterly I fought my lonely fight against the initial ignorance of gold, and how in the mid-60s, a flicker of national awareness emerged. Yet, as late as 1965, no one did anything substantive. No President stepped forward and urged us to chuck national dreams of everlasting and total prosperity, or asked us to live within our means. No President, no leader, demanded that other nations stop begging from us; instead, they encouraged it in order to enforce our occasionally unsound foreign policies with threats that we would disinherit them like recalcitrant heirs. Nineteen seventy-four saw massive economic conferences to decide what the causes of inflation were. No conclusions were reached. I have clearly outlined what inflation is, how it began, and what its long-term solutions are, nailing the hides of many politicians to the barn door with it. I have blown out of the water the idea that government deficit spending creates full employment, prosperity, and rapid growth for all, and the idea (for that is all it is) that the less we work and the faster we obsolesce material possessions, the more we all prosper. Few protested when the richest city on earth, New York City, visibly began to go downhill. What protest did you make when government, whose primary and original function was to preserve the safety of its citizens, began to give you everything but that safety?

The depression I foresee has not yet even arrived, yet I know that many will be so scarred by it they will have great emotional difficulty following me back into the stock market when this is all over. I offer the "Odyssey" chapter as an inspirational calmer of fears, and if it turns out that I am only with you in spirit, then another must use this book to pick up the baton.

I hope that after reading this book, you experience the same kind of righteous indignation I have felt for over a decade; then the time taken from my life to write it will have been more than worth the enjoyment I could have had otherwise.

I would like to thank my loyal staff and subscribers who supported me over the years while I was thinking out these weighty problems and translating the solutions into simple language.

I also wish to thank the many "goldbugs," only a few of whom are mentioned in this book, for being there, and stimulating my thinking. I am especially appreciative of Ira Cobleigh, an author of many books on gold, for the final reading of this book and his helpful comments. My appreciation also to Jacques Rueff, adviser to Charles de Gaulle, for his personal encouragement to me over the years.

The Invisible Crash

Introduction

The following charts of the NYSE Composite Index and the Composite Value Line Index tell very different stories. The NYSE Composite Index is heavily weighted in favor of blue-chips, and is quite similar to the Dow-Jones Industrial Average. Since 1968, the NYSE Composite Index has dropped 46% from a high of 61 to 33. However, the Value Line Index is an unweighted Average, which means it shows a much broader view of what stocks have really done, as opposed to what the blue-chips have done. Since late 1968, the Value Line Composite Index of 1,538 stocks has dropped 74% from 188 to as low as 48.

Since most investors watch either the DJI or the NYSE Composite Index, they probably thought that their catastrophic losses were peculiar to themselves and that most other investors were doing better. The sad truth is that because of the lulling effect of these Averages, investors did not realize they were in a crash. Partially for this reason my book is entitled "The Invisible Crash."

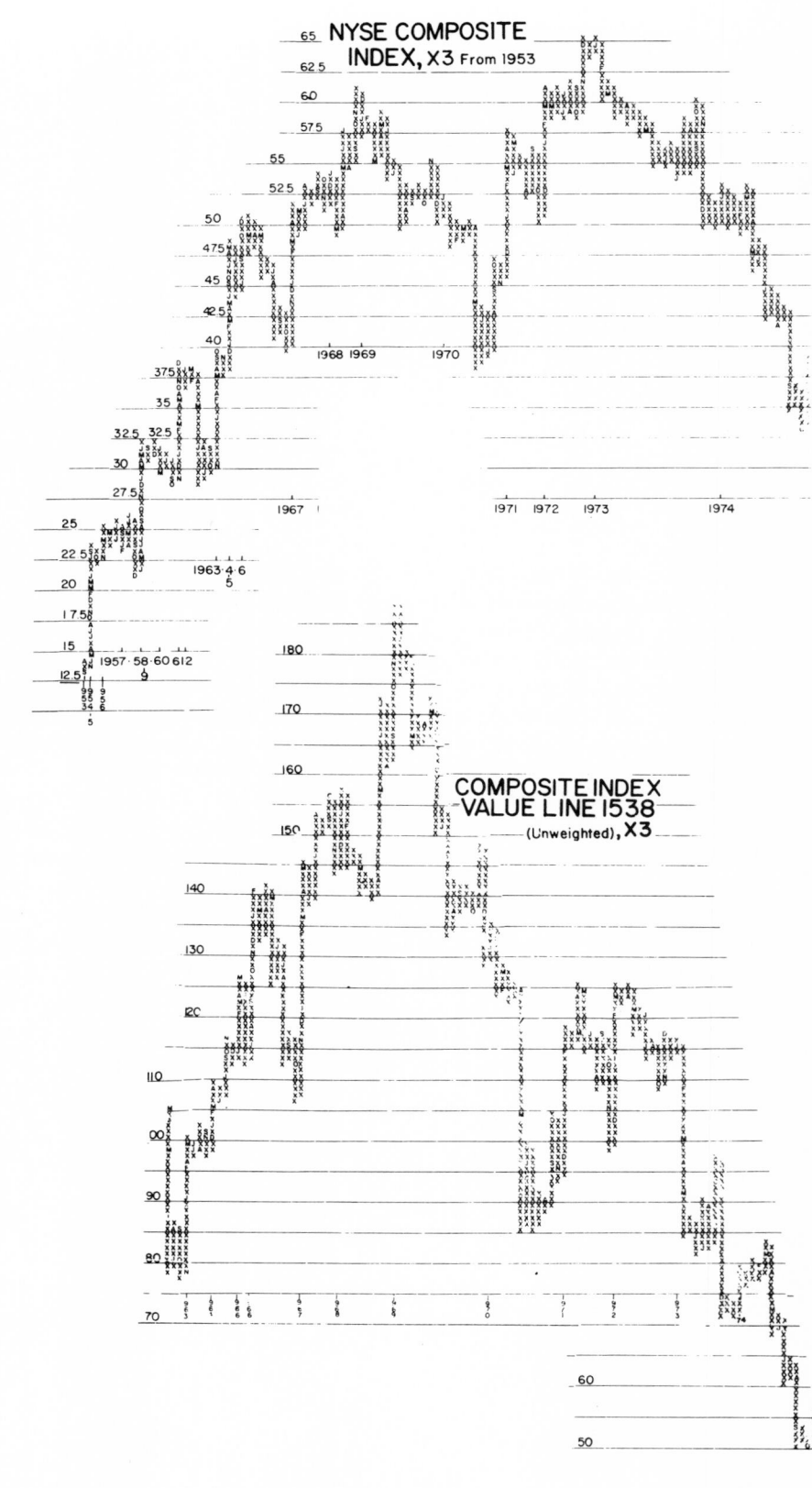

1

The Present Malaise

There is obviously something very wrong with the world today. The cost of living has moved irregularly higher for many years.

The "almighty" dollar already has been devalued twice this decade, and foreigners recently committed the unheard-of act of refusing to accept more paper dollars, except at steep discounts.

Personal, corporate, and federal debts are at unbelievably high levels, and there is no way of ever paying them off, other than through bankruptcy or runaway inflation.

The world's banking system is so overextended that it is at its most vulnerable point since the 1929 Crash. Who, as recently as 1973, would have believed large banks would fail all around the world? The coming banking collapse has probably already begun.

Interest rates fluctuate near their all-time highs as lenders weary of being gutted by inflation. A 1974 prime interest rate as high as 12% suggested an economic crisis was under way and that its tentacles reached far and wide.

Wage and price controls and material shortages and dislocations, both unheard-of in peacetime America, began to appear throughout the economy in 1973.

There has been an invisible stock market crash. From 1968 to 1974, the Value Line Average, far more representative of the majority of stocks than the Dow-Jones Industrial Average, dropped over 70%. Add the effects of inflation to this already staggering loss, and there can be no denying that nearly all investors of recent decades have been wiped out in terms of purchasing power.

The energy crisis is threatening the actual form of our economic and, therefore, political system as we know it. It did not cause the current malaise; ignorance of gold did. By sucking dollars from all over the

Western world to the previously destitute Arabs, the energy crisis has hastened the collapse I have so long envisioned. And I stress the word hastened. The energy crisis was an aggravant, a precipitant, a catalyst, but not the cause, despite what will be said in the future.

Americans must understand that it is not only the quantity of dollars they get in their pay checks that counts, but the quality of those dollars. The threat of a world-wide crash has instilled a doomsday attitude into many otherwise optimistic people, and even small investors are becoming gold speculators. Ordinary people watch in dismay as the buying power of their bank accounts, retirement funds, mutual funds and insurance policies melts away like ice in the Sahara. Many will say that the Federal Reserve System must come to the rescue, yet I take the daring position that the Fed was one of the prime causes of this desperate situation. There are some who want the Federal Reserve System expanded to an international basis, thus changing the disaster from an American to an international phenomenon. I hope my work will be of some help in blocking this.

The roots of all these problems come from one source and one source only: the world has too much government. Basically, our economic system as it now stands is rotten. That is why our leaders appear to be milling about in confusion. The system is at a dead end. Our system is certainly not capitalism, and certainly not free enterprise, although the façade contains vestiges of both. When this system collapses, some will say it was a failure of capitalism, and those who believe that will be misled once again. Our society is so corrupt and immoral that when an honest man actually stands up and fights corruption, they make a movie about it like *Serpico* and *Walking Tall*, that the crowds may gape at him like some freak, or weird creature in a zoo. Imagine, an honest man! Who will turn the moneychangers out of the temple?

Long before 1974's crashing stock markets I envisioned the world living under the growing shadow of an onrushing economic cataclysm. The crisis has been building since 1933, and there really has been no precedent for such monetary instability in the financial history of the world. The man in the street has no concept of the financial panic and the liquidity crisis which could lie just before us. The 1929 Crash was blamed on vague notions of "beggar-thy-neighbor" nationalism, floating exchange rates, gold, or a myriad of other excuses and, therefore, our leaders have been unable to avoid another such catastrophe. Presenting this review of history is my contribution to forestalling another depression around the end of this century.

Some think "a little inflation," usually admitted with a small wink, is "good," meaning they think they are benefiting from it somehow. But it must be at someone else's expense, and therefore it is inherently immoral. To accept it at all makes such people perhaps as guilty as those in Nazi Germany who knew of the atrocities but said nothing. All of us are drenched in the inherent immorality of inflation.

Inflation actually represents a federal embezzlement of bank savings, life insurance, and pensions at the expense of the poor, the aged, the little, and the weak. It is the most insidious and hideous form of economic imperialism. Even now, no government leader dares to stop it; in their speeches they merely want to slow inflation down. You cannot slow pregnancy down, either. Is capital so cheap and greed so abundant that the little man will be sacrificed on a cross of socialism?

Inflation has led to the spectacle of those who are on relief conducting mass protests for higher benefits. There are no jobs for these people because the wages they seek are too high and inflexible, and rising wages over the years have led to built-in inflation and non-productivity which must lead to a crash. Every effort to avoid another 1929 has inexorably led right back to it.

The long bull market from 1949 to 1968 created its own excesses, which needed to be corrected, but the stock market decline since then has gone beyond a mere correction. I firmly believe this is due to the Keynesian* practice of printing unlimited paper money and encouraging domestic and international deficits. People are overextended in debt hoping inflation will bail them out at someone else's expense. As a result, exorbitant interest rates are necessary to compensate lenders against inflation. These rates are ruinous for business. Consequently, the world will suffer a terrible recession, replete with radical political shifts and malinvestments in world trade. Add to this the oil crisis, and you can see why I fear social and political changes ahead that could alter the history of the world.

The high volatility of international money flows, unbelievably low Price/Earnings Ratios on stocks in 1974, unheard-of levels of currency speculation, skyrocketing gold prices, people actually pulling money out of banks to buy gold coins, big changes in the parities of currencies which are suddenly floating, the lack of convertibility of the U.S. dollar into gold, the imposition of exchange controls, Mafia-high international interest rates, the smashing of the Nairobi and Smithsonian international monetary agreements, massive deterioration in many countries' trade balances, food and material shortages, unrestrained government spending, swelling of domestic and international relief rolls and terrifyingly low bank liquidity are all part of an ominous pattern signaling the disintegration of the international monetary system as we know it, and the

---

* John Maynard Keynes was an economist who greatly influenced all modern economics, and he will be discussed later on.

end of an era. I predict the next few years will be the watershed of a "new era," born in grief and perhaps blood. It will be difficult to avoid despairing for capitalism and freedom; but we simply cannot give up without trying.

The present malaise, including environmental pollution, has its roots in a gold crisis of which the public is generally unaware. There are two related problems: the relationship of all paper currencies to each other, and the relationship of all paper currencies to gold. The first problem has already been solved by floating exchange rates, but the second crisis is yet ahead—and it is the crusher.

However, it is not too late. The spirit of 1776 could be revived in **1976.**

2

The History of Gold

How Did We Get Here: A Devastating Historical Exposé. The real economic history which inexorably led us to the Crash of 1929, and which could lead us to much worse ahead: Not to be found in economics textbooks.

> *There is nothing inherently wrong in effect with fiat money, provided we get perfect authority and God-like intelligence for kings.*
>
> *Aristotle*

There is no way to know where we are headed without carefully delineating our trajectory. That is why I expect you to read this chapter carefully. Some have a natural aversion to history, but it is precisely that ignorance of history that has led this nation to repeat its errors.

What Is Gold?

Gold is a very special and completely unique metal. This yellow element has spurred emotions, ignited passions, and spilled blood like nothing else in the history of the world, with the possible exceptions of women and land.

For more than six millennia, gold has been mankind's number-one portable asset. Aside from silver, it is the only metal universally accepted as money, or backing for paper currencies. Gold has a reputation which has proved to be the best hedge against financial busts, currency debasements, war and mass manias. In fact, all of the great monies of the world were convertible into gold—the daric, dinar, bezant, ducat, florin, sovereign, French Napoleon, and the U. S. dollar (which was convertible into gold from 1900 to 1933). Caesar's gold coins still have value, though his written promises are worthless. Yet all the gold mined in the last 6,000 years comes to only 80,000 metric tons. (One metric ton equals 2,204.6 pounds.) Purified, this gold would fill 30 large railroad boxcars, or a cube with 53-foot sides, or a 2½-foot thick slab the size of a football field. The largest single national gold hoard belongs to the United States, but it has dwindled to 8,584 tons from 21,530 tons in 1950. There are around 12,600 metric tons in a New York vault of the Federal Reserve Bank. The U. S. Treasury values its share at approximately $11 billion (at $42.22 an ounce) although the free market price is much higher.

Alloying with silver makes gold whitish; with copper, reddish; with cadmium, greenish; and with iron, bluish. Jewelers rate it by degree of purity. Refined to 99.5%, it is called 24-karat gold. Goldsmiths add a drop of nitric acid and what is not gold bubbles away.

Jewelry manufactured in the United States is usually 58.33% gold, or 14-karat. In most of Europe it is 18-karat, or 75% gold; in Cairo, 21-karat or 87.5%, and in India, 91.66% or 22-karat, which is almost too soft to wear daily as jewelry. The value of gold in U.S.-made jewelry is only about 20%, so if you spent $100 for a piece of machine-made jewelry in New York City, it would contain around $20 worth of gold, on the average.

Gold has long been appreciated for its aesthetic beauty and recognized for its economic power when handled by people who knew its value. Yet only about 2.4 billion ounces of gold have been mined since 1500 A.D. (more than half of that since 1900). The amount of gold mined in the 6,000 years since Egypt was the leading gold producer in 4000 B.C. is estimated at around 3 billion ounces.

The world's remaining unmined gold reserves are estimated at less than 1 billion ounces, 80% of which exist in South Africa. Nearly all of it will be mined by the end of this century. Russia's reserves are estimated at between 200 to 300 million ounces. Gold is valuable precisely because its availability is so strictly limited.

The Greek Disease and Frankenstein's Birth

When venereal disease initially appeared in England, it was promptly labeled "The French Disease." Of course, the French, in turn, called it "The English Disease." The ancient Greeks are apparently responsible for the inflation disease, and in my opinion, it is the Frankenstein they created that destroyed the Greek Empire.

The argument over the merits of paper and gold money goes back many centuries. Over 2,000 years ago, Aristotle favored metallic money, while Plato advocated fiat, or paper money. So this is an old and recurring problem, much like the age-old struggle to determine whether the individual is more important than the state or the state more important than the individual. Over the centuries, man refined the concept of money through painful trial and error. Those in power realized the importance of money and attempted to seize control of it as a source of power. The question leaps to mind whether governments should be trusted to create money or whether this should be the responsibility of private enterprise.

The ancient city-states created the first commercial middle class, and their voice in government was democracy. By 560 B.C., the Greeks were using a gold coin called the dinar at a time when Greece was the paragon of the world. They elected leaders who could properly be described as the first politicians (another invention for which I might not forgive the ancient Greeks). These politicians discovered that the way to ensure their reelection was to give the people more in services than was demanded from them by way of taxes. This should sound familiar.

Unfortunately, since politicians could not create gold, they soon resorted to adding some copper to the gold dinar, and eventually eliminated gold entirely, thereby inventing inflation, which I call "The Greek Disease." Charles I. Black, author of a gem of a book vital to every complete financial library, entitled *Inflation: The Tragic Economic and Investment Consequences,* defines inflation as "the excessive creation of money or credit in relation to goods and services available." In 400 B.C., Greece was struck by an inflation-induced depression from which it never recovered, because it had lost its commercial and military advantages. Frankenstein, created by politicians, had claimed its first victim.

Ancient Rome prospered when it had a fixed relationship to gold and silver coins in its currency. Roman currency degenerated when Nero demanded that taxes be paid in gold and silver, spent the money and issued base metal currency back to the people, just as Hitler did in the 1930s, and the United States in the 1960s. While inflation ruined the Roman middle class, the lower classes became totally dependent on the government and demanded circuses to occupy their time—reminiscent of the welfare states nearly two millennia later.

In stark contrast, the Eastern Roman Empire founded by Constantine lasted 1,000 years beyond the fall of Rome, probably because the bezant maintained the same value for over 800 years without inflation. Destruction of this currency in 1282 A.D. coincided with the Eastern Roman Empire's downfall.

With the fall of Rome came the Dark Ages, and for nearly eight centuries barter dominated the feudal scene. By the twelfth century, the increasing development of cities created the need for money and a system beyond barter. With the creation of the florin in the mid-thirteenth century, the Italians flourished once again. By the fifteenth century, the nation-state was viable because commerce could be carried on. This led to the emergence of uniform taxes, laws, weights and measures, military forces, and transportation facilities. By 1600, modern banking practices operated throughout the Mediterranean. As Black writes, "This unity of a strong self-sufficient economic state as a natural partner of a political state is known as mercantilism." Mercantilism stressed silver and gold as indispensable to a nation's wealth, and the subsequent expansion of commerce led to the colonization of the New World, Asia, and Africa. Mercantilism furnished the drive behind the need for gold which drew early explorers to the four corners of the earth.

By the seventeenth century, precious metals had become dominant, culture flourished, and man's standard of living rose toward modernity.

Gold was the only medium of exchange, but people did not want to carry all that heavy metal around with them, so they left it with goldsmiths. In return they received gold receipts. People learned to sign these receipts over to others, so this paper became the forerunner of checks and paper money. Anyone could take a gold receipt and receive gold for it on demand. However, goldsmiths became aware that people left their gold undisturbed for long periods of time, so they lent out 10% more than the gold they held, thus beginning the "fractional reserve system," a source of mischief for centuries to come. For the first time there was more paper in existence than gold backing it. Part of the enormous profits goldsmiths made on lending this money out was paid to these depositors in the form of interest. This was the forerunner of the modern bank. Modern banks are middlemen between borrowers and lenders, but their forerunners were goldsmiths.

Unfortunately, this system was vulnerable to panics and runs. If all the depositors were to demand their gold at once, the last 10% could not possibly withdraw their gold. To avoid runs, banks developed reserves for one another, but maintained no protection against a wave of bank failures. Furthermore, over the years goldsmiths increased their lending to 20%, and then 30%, and more.

In 1718 John Law, a Scottish financier and speculator, practically ruined France by his over-issuance of paper currency. The story of "The Greek Disease" is well-documented in Charles MacKay's great classic, *Extraordinary Popular Delusions and the Madness of Crowds* (another essential book for your financial library). The paper money Law created was first backed by confiscated church lands, later by French land in Louisiana, and finally by hot air from politicians' lungs. People felt they could trust this land-backed paper, but when the King needed more money, he had Law print more. Louis XV thus impoverished the French middle class, which eventually ended his family's reign during the subsequent Revolution. This mishandling of gold was washed away by blood.

Having learned nothing and forgotten nothing, the newly organized National Assembly issued paper money called *assignats* in 1790. Andrew Dickson White's great book *Fiat Money Inflation in France* describes why this paper system supposedly would work despite the fact that John Law's had failed: "Paper money under despotism is dangerous, it favors corruption; but in a nation constitutionally governed, which itself takes care in the issuance of its notes, that danger no longer exists."

When this new paper money was first issued, trade and investment credit increased, and a boom ensued which government economists declared would never end—it was like the United States in the 1920s and 1960s. Unfortunately, as the people began to doubt the *assignat* currency, they began to get rid of it by buying things they did not really want or need, but which they thought would retain value. Wage and price controls were established, while rationing and the black market proliferated. Cheaper foreign products proved to be unbeatable competition, and commerce began to slow down. Loans were not made because moneylenders could not be recompensed properly for inflation, and speculation became more attractive than production. A period of corruption and tax cheating followed, and by 1796 prices of nearly everything in France were a thousand times higher than in 1789. Amidst this terrible misery, unemployment, moribund financial institutions, and a failing government, a man on a white horse named Napoleon appeared and re-enthroned gold.*

What Is the Gold Standard?

> *Of all the contrivances for cheating the laboring classes of mankind, none has been more effective than that which deludes them with paper money.*
> Daniel Webster

Let me digress for a moment to describe the gold standard.

Over the centuries, countries were said to be on the gold standard when their currencies were exchangeable for gold at a fixed price. Thus, when a country ran a balance-of-payments deficit, the difference had to be paid in gold (unless loans or credits were arranged). The amount of

---

* There are many interesting parallels between these two French inflations and the United States in the '20s and '60s. In each case, inflation involved a man on a white horse: Napoleon and FDR. Other similarities suggest that all inflations are alike and involve the failure to grasp some basic economic truth. This book suggests it is an understanding of the function of gold.

domestic currency in circulation was linked directly and precisely to the amount of gold a particular country's central banks owned. When a country had a payments deficit, gold flowed out in settlement, and the subsequent reduction in the domestic money supply caused a contraction in internal demand for goods, which in turn caused prices of goods to fall. Imports declined as economic distress mounted, and exports rose as goods became cheaper. In those days, wages were flexible and fluctuated depending on the economic situation.

This sounds complex, but all the gold standard did was to make nations give to the world as much as they took, using gold as an international common denominator of all currencies.

A return to some form of gold standard is the only way the world is going to develop a sound currency again, especially after two colossal disasters in a row (the 1930s and the 1970s). Yet the true-blue gold standard has been so bitterly criticized by modern economists that I doubt it will return in its old form. The most serious criticism of the gold standard is that it led to booms and slumps. Unfortunately, the alternatives are much worse. It is better to have mild booms and slumps than speculative orgies and devastating depressions.

The Nineteenth Century—A Brief Moment of Bliss

The nineteenth century saw an amazingly sound and self-regulating system based on gold, in which currencies were redeemable in gold at fixed exchange rates. A balance-of-payments deficit was settled quickly in gold, and since all paper was immediately convertible into gold, no government could get away with unlimited creation of new money.

The nineteenth century was a golden era, as all major industrial nations supported the gold standard, giving the world a stability it had never known before—or since. From 1814 to 1914, Europe saw no depressions, and no prolonged wars. Business flourished and there was great scientific and cultural progress. The word inflation almost ceased to exist. The United States went on the gold standard in 1900. The automobile, airplane, electricity, and steamship were developed during the gold standard's reign.

England went on the gold standard in 1816. Favorable conditions such as wage and price flexibility, untaxed goods, public acceptance of unemployment (which weeded out the inefficient), and relatively high political stability (probably partially due to the gold standard), allowed the gold standard to work well. In addition, gold was discovered in the nineteenth century in California, Australia, and South Africa, and this, along with the invention of the cyanide process, made possible the use of lower-grade ores. This new gold led to an increase in the money supply—but based on the solid finding of a rare element, rather than on the limitless fantasy of political greed.

World War I Guaranteed the 1929 Crash

> *Government is the only agency which can take a useful commodity like paper, slap some ink on it and make it totally worthless.*
>
> *Ludwig von Mises*

The mistakes made around World War I haunt us half a century later, and are at the root of most of the evils in the world today, in my opinion. I do not expect many people to agree with me, so let us calmly look at the facts.

World War I was extremely expensive. Far more expensive than the governments involved could afford. Therefore, they abandoned the gold standard and resorted to inflation to pay for that war. Thus, Frankenstein was resurrected. England valiantly tried to return to the gold standard in the 1920s, and to the relative peace and progress of the nineteenth century. Unfortunately, England tried to do so with an overvalued exchange rate instead of hard work, and this was a major cause of the 1929 disaster.

Beyond the printing press mania used to finance World War I, an incredibly unreasonable "peace settlement" on Germany guaranteed another world war, and ruined Germany by rendering her incapable of paying her war debts in any other way than by the legendary inflation she suffered in the early 1920s. This was followed by the complete repudiation of the mark. Germany was forced to ignore gold to meet her impossible debts, and the results were inevitable. The peace and happiness of the nineteenth century had lulled people into trusting paper money and paved the way for the great inflations of the twentieth century.

World War I was the watershed which unleashed inflation and other forces which still exist. By the time De Gaulle came to power, the franc was worth 1/200th of its value in 1914. However, England did not suffer nearly as badly, and the United States lost very little in purchasing power. But the movement from a strict gold standard to various non-gold standards gave the printing press a foothold which was relentlessly expanded and will eventually lead to ruin.

It can be seen how the rapid deterioration of the quality of paper money snowballed and led more nations to resist the acceptance of foreign paper money. The number of anti-capitalist international tariff laws in the 1920s led to a virtual halt of world trade in the next decade. Like the competitive devaluations of the '30s and the stock market crash itself, this was blamed on the stock market, gold, and anyone or anything but the real culprits: the anti-capitalistic inflationists themselves.

In the 1920s, the discipline of gold began a long decline (not to be arrested until the 1970s). As conservative financiers wrestled with inflation, problems of protectionism, high tariff walls, and competitive money devaluations, they realized they had to return to the classic gold standard extant before World War I. Tragically for the world, the method they chose in Genoa in 1922 only aggravated the problem by creating a "gold exchange standard" which led to the inflationary boom of the '20s. Even though that boom ended in the Crash of 1929, the same mistakes were made after World War II at Bretton Woods.

A truth of financial reality is that any prosperity built on paper has usually been fun for a while but has always ended in catastrophe. In the 1920s (as in the 1960s) the stock market was fueled by surplus paper spilling over from banks, which borrowed it from the U. S. Treasury, which had borrowed it from itself. It is inconceivable to me that the same unwise mistakes made in the '20s were repeated in the 1960s. The question which remains is will I be able, in some slight way, to help prevent it in the year 2008? I am hopeful, but sometimes hope can be a sophisticated and diabolically clever form of despair.

The United States—A New Factor in the Equation

Since the United States dominated international monetary finance beginning in the '40s, and since it is vital to today's monetary picture, let us go back to the beginning of America's financial development and see how it began to intertwine with the European picture during the critical years following World War I.

The future of the U.S. dollar is especially bleak in view of the fact that the United States has one of the world's worst financial histories. The American Revolution induced the ruling Continental Congress to print paper money unbacked by precious metals. When this currency became worthless, the expression "not worth a Continental" evolved. For this reason the new American Constitution declared

> No state shall enter into any Treaty, Alliance, or Confederation; grant Letters of Marque and Reprisal; coin Money; emit Bills of Credit; make any Thing but gold and silver Coin a Tender in Payment of Debts; pass any Bill of Attainder, *ex post facto* Law impairing the obligation of contracts or grant any Title of Nobility.

Since the Constitution prohibited the states and the federal government from issuing paper money, our founding fathers unquestionably felt they had nothing to worry about from fiat money in the future. The Coinage Act of 1792 made our currency a "hard money," that is, backed by precious metals with a ratio of 15 ounces of silver to 1 ounce of gold.

Bi-metallism did not work well because the market price of silver declined, and gold became undervalued. According to Gresham's Law, the bad money drove out the good and gold became scarce. The American people have not learned that strict and artificial ratios applied to precious metals cannot withstand the test of time. Thus, in the early 1800s, half the currency base was out of circulation, and private bank notes began to circulate freely to fill the vacuum. Frequent bank failures led many to call the defaulted currencies "shinplasters" and these silly credit instruments also became discredited. This was the second group of people ruined by a phony U. S. currency. There were more to come.

In an effort to make the U. S. currency coherent, the Second Bank of the United States was chartered in 1816. Expansionary policies were favored by government as capital moved from the relatively developed East to the pioneer West, so the bank ran into liquidity problems. To pander to the debtor class, Andrew Jackson vetoed renewal of the bank's charter in 1832. Jackson won reelection that year and permitted private banks to issue their own money, sowing the seeds for a great speculative financial orgy in 1834. In July 1836, inflation was so wild that Jackson demanded that all down-payments for land be made in precious metals. In the ensuing panic, land values collapsed overnight. Government distribution of paper money to banks did not help, because people had lost faith both in land and in paper money. In 1839 we had the worst financial panic in American history. Counterfeit paper money began to circulate. By 1841 the devastation was complete and the country was virtually at a standstill. The United States did not fully recover until just before the Civil War, and many land values did not reach their former heights for half a century. (Doesn't this sound like the 1930s?)

Greenbacks were issued in 1862 because Lincoln needed the money to finance the Civil War. He promised that this paper would be retired right after the war. Naturally, by the time they were redeemed in 1879, greenbacks were worth only a small portion of what they were worth in 1862. Debtors who fought this redemption formed a political party called the Greenback Party which advocated fiat money and social security—radical ideas denounced in those days. However, all forms of money were to be redeemable in gold, and from the Civil War until the 1920s America had no major economic setbacks. I think it was gold which, by blocking politicians from inflationary policies, led to that economic security.

In the nineteenth century, the constant creation of paper led to malproduction and malinvestments, and financial crises were needed to get back to reality. Thus came the banking crises of 1873, 1884, 1893, and

1907. In 1907, people demanded a solution. The National Monetary Commission wrote an act in 1908 which sought a flexible money supply to "meet the needs of business." They set up the Federal Reserve System to act as a fire department, a decision which would bring trouble and destruction for the entire modern world, almost as efficiently as if it had been intentional. When banks ran into trouble, the Fed printed money and loaned it at a penalty rate and bailed them out. This law, designed to stabilize prices and eliminate depressions, went into effect in 1914 after war broke out. People were so excited by the war that they barely considered the tremendous significance of this law. The Fed's purpose was to provide more money when business needed to expand, and contract the money supply when it needed less. It was thought that the dollar would be sound because it was backed 40% by gold and 100% cash reserves so that it was 140% secured. Federal Reserve Notes were sold to a bank to help business. The bank could then sell the Note at a discount and it was then rediscounted. The Note was canceled when paid in full. In theory, it sounded great.

Elihu Root, then an opponent of Federal Reserve Notes, said that the United States is basically optimistic, so this power would be abused and there would be a bias toward inflation. He was so right.*

Prior to the creation of the Fed, banks could theoretically lend whatever they wanted. In actuality, they were limited by the needs of business and by the amount of deposits on hand. By artificially lowering interest rates, more deals became profitable; thus increasing lending. A banker would lower his rates to artificially create a demand for loans. This could be done by juggling the books, that is, placing the borrower's note opposite his balance. However, the marketplace was not quite the same. Suddenly, there were more buyers for the same amount of goods, which led to higher prices and misdirection of production. Many years later, the Fed went to 25% gold backing of the dollar, and after World War I it went down to 13%.

The Fed's original purpose was changed from meeting the needs of business to meeting government's need for war borrowing. At that time, they sold $21 billion in Liberty Bonds. The banks created money via the Fed while their own depositors were being embezzled. On June 30, 1914, the United States had $16 billion in money supply and zero national debt. We loaned $7 billion to the Allies and an additional $3 billion after the war. By April 1919 we needed to sell $5 billion in Victory Loans, so it was decided to keep interest rates down until after

---

* Those who wish to know more about this period are strongly urged to obtain *Monetary History of the '29 Depression* by Prof. Percy L. Greaves, Jr., 422 First St. S.E., Washington, D.C. 20003.

their sale. By June 1919, cash reserves were $29.7 billion, with a $292 million national debt. This was not free enterprise, and it was a fraud perpetrated on people who trusted their own government.

Eventually, a sharp boost in the discount rate from 4% to 7% stopped the expansion because inflation was becoming a problem. The rise led to a recession after World War I that hurt farmers and caused the money supply to contract. This pain was a result of paying for World War I by inflation. At that time wages were dropping, since they were flexible and the economy was correcting naturally. U. S. gold reserves were rising and cash reserves were dropping. This made it harder for Europe to pay her debts, but of course she always had the recourse of paying by selling goods to America. Short-sighted people elected tariff-erecting Republicans, thus blocking Europe's ability to pay.

The Federal Reserve Board was supposed to have been designed to prevent it from becoming political or partisan, but it had naturally become so. Europe's inability to pay led to the catastrophic German inflation in the early 1920s which paved the way for Hitler and World War II. Yes, I even blame that war on the mishandling of gold. Unbacked paper money is truly the root of all evil.

The Genoa Convention of 1922

Meanwhile, some politicians were desperately trying to get back to a pre-World War I situation. Unfortunately, they chose the gold *exchange* standard. Lloyd George convened a group of international bankers at Genoa, Italy, in 1922. At this little-known convention, it was decided that each nation could issue its domestic currency against its gold reserves *plus* its holding of British pounds. (U. S. dollars which were convertible into gold were later included.) It was argued that these pound and dollar reserves were superior to gold because they could earn an income when deposited in banks. This concept was the forerunner of present-day SDRs, otherwise known as "paper gold"—an oxymoronic experiment, in my opinion, doomed to failure. Gold is gold and paper gold is paper, and the twain will never meet.

Unfortunately, what was needed in 1922 was a deflation to get back to reality. Since those politicians found it politically unacceptable to experience a mild deflation, they favored the gold exchange standard, which led to a new boom and the catastrophic deflation of 1929. By avoiding a spoonful of castor oil they wound up with a pailful of arsenic.

The fallacy here is that the gold exchange standard counted the gold twice—once where it was stored, and again in any country which held paper dollars or pounds. Thus was sound banking abandoned. Whereas the gold standard was automatic, the gold exchange standard did not force immediate payment, or export of gold which forced deflation. Nations printed more paper money instead of shipping gold, thereby postponing the awful day of reckoning. Thus the present system allows the poison to accumulate to lethal levels.

Currency should be no more than a "claim check" on all the goods and services of a society; no new goods being produced, no new paper money. The elimination of Keynesian perpetual inflating is the key to permanent long-term prosperity. In no way is avoiding a recession worth undergoing a depression later. Yet, that is the path these people chose in this little-publicized meeting in Genoa in 1922. They stuck the next generation.

In the 1920s a new boom started, but this time the Fed would not let things get out of hand. Since political parties seek booms during elections, the Fed inflated to make the economy look rosy. Any political connection between the Federal Reserve Board and a government in power must be corrupt. In 1923–24, the money supply rose again, but our gold reserves did not. Interest rates in 1924, an election year, were held down, and this led to a false sense of prosperity. The Fed bought government bonds and printed paper money, so that 1924 really marks the Fed's first aggressive and intentional act of inflating. The Fed was used to manipulate the economy. Frankenstein was reincarnated.

In June 1924, in an effort to return to the pre-World War I gold standard, the Bank of England colluded with the U. S. Federal Reserve to push interest rates lower in the United States and maintain higher rates in England—a violation of the free marketplace and free capitalism. In December 1924, the Fed gave $200 million in depositors' money to England and received nothing in return. The free market price of the pound was $4.40 at that time. All of this was kept a secret from both Parliament and Congress. Accordingly, on April 28, 1925, England went back on the gold standard and the bank rate was raised from 4% to 5%. England tried to get the pound up 10% to its pre-war level of $4.86, but the currency manipulators feared the United States would attract English gold unless interest rates were higher in America. Therefore, the entire U. S. economy and money supply was manipulated to help England at a time when the free market said the pound was only worth $4.40. These individuals thought they were smarter than the marketplace. Fight such intellectual arrogance when it reappears in the future.

Artificially low interest rates in the United States enabled banks to borrow from the Fed at the low rate of 4% and lend it on Wall Street at 5% or 6%. Bankers were thrilled. But when the Fed buys bonds as they did in 1925, there is more paper in the marketplace competing with the depositors' original deposits. Their own money was coming back at them. The Fed did it to help the stock market, which worked, and the surplus paper money spilled over into the legendary Florida land boom.

On April 26, 1926, the United States lowered its discount rate from 4% to 3⅓% with all discussions kept secret. At this time the pound was valued at $4.86. English prices were up 10% and lower costs were needed, but an overvalued pound and intractable labor unions blocked that path. The refusal to accept lower wages is a source of unemployment. Instead of concentrating on how high or low wages are, what is significant is purchasing power. A lower wage in a sound economy could mean more purchasing power than is possible from a higher wage in a runaway inflationary period, but pusillanimous politicians do not explain this. Any built-in inflexibility of wages makes higher prices inevitable. Seniority is another unfair practice because it gives old union members a guaranteed income at the expense of new members. Unemployment compensation then shifts the burden from the unions to the taxpayers. All of this is unfair because it is an improper allocation of costs.

In May 1927 it was announced that the 4% Liberty Loan would be refinanced at 3½% the following November. The Fed was determined to hold interest rates below the free market price until then. This was not capitalism, and this action forced prices to rise artificially. Easy money was the answer then, as now, and an artificially-induced boom spread. That year, France withdrew gold from the Bank of England and sent it to the United States for safekeeping, causing England some economic heartache. And, of course, the United States printed money to make up for it. This was not a free market, and the Fed used depositors' money to support the English pound. (All this has an astonishing similarity to events occurring nearly forty years later when De Gaulle was roundly condemned for being suspicious of the U. S. dollar.) The United States lowered its discount rate to 3½% to aid England on the theory that England would then buy our food and other exports. It was even said this was done to help U. S. farmers, without mentioning England. Nonetheless, 1927 saw an expanding money supply with remarkably stable prices. This early stage of inflation was masked by radically increased production, and it was not seen that inflation was distorting the economy.

There is always a new and good excuse for immorality. The government could not tighten up in 1928 because it was a Presidential election year, thus exposing the fallacy of ever allowing politicians to get near the money supply. In January 1928, brokers' loans were a record $3.8 billion. From 1922 to 1926, the money supply increased from $33 to $43 billion and more and more of it flowed into the stock market. Each time the market dipped, politicians would "jawbone" it higher, as Nixon was to do in the '70s. Banks were borrowing at 5% from the Fed and lend-ing it at 8.6% to stock speculators. Who dared kill the golden goose? The Federal Reserve Board tried to raise rates to 6% as the money sup-ply rose to $45.7 billion, but Washington refused. Brokers' loans were $8.5 billion by the time of the Crash, which gives some idea of the incredible rate of increase. All this was done with the money of innocent depositors, and the subsequent collapse of the banks (which wiped out the savings of those depositors) meant they were being hoisted by their own petards. It is astonishing that there are still people who favor "easy money." I firmly believe politicians should not be given the power to manipulate the money supply at all.

Conclusions

There should have been a recession in the early 1920s to wipe out the excesses of World War I in Europe. This was postponed by the addi-tion of the gold exchange standard at Genoa in 1922 which took the rotten structure to yet higher levels by 1929, where the collapse was to be even more devastating. Even the 1929 Crash could have, at that point, wiped out the mistakes of World War I's inflation and the errors pro-liferated by the gold exchange standard of the 1920s. Ordinarily it would have, which is the really important point to grasp here, since it ties everything together.

While the situation in Europe was going through its natural correc-tion, a new monster was unfortunately evolving in Washington, D.C., in the form of the Federal Reserve System. When the United States as-sumed world financial leadership after World War II, the Fed became an engine of inflation for the entire world; a Frankenstein monster bigger than anything which had yet been created by the imaginative folly of man. Grasping the intertwining of the American and European histories is an underlying cause of the long-term pessimism of my stock market publication, *The Dines Letter*.

The gold exchange standard initiated in 1922 led to a collapse in land values in 1926, while stocks continued to rise for three more years. The lack of real estate opportunities caused speculators to look for greener pastures on Wall Street. (In the 1970s the sequence was ex-actly the opposite. Since there was very little borrowed money in stocks compared with 1929, stocks began declining at the end of 1968. How-ever, real estate, propelled by 90% loans in some cases, continued to rise for several more years, particularly as more people became disillusioned with the stock market. Thus does history tirelessly repeat itself.)

It was not capitalism that failed, but political tampering with the money supply for the purpose of pandering to the public. The gut cause of the misery of the little man is the desire to get something for nothing. It is this immorality in an inflationary system which must lead to catastrophe each time, as inevitably as the earth revolves on its axis.

So, World War I's inflation temporarily derailed the gold standard, which strictly limited the money supply, and the Genoa Convention legitimized governmental tampering with the creation of paper money. No government has ever resisted abusing that right, and the direct result was a runaway inflation in the 1920s and the inevitable corrective Great Depression.

The Great Depression

No one needs the 1929 Crash and 1932 Depression described again. It was bad. It was *the* Depression. Yet, I fear we might see an even worse depression in coming years.

Interestingly enough, the word "depression" when first used by Herbert Hoover in 1929, was actually a euphemism for what used to be known as "hard times." Hoover treated the market crash as a psychological phenomenon. He used the word depression because it sounded less frightening than panic or crisis. By the 1950s the use of the word depression was unofficially prohibited from all Wall Street market letters, and the word "recession" (especially "extreme recession") became the euphemism. For the record, in mid-1974, the word depression began to be used publicly again. What it means in the larger scheme of things remains to be seen.

As I pointed out in my first book,* the stock market "discounts" the future, so that when people refer to "1929" they mean not so much the market crash of that year, but the Great Depression which the barometric stock market forecasted for 1932.

While the stock market anticipated the future, crashes might be something of an exception in that such market plunges might be self-fulfilling prophecies. This is an insightful sidelight. At the bottom of the Great Depression people began to look for scapegoats. Among others, they accused Wall Street of causing the Depression. This accusation perhaps has some merit—Wall Street can be considered a contributor—but of course inflation was the real cause. Perhaps stocks should be included in the definition of "money supply."

Something as violent and drastic as a stock market crash might actually aggravate, and conceivably cause, an economic downturn. For example, in 1974, IBM dropped 10 points in one week, equal to $1.5 billion, and Eastman Kodak declined 6 points, equal to about $1 billion. Exxon

* *How the Average Investor Can Use Technical Analysis for Stock Profits.*

lost $1 billion in one week, and Polaroid, down from 143½ to 18½, lost around $4 billion in value. Where did this money go and what happened to it? Paradoxically, the money really never existed. Yet, when those shares were still high, shareholders could not only borrow on them, but they were also considered psychological assets. When most of us have a stock rising steadily we feel richer, and are more willing to spend money. Furthermore, when a stock drops, we naturally tend to pull back and cut back on our spending. If that stock is pledged as collateral, one can actually borrow less money on it, so it is a real loss in every sense. Pension plans which appeared to be "fully funded" (that is they seemed to have enough assets to meet the income requirements of retiring employees), suddenly find that there is insufficient money. Money must then be pulled out of the corporate income stream to make up the difference, which must be an unpleasant surprise to corporate treasurers. A rising stock in a strong market makes it easy for a company to raise money through the sale of new common shares, while in the opposite condition corporations need to borrow from banks and might become overburdened by debt.

Another point to remember is that appreciated stock is frequently given to charity, which can be donated at market value for tax advantages in the gift. Thus, for all these reasons, a stock market crash can make its negative prophecies to some extent self-fulfilling. Secondary effects from a decline in security values might conceivably bring about, or at least accentuate, a business slowdown.

Some who seek scapegoats will say later that gold caused what I predict will be called "the Crash of 1969." They will say that since gold is money, any increase in the gold price is inflationary. This is untrue. When Roosevelt raised the gold price from $20.67 to $35 in January 1934, there was no runaway inflation in our economy. The gold price-hike wiped out the previous paper money creation binge of the 1920s. Others declared that declining world gold output and a reduction in the quality of South African gold reserves caused the Crash. I mention these points not to dredge up ancient history, but to forestall criticism I predict could return within the next few years. Be ready to confront and reject it.

Everything got blamed for the '29 Crash sooner or later, including stockbrokers, politicians, Hoover, England, sun spots, anything but politicians' own greedy refusal to accept small economic downturns. Recessions are a necessary Darwinian evil to root out inefficiencies and to keep prices reasonable by competition.

In 1933, Professors G. E. Warren and F. A. Pearson wrote *Gold and Prices,* which declared there was a precise relationship between new gold production and the supply of old gold reserves on the one hand and the level of wholesale commodity prices on the other. Roosevelt reasoned that by reducing the gold content of the dollar (or raising the price of gold) domestic commodity prices would turn up and exports would be stimulated. Thereby the gold content of the dollar was reduced from 25.8 grains to $15\frac{5}{21}$ grains of gold $\frac{9}{10}$th-fine ounce, raising the price of gold from $20.67 to $35 per ounce. The devaluation about equalled the shrinkage of wholesale commodity prices which had taken place during the Great Depression. Commodity prices, as it turned out between 1932 and 1937, moved up around 20% and did not equal the 1928–1929 range until 1941. To that extent, Warren and Pearson were proved wrong.

C. O. Hardy's influential 1936 book *Is There Enough Gold?* answered the question "yes." Hardy rejected the argument that the inadequacy of the gold supply did not contribute to price declines between 1924 and 1929 and that there was no immediate prospect of an important decline in gold production before the Great Depression struck. Hardy also said that had gold reserves been larger or smaller, central bank policy would have been the same.

Each economic cycle in history shows itself to be the same animal every time; the "Odyssey" chapter will present a microscopic examination of one of those cycles. Only through rigorous self-training can you become immune enough during the emotions of such a cycle to remain the master of your own destiny. Many of those who were wiped out in the Crash could have saved themselves if they had mastered the "Odyssey" chapter. Thus, there is hope, if this book is brought to the attention of enough people, that they will be able to detect these cycles for themselves.

Oh No! Not Again!

At the bottom of the 1932 Depression, a unique and undoubtedly well-intentioned man arrived on the scene who was to start the same dreary cycle over again. Almost half a century later we are still in that cycle, one which has been repeated time and again, a cycle which man stubbornly refuses to confront and master.

Franklin Delano Roosevelt was elected in November 1932 when the economic world was prostrate. One can see how anyone with feeling for the little man would try to take direct action to help him, and in that sense he cannot be blamed. Yet all the excesses derived from the inflation to pay for World War I, all the foolishness of the Federal Reserve Board had been liquidated violently, and FDR was presented with a clean slate. A really sound currency established at that time would almost certainly have prevented World War II and surely would have changed the economic history of the rest of this century.

FDR's "New Deal" has always been controversial. Many have sworn by it or at it. A century from now, I predict there will be a consensus which will declare that both sides were right. I flatly predict the New Deal will go down as the greatest short-term success (four decades) and the biggest long-term failure of any economic system the world has yet seen.

Here was a God-given opportunity to liquidate the Federal Reserve, reduce taxes to virtually nonexistent levels and establish a really sound currency. The recovery would have been slow at first, but it would have been as profound and long-lasting as the halcyon days of the nineteenth century. Instead, because he was influenced by British economist Lord John Maynard Keynes, Roosevelt took the opposite tack. Santayana said, "Those who cannot remember the past are condemned to repeat it."

Let us carefully dissect the economic policies of this period so that we can demonstrate where we went wrong to future generations who might be tempted to abandon gold again. To oversimplify Lord Keynes' theories greatly—which perhaps means to look at them realistically— unless the government spent a lot of money, even more than it had, the depression would never end. He convinced Roosevelt of this, even though all of our previous depressions had ended without government intervention. Keynes believed that during bad times the government should borrow money from itself by creating money, and funnel it into the economy to stimulate business; then, when the economy recovered, the government should tax heavily and repay those deficits.

This practice was like leaning into the wind, with the government becoming a balancing mechanism. In fact, Keynesian economics, as it came to be called, became the dominant economic theory of the world from then on, and in the 1960s it actually became the "new economics." Actually, if one knows the history of these cycles, it can be seen that there was nothing "new" about Keynes at all. In fact, as Charles Black points out in his excellent book, there were great parallels between Keynes and John Law, in that both were speculators, gamblers, philosophers, intellectuals, and neither conservative nor cautious. Both preferred sudden wealth over patience and frugality, and where John Law had Louis XV, Keynes had Roosevelt.

Unfortunately, Keynes' theories had the same weakness as Law's: that of human nature itself. Politicians did not like cutting back in good times, so the second half of Keynes' philosophy was disguised, suppressed, forgotten, and conveniently overlooked at every opportunity.

Keynes was not necessarily wrong, but his ideas were unrealistic in view of human nature. Can men get along without gold? Can there be a sound currency without gold? Theoretically, of course. But until men become saints, we never will. So, the acceptance of Keynes' ideas in 1932 became a one-way ticket toward a guaranteed inflation which would eventually bring ruin in the form of another depression. The little guy took a beating from the Great Depression, and instead of being cured by it, went right back to the heroin needle.

Perhaps Keynes, a socialist, understood that his theories would eventually fail. When asked about the long-term problems involved in his philosophies, Keynes answered, "In the long run we are all dead." Did he mean that by the time he was brought to account for his disastrous policies he would be dead, or that the generation who suffered the Great Depression would be dead? Did that rascal know the point I am making in this book? If he did, you can bet he is not in heaven.

Thus it was that the government wholeheartedly accepted the concept of perpetual deficits without ever thinking about how to repay the ensuing debt, a debt which was to slowly wind its way around our collective necks.

The Son of Frankenstein

It is my deepest conviction that gold is the fulcrum, the very central lever within the deepest inner recesses of capitalism itself. Touch the golden lever, and its effects are so profound they can surface half a century later. It is the very pituitary gland of our economic organism, if you will. The more I study economic history, the more I become convinced that this is true, and no politician should be permitted to get near this lever unless he really understands what he is doing, and the public willingly accepts the consequences.

Roosevelt took office on March 4, 1933, and on March 6, 1933, blazing *New York Times* headlines declared: "Roosevelt Orders 4-Day Bank Holiday, Puts Embargo on Gold, Calls Congress." Other headlines included: "Financiers Look for Little Interruption in Business"; "Emergency Step Praised—President Takes Steps Under Sweeping Law of Wartime"; and "Prison for Gold Hoarder." Roosevelt's actual proclamation begins

> WHEREAS there have been heavy and unwarranted withdrawals of gold and currency from our banking institutions for the purposes of hoarding; and WHEREAS continuous and increasingly extensive speculative activity abroad in foreign exchange has resulted in severe drains on the nation's stocks of gold; and WHEREAS these conditions have created a national emergency . . .

And so it went. Gold was somehow blamed for what was wrong with the world, rather than blaming the greedy people who abused it. It was at this instant that the Son of Frankenstein was reborn for the umpteenth time.

On March 9 all gold and gold certificates were called in and the die was cast for a fiat money currency inflation which is now near its climax. Roosevelt accepted this long-term problem probably because he felt traditional methods of resurrecting the economy would be too slow. Perhaps the nation would have turned to communism if it had had to wait that long, and in that respect perhaps he was right. As hard as it was to follow or even envision at the time, the sound currency path was the only way to break the vicious cycle of inflation and depression.

Instead of dissolving the Federal Reserve Board, Roosevelt made it a more powerful tool of government to implement his inflationary plans. FDR castrated gold by nationalizing it and gave it as a gift to the Federal Reserve Banks. With the United States severing an important link between paper money and gold, a chain reaction started which reached its climax on August 15, 1971, when President Nixon axed that final link between gold and paper—unleashing a tidal wave of inflation. On that day, Nixon reneged on all our nation's solemn promises to exchange gold for dollars held by foreigners. Nixon, who had proudly and defiantly declared after his election, "I am a Keynesian," once again permitted the United States to print unlimited amounts of paper money. Inflation subsequently accelerated. Gold did not cause anything: a penny was stuck in the fuse box so that gold could no longer function as policeman. The sad part of it is, FDR's act not only flew in the face of history, but in all probability was illegal. He might have done it because public pressure on him to do something, anything, to get out of that Depression was so immense. People always get the government they demand sooner or later, unfortunately, which suggests a failing of democracy itself.

FDR's outrageous act of withholding gold from the American public was brilliantly discussed in a Winter 1973 *Brooklyn Law Review* article by Henry Mark Holzer entitled, "How Americans Lost Their Right to Own Gold—And Became Criminals in the Process." Many knew this act was illegal at the time, but winked at it. Thus, Frankenstein was born. I cannot improve on Holzer's story of how Roosevelt rammed this legislation through the Courts, and if this particular aspect of history interests you, then obtain this article.

Legislation restored this right to the American people on December 31, 1974. Consequently, the problem is moot now, except that it might arise again in the future. If you let them take gold away again then you really do deserve what will happen to you.

Only two centuries ago, gold and silver coinage were the principal means of paying for goods and services. I have described the gradual emergence of a credit economy which changed the function of gold from a medium of exchange to a leash on the banking system ostensibly to limit the creation of credit. Credit was Frankenstein on a golden leash. Since scarcity is the source of all value, a golden leash affords the discipline which maintains the integrity of a currency and prevents Frankenstein from getting out of control.

The Federal Reserve System was required to set up a gold reserve of 35% against currency notes issued, and 40% against member bank deposits. In addition, all parties entering into contracts had the absolute right to insert a gold clause guaranteeing payment in gold, which protected the lender against inflation and repayment in a debased currency. Roosevelt continued the chipping away at the power of gold by eliminating gold coinage on January 31, 1934, and the private ownership of gold by American citizens. One of the controls the individual had over his government was eliminated, and Americans no longer had any realistic domestic alternatives to paper money or defense against inflation. Roosevelt also raised the price of gold to $35 per fine ounce. To quiet critics, legislation was passed demanding that U.S. currency be backed by at least 25% gold. Since the United States owned a large percentage of the world's gold at the time, and since foreigners (but not U.S. citizens) could still convert U. S dollars into gold, the assumption developed that the U. S. dollar was still "as good as gold." For a while it really was. But Frankenstein was just an infant.

Roosevelt's inflationary policies seemed to work at first, and by 1936 recovery seemed complete. Unfortunately, a relapse in 1937 led to a stock market collapse and the return of massive layoffs. When Roosevelt took office there was an army of 14 million unemployed, and as late as 1939 the unemployment rate was around the 10-million mark. So the New Deal never really succeeded in bringing back prosperity or ending unemployment. All Roosevelt had done was to raise the national debt from $16 billion to $56 billion by 1941. Roosevelt did not blame his own policies for the failure, but business itself, just as Presidents Eisenhower, Kennedy, Johnson, and Nixon were to do at later dates. Business was saved from Roosevelt's scapegoatism by the outbreak of World War II. In fact, there are some who accuse Roosevelt of actually engineering World War II to save his own skin. The war devastated all the other major industrial nations while we were building up our industrial capacity, and accumulating 60% of the world's gold.

At the end of the war, the United States was at the absolute summit of the world economically, morally, politically, militarily, and financially. In 1945 the United States called the shots. This was another

God-given opportunity to set the world's monetary system straight. Instead, at the Bretton Woods Conference in 1944, our leaders again had locked us onto a path ineluctably aimed at another depression. At Bretton Woods, England's Lord Keynes and America's Harry Dexter White tried to eliminate gold from our currency, which probably would have been for the best because that system would have failed much sooner than the present one will. Instead, the United States went on a dollar exchange standard, just as the gold exchange standard was adopted in Genoa in 1922. Other nations were hungry for dollars which were freely convertible into gold. But given human nature, paper money which is not backed by gold will eventually be corrupted. The United States pumped dollars into Europe, which Europeans used to buy American materials, and the dollar became the universal medium of exchange. This could have lasted until the end of recorded time had our money been managed wisely, since the world had actually gotten used to the paper dollar. But U.S. leadership snatched defeat from the jaws of victory.

The United States had gold holdings of $10 billion in 1935. This swelled to $18 billion in 1939 as a consequence of gold fleeing Hitler, and another $5 billion was added before our entry into World War II. This build-up did not stop until 1949, when we had accumulated $24.6 billion in gold at a time when the entire gold reserves of all governments in the free world totalled only $35 billion.

Bretton Woods—Frankenstein's Baptism

In 1942 Harry Dexter White of the United States Treasury and Lord John Maynard Keynes of the British Treasury developed proposals for a new international monetary system. This fountainhead of financial troubles was convened at Bretton Woods, New Hampshire, in 1944 with the launching of the International Monetary Fund and the World Bank. Since the United States was the strongest country in the world, the opportunity was seized to change the old gold exchange standard to a *dollar* exchange standard. Thus, the New Deal's direction was internationally legitimized, and firmly locked the world's economies on their tragic rendezvous with destiny.

The IMF required countries to define the parities of their currencies in gold or U. S. dollars and to keep exchange rates firmly locked within one percent of "parity." To give countries time to adjust imbalances in their international payments, the Fund's resources extended credit. This credit emasculated the power of gold to force countries to sell internationally in roughly equal amounts against what they were buying. This fatal flaw threw the international balance wheel out of whack. The gold lever had been tampered with and now was being strapped down out of position. If a country could not restore its balance-of-payments without deflating its economy or imposing restrictions on current trade and payments, the country could then change the parity of its currency, thereby eliminating all deflations and creating a perpetual "high." So, all the world's currencies were expressed in terms of, and closely tied to, the U. S. dollar, but in turn the dollar was still fixed to gold. The dollar was the yardstick. Only the United States could change the price of gold, and all other nations were forced to upvalue or devalue in terms of dollars. This incredible power given to the United States would eventually be devastatingly abused, but Lord Keynes and other engineers at Bretton Woods were too busy constructing Frankenstein to worry about such mundane matters.

Bretton Woods also ruled that a nation's reserves could be composed either of gold or any currency convertible into gold. This last one was the killer because it included the dollar, and later the pound, when it was strong and freely convertible into gold, and in no way anticipated the way the dollar would be run into the ground. This subverted virtually every other currency in the world and launched an international inflation the likes of which had never before been seen. Meanwhile, the United States solemnly pledged to maintain the value of the dollar by buying or selling unlimited quantities of gold at $35 per fine ounce.

Since other countries did not have to make their currencies convertible into gold but merely into dollars, their governments expanded the total amount of their paper currencies freely. Devaluations came casually, since the United States never retaliated, and nations began to develop trade advantages with the United States, leading to, among other things, the invasion of the United States auto market by Volkswagen. To pay for these losses, the U.S. government inflated even faster than the others, which is why the United States began to lose gold after World War II, falling from $24.6 billion in 1949 to around $10 billion in 1971.

The United States should have lost more gold than it did, should have in fact been cleaned out, as it would have been under the classic gold standard. This would have stopped the inflationary mess long ago, illustrating the importance and function of gold. However, other nations did not demand gold, since they could acquire and use paper dollars in their reserves. While dollars were technically convertible into gold, the United States did not mind foreigners collecting them because the Washington Economic Establishment could always print more.

Thus it was that the International Monetary Fund proved to be one of the greatest engines of inflation in the world's monetary history. So it was that no nation except the United States was limited by gold. Even the United States did not have that constraint because it could simply print more paper. When U.S. inflation began to accelerate toward its terminal stages, the IMF forced other countries to keep their currencies from going to a premium against the dollar, so the United States forced the world to import inflation. Germany, for example, was forced to buy billions of dollars to support the dollar in relation to the mark and, in order to do so, issued billions of new marks. This increased Germany's money supply, and the dollars it absorbed became its "reserves."

There is no evidence that the IMF, from 1946 to 1972, did anything to stop or reduce inflation, and the IMF was obsolesced when the system of fixed exchange rates was finally abandoned in 1971 and all currencies were allowed to float. Floating calls much more attention to the need for each country to maintain a sound currency, and immediately exposes individual weaknesses. By a sound currency I mean strict limitation on the issuance of paper, especially by maintaining its convertibility to a universal standard of value like gold or silver. Floating reveals true weakness which cannot be wall-papered over by "international cooperation," or using strong currencies to buoy up weak currencies. This only allows countries with weak currencies to prolong their reckless policies, and in the end makes it worse than it would have been had the problem been immediately confronted. Attempts to eliminate Darwin's law of the jungle from currencies only results in the "survival of the´unfittest," and requires a depression to eliminate them.

The IMF is simply a pool of currencies and gold, originally worth $8 billion, designed to help countries over near-term imbalances in their international accounts; an "international Fed," in a way. The IMF helps avoid the rigorous discipline of gold. It is dedicated to the idea that currencies should have stable values and nations should not be subjected to severe deflation in order to adjust to short-term imbalances in their flow of funds across their borders. It was a desperate effort to legislate against the Great Depression. In a sense, we were still fighting dinosaurs long after their extinction. Of course, the founding fathers of this useless organization ignored the consequences of what happens when currency values get fundamentally misaligned. Three decades later when this fixed parity system collapsed (as I had long predicted it would), it was followed by the "floating" of all currencies in relationship to one another. To me, this is a cheaper and more sensible solution than legislating, by dictatorial fiat, what currencies ostensibly are worth.

Handmaiden to the IMF is the World Bank. This is a multinational mechanism by which developed countries can lend long-term money to underdeveloped areas of the world. The bank sells bonds in the United States and Europe and then lends the proceeds to finance projects such as agriculture, public utilities, education, transportation and, to some extent, industry. It is really a form of charity. I do not criticize charity, but merely point out that if we spend more on charity than we should, everyone will suffer a depression. The choice should be up to the electorate.

Briefly, here is the sad story of the severing of the link between paper and reality. In 1945 the Federal Reserve Bank, having maintained 35% in gold against currency notes and 40% in gold against bank deposits, lowered the requirements on both to 25%. In 1964, the gold reserve requirement behind the Federal Reserve deposits was abandoned, leaving only the 25% gold backing for Federal Reserve deposits of member banks until March 14, 1948. At the time *The Dines Letter* objected vociferously, but few cared in those days. In 1965 silver (silver certificates) was eliminated from backing U. S. dollars. Since all silver certificates were called in and paid off in silver, all links with precious metals had been severed and there were no limits to the printing press, and a horrendous inflation became inevitable. By calling in our gold coins in 1933 and our silver in the 1960s, real value was taken from citizens and counterfeit money was issued in its place, just as in ancient Rome. The government made a 100% profit on the transaction, which is nothing more than a sophisticated form of embezzlement. The purpose of downgrading gold can be seen to have been a deliberately immoral attempt to inflate, knowing it would be a secret tax levied on the public. No state or bank has ever had the right to print unlimited money and resisted the temptation to abuse that right.

These are chilling events because there will ultimately be a terrifying backlash against inflation. Undoubtedly, some "genius" will point out that there is no inflation in China or Russia, to which I add, "Nor was there any in Nazi Germany." Unfortunately, I doubt there will be any leader to stand up and say there is no freedom for their citizens either. Freedom cannot exist for long in the absence of a sound currency.

In the '30s, '40s, and '50s, deficit after deficit was piled up in the U. S. international balance-of-payments. Our balance-of-trade was positive, but the government gave away so much money in foreign aid, and blew more on wars and other costly projects, that an accumulation of dollars began to develop overseas. Some foreigners began to exchange these paper dollars for gold, as rising U. S. prices made it increasingly unattractive for foreigners to exchange them for goods. U. S. prices were rising rapidly because of an internal inflation based on internal deficits. Thus dovetails the internal and external U. S. currency picture.

Frankenstein's Coronation

Still overreacting to the staggering unemployment of the Great Depression, the Employment Act of 1946 was dedicated to the idea that the government had an obligation to stabilize employment at a high level, even though this might increase inflationary pressures. Since wages consequently lacked the flexibility to decline, this cost to business could never decline and, therefore, long-term price pressures had to go up. To a great extent, business did attempt to offset the rising cost of labor by innovations, inventions, higher productivity, better machines, labor-saving devices, and electronic computers. It also forced commodities to be cheaper and less important than labor, since commodity prices had the flexibility to decline, and something "had to give" during economic downturns. But in the end, nothing could really help and ineluctably rising wages had to come out of profits, then capital itself, eventually killing the golden goose. Another inflation and depression thus became inevitable and institutionalized by law.

To bring us to the present, I wish to call to your attention an interesting editorial *Barron's* published on August 26, 1974, entitled "Bugaboo of Joblessness. The Unemployment Rate Is a Poor Guide to Public Policy." Aside from compassion, the real question is, Which helps the workers more: temporarily suppressing unemployment leading to another Great Depression, or accepting certain levels of unemployment as the price of an incredibly productive and free society? Opportunistic or compassionate politicians have answered this question incorrectly for too long. In this editorial, Robert Bleiberg writes, ". . . unemployment, properly analyzed, is not a major problem in this country. Furthermore, the unemployment rate could rise substantially without causing significant economic hardship. Full employment is usually defined as unemployment which does not exceed 3%–4.5%. As of last month (July 1974), the unemployment rate was 5.3%." Bleiberg then shows a table indicating that for males, aged twenty years and over, the unemployment rate is significantly below the total rate, and has been for the last decade. This group still constitutes the nation's top primary breadwinners, so 5.3% overstates the case. Current figures take into account the number of unemployed persons per one hundred in the labor force, but take no account of how long they have been unemployed, or how long it takes them to find a job. The duration of the unemployment has a great deal to do with the seriousness of the situation, from the individual's standpoint.

The addition of women and teen-agers entering the labor force, not usually primary wage earners, makes the figures look much worse than they are. Distorting unemployment rates in an upward direction prompts government to overreact, thereby distorting economic planning. Policy makers react to this distortion by creating more inflation. Bleiberg therefore feels that unemployment "is not the proper yardstick to use for policy decisions." Unemployment is not caused entirely by people being thrown out of work, since it can rise if more people quit or try to enter the labor force suddenly. Therefore, our fiscal and monetary policies have been frozen into an expansionary mode, feeding an inflation and disrupting the marketplace. Bleiberg points out that "the unemployment rate, as traditionally viewed, can probably double without significant damage being done to the economic structure." He goes on to write, "For forty years, it has been this country's policy to react whenever the unemployment rate rises. It is time that horizons be expanded to include new data, or fresh interpretations of old ones. Since the Great Depression, major programs have been established which serve to cushion the shock of unemployment. The basic necessities of life are not denied if one loses his job. An income still flows in. Thus policymakers should look past the unemployment rate and focus instead upon the havoc that is being wrought by an inflation that is not far from runaway."

Nixon Comes Up with a Weird Theory Which Fails

The basic idea of full employment was developed by Charles L. Schultz and implemented by others well before President Nixon took office. Seeking to cope with the fiscal problems posed by an economy operating well under capacity with sluggish revenues (in other words, short of full employment) the budget was likely to be in deficit without being stimulative for the economy. So went the theory. In 1971 Nixon declared, "The full employment budget is in the nature of a self-fulfilling prophecy: By operating as if we were at full employment, we will help bring about that full employment."

If, under this concept, it could be shown that the existing taxes and spending would produce a large surplus if the economy were operating at capacity, the remedy would be a tax cut, or more spending to stimulate the economy toward full employment. Thus, a tax cut or more spending became respectable even though the budget was already in deficit. The theory was that a tax cut could actually produce more revenues and reduce the deficit by stimulating the economy. In other words, if you are going broke, spend more. This excuse for an inflationary budget policy rationalizes spending by assuming you have it. Amazing!

The idea was discredited in 1974 not only because double-digit inflation outraged the American public, but also because a Brookings Institution budget study pointed out that government revenues are much more sensitive to inflation than expenditures, at least in the short run.

If the above is unclear, it is probably because of the illogical nature of the program. That is why Nixon junked it in mid-1974 and began talking about a huge full-employment surplus on the old basis. At the time of his resignation, Nixon still aimed for a balanced budget in the 1974–75 fiscal year on an actual receipts and expenditures basis, even though unemployment was likely to be well above 5%. As these U. S. budgets become less "stimulative" they will presumably become less inflationary. But unless there are some substantial and sustained federal budget surpluses, I still see no hope.

The '50s and '60s Were Only the '20s All Over Again

In 1949, many European currencies were devalued on a broad scale to adjust for wartime credit inflation used to pay war debts. The United States magnanimously did not devalue the dollar, and because of the resulting competitive trade disadvantage and a generally resurgent world-wide economy, the U. S. dollar entered the 1950s with one hand tied behind its back. The U. S. dollar was overvalued at that time, just as the English pound was in the 1920s. History repeats itself, albeit imperfectly, and that is why I have devoted so much time to it.

In 1950, the U. S. balance-of-payments surplus slipped into red ink and, with rare exceptions, it has remained in deficit ever since. Since most European currencies were inconvertible, the dollar was widely sought as an international medium of exchange. Because of the great demand for dollars at that time, the United States was permitted to run a large balance-of-payments deficit. From 1950 to 1957, only $1.7 billion of our total accumulated deficit of $12.5 billion was exchanged by Europeans into gold. The balance was financed by foreign central banks acquiring those dollars. This blunder was the beginning of the end for Europe (as Frankenstein went international), for dollars owned by foreigners are debts which the United States must eventually pay, but which Washington apparently never intended to pay.

In 1958, leading European currencies became convertible again, offering alternatives to the dollar for the first time since World War II. Around this time, the U. S. government became more reckless in the amount of money it spent and gave away overseas. In 1958, the U. S. balance-of-payments deficit of $3.4 billion was financed when 68%, or $2.3 billion, was converted into gold. In 1960, the U. S. gold supply declined to $17.8 billion, while liabilities against that hoard ballooned to $20.9 billion. It was around this time that I began to understand, albeit dimly since I was alone in those days, that the dollar was bankrupt. I had stumbled on Frankenstein but he had laid the groundwork well and my alarms went unheeded by most people.

On October 11, 1960, fears about the dollar drove the London gold bullion price to a new all-time high of $40.50. It was during this first gold crisis since the Great Depression that *The Dines Letter* tested out its new theories on gold and silver, and their mining shares.

In 1960 I also realized that trouble was brewing when I compared the amount of foreign claims overseas to the dwindling U. S. gold supply. To my horror, I discovered that they were almost in balance. I then began to make statements in the early 1960s to the effect that the U. S. dollar was approaching bankruptcy. Never in my wildest dreams did I think that the financial wastrels in Washington could postpone the gold crisis for over a decade. They succeeded in delaying it, but they only permitted the situation to worsen in the process. Had bankruptcy of the dollar been filed then, the price of gold increased, and a sound currency developed, many of the ills of the '60s and '70s could have been eliminated.

The fraudulent inflation of the 1960s could never have continued if the United States had recognized the power of gold, and perhaps why gold was so suppressed. John F. Kennedy slapped an Interest Equalization Tax on Americans that discouraged foreign investments. This helped prevent dollars from leaving the United States, so that citizens partially lost their freedom to invest or spend money overseas, thus allowing the U. S. government to do so instead. This is not free enterprise. The removal of the 25% legal backing of the dollar, so that all of our gold holdings could be made available to foreigners who wanted gold, was strictly to reassure and ensnare them into trusting us for a few more years. The immorality of this, again, is not free enterprise. It was an open secret that Lyndon Johnson discouraged Germany, for example, from converting their dollars into gold by reminding them that our troops stood between them and Russia. This was not very neighborly. The amount of goods that American tourists in Europe brought back to the United States was reduced, and new tourist taxes sprang up everywhere. This chipping away at the freedoms of Americans was met without complaints. Pressure was placed on U. S. corporations to repatriate U. S. funds under the lure of patriotism. Actually this was a fascist act of expropriating the assets of U. S. stockholders so that the government could spend that money instead. This is not capitalism. Such chicanery by people in high office is not the kind of free government which can support capitalism. Economic freedom is the historic enemy of the fascist state.

Economic imperialism was a natural consequence of the type of mentality dominating Washington in the 1930s. The fascism which spawned Hitler apparently had more subtle manifestations in other countries, not all of which have been extirpated to this very day. The people in Washington stopped at nothing to show that gold was an unsatisfactory investment, so great was their fear of this metal. Their anti-gold hysteria became a self-fulfilling prophecy from 1933 to 1968. But as great as was the power of these politicians, even greater was the power of gold. That anti-gold policies would virtually drive the entire United States gold mining industry out of business, Canada to a subsidy system for their gold mines, and South Africa to paying pennies to their workers because it was all they could afford, was blithely ignored by the powers in Washington. I wonder if the people responsible for these outrages will ever be brought to justice. Currency manipulators are not limited to the U. S. Treasury, but include a long procession of bankers, economists, politicians, and college economics professors, who have dominated the Washington scene far too long.

The London Gold Pool was created in 1961 after the 1960 gold crisis. The government, completely misreading the gold crisis, immediately blamed it on the "gnomes of Zurich" and "currency speculators."*

Thus, the guilty parties, the creators of inflation, immediately sought a scapegoat. Their reaction would prolong the crisis because it concealed the true culprits. The Pool was designed to dump gold on the gold market whenever it began to rise. This suppression of free market prices is not free enterprise. This Pool actually enlisted the central banks of the United States, England, West Germany, France, Italy, the Netherlands, Switzerland, and Belgium as their accomplices. Their specific intent to suppress the gold price at the $35 official level succeeded for

---

* Much of the blame for the gold crisis has been placed on "speculators" in the most negative, modern sense of the word. Irresponsible, or worse, ignorant, journalism has heaped great blame on speculators for "causing" most of the dollar's problems. This is hardly surprising when most Americans have been raised and educated under Keynesian economics either to disregard gold completely or to despise it. Actually, the word speculation derives from the Latin word *speculari* which means "to spy out or observe." Bernard Baruch defined himself as a speculator who observes the future and acts before it occurs. Why put people down when they see trouble coming and try to protect themselves from that trouble? Is it somehow more commendable to understand that your government is trying to destroy you, and then make no effort to protect yourself? As a matter of fact, the Bank for International Settlements once wrote, "Most of the component of private gold offtake is accounted for by fairly firmly held savings in gold rather than by a large block of speculative holdings awaiting a short-term capital gain." Finally, let me add that there is nothing in the U. S. Constitution prohibiting speculation. Or jealous morons either. So far detached from reality were our leaders that John Connally, then Secretary of the Treasury, Paul Volcker, then Under Secretary for Monetary Affairs, and Rep. Henry Reuss, Chairman, Subcommittee of the Congressional Joint Economic Committee, favored selling off gold just to push the price down artificially, to punish speculators and prove that gold is "just another commodity, like pork bellies."

seven long years. Money was taken, via printing press inflation, from bank depositors and used to suppress a natural market force. The losses eventually came out of the hides of all those innocent bank depositors who never seem to anger.

The United States provided 50% of the total net gold sales by the Pool, which tends to show the geographical location of the true culprit. William H. Tehan of P. R. Herzig and Company (an outstanding gold observer, in my opinion) has done some original research suggesting that the Pool was a net buyer of gold until 1966, but for the first time in modern monetary history all of the newly mined gold went into private hands mostly for investment and speculative hedging against a possible U. S. dollar devaluation. In 1966, the Pool lost $991 million in gold. In June 1967, France wisely withdrew from the Pool amidst heavy criticism and pressure (details of which you will find discussed in the "Odyssey" chapter).

However, in November 1967 the situation changed when the English pound was devalued. Led by Lyndon Johnson's brave announcements about the determination to maintain the official price, England was secretly preparing a fallback position. On Thanksgiving Day 1967 at a central bankers' meeting in Frankfurt, Italy's Guido Carli was busily finalizing that fallback position. After losing $3.2 billion in gold between November 1967 and March 1968 in London Gold Pool operations, then Secretary of the Treasury Henry Fowler decided to call it quits. Europeans refused to accept more losses in the Gold Pool, so by March 1968 nearly all of the increasingly large gold sales were supplied from the U. S. gold stock. By March 1968 the U. S. gold stock had plunged to around $10.5 billion, due to the Treasury's unwise insistence on suppressing gold at the unnatural $35 gold price. On March 14, 1968, Fowler and Chairman of the Federal Reserve William McChesney Martin, went to the White House at 2 P.M. and persuaded the President to ask the Bank of England to close the gold market and convene an emergency weekend conference in Washington of the seven remaining Gold Pool partners. It was decided finally to accept Carli's Thanksgiving proposal of a two-tier gold price system.

These gentlemen were telling the free market that even though gold was rising on its own, they knew better than the free market and would continue to attempt to suppress the price of gold. Their failure to raise the price of gold at this point, and the staggering losses already taken, were concealed from the American public and were paid for by money embezzled via inflation from American bank depositors. This was not capitalism.

At this point, I cannot resist showing you a quote from a *Barron's* editorial on May 27, 1968 by Robert Bleiberg:

. . . the files fairly bulge with the repeated assurances of Secretary Henry Fowler and his predecessor on the imminent improvement which somehow never came in the nation's balance of payments. Again, just a few days before sterling was devalued, Secretary Fowler went out of his way to give the pound a rousing vote of confidence . . . The "two-tier" system, so the master financial craftsman recently boasted about his dubious handiwork, will endure for decades. Frederic L. Deming, Under Secretary of the Treasury, outdid his boss by predicting that it could last "till hell freezes over."

The ending of the Gold Pool, as Tehan aptly points out, was the beginning of the end of monetary cooperation among the central banks. Washington was unable to suppress the price of gold starting in 1961, and actually wheedled seven other countries to add their resources in future attempts to suppress the free market. The gold, a national treasure used to defend a non-market price, was squandered by those who refused to understand the function of price as a device for rationing scarce resources in a free marketplace. Sadly, charlatans had legal possession of the gold lever, and were using it to selfishly reelect themselves.

With the ending of the Gold Pool, Washington established a two-tier gold price, an obvious last-ditch attempt by the currency manipulators of the Washington Economic Establishment to suppress the gold price.

The generals were ordering yet another attack while their troops were already in full retreat. The United States was attempting to suppress an ever-stronger gold demand, increasingly weakening itself in the process. The two-tier system ordered all gold frozen at the $35 level, with bankers forbidden to use the free market price. Thus, there was the irrational government price of $35, and a much higher free market price. This was the beginning of the end for the currency manipulators. Gold can be pushed only so far.

In the 1960s Washington whimpered its pathetic insistence that gold needed to be demonetized, even while crisis after crisis, devaluation after devaluation, and floating currency after floating currency began to crop up in the financial news. People began to ignore Washington. More and more people around the world learned that paper money was not to be trusted, that it did not maintain its store of value, nor could it be used as a measure of value. This left gold as the unquestioned financial master of the world, and the world's ultimate money to those who would see.

The Gold Crisis Becomes Conceivable

Few understood what was happening as dollars began accumulating overseas, becoming a sort of Damoclean Sword hanging over our gold reserves, our national patrimony itself. We have seen how the first run on the U. S. Treasury (akin to a run on a bank), occurring in the fall of 1960, boosted the price of gold to over $40 in London.

Lord Keynes called gold a "barbarous relic of the past" and decades later the U. S. government blamed gold problems on "speculators." Because the free market price of gold was suppressed, dollars were permitted to accumulate overseas in vastly larger quantities than could have been possible otherwise, and this made the ensuing potential disaster much greater than it should have been. These efforts are alien to capitalism. Under a true capitalist gold standard, the United States would have lost gold, and the U. S. economy would have contracted to keep our costs competitive with the rest of the world as their industrial bases began to improve and Europeans became more effective competitors. In our pathetic refusal to accept the downs in life as well as the ups, the attempt to block this capitalist adjustment led to an ever-increasing rise up to an unsustainable height which necessarily had to come crashing down to reality. Instead of being allowed to correct itself naturally and moderately, the problem grew into a more and more threatening monster. Presidents Kennedy and Johnson were so deeply imbued with Keynesian economics that they did not grasp what was really going on, thus permitting an ever-increasing rate of inflation, steadily higher interest rates, deficit after deficit in our balance-of-payments, a steady loss of gold, and an increasingly overextended stock market.

Throughout the '40s, '50s, and '60s, the hemorrhage of dollars gushed into Europe. What started out as an effort to help a ravaged Europe recover from World War II soon became a habit impossible to kick. U. S. gold reserves were lost due to foreign aid, wars, other military commitments abroad, and tourism bloated by distorted currency relationships. We spent it, loaned it, and gave it away in ingeniously devised ways. In this chart you can see why in 1960, when dollar liabilities to foreigners moved above our reserve assets, I realized the dollar was then bankrupt. The problem has grown steadily worse since then, and has now approached terrifying proportions.

In the 1960s, people were "taught" that frugality, hard work and the avoidance of debt were not smart qualities. Borrowers who paid back in cheap dollars and speculators making quick profits were all highlighted in the financial publications of the times and it led inevi-

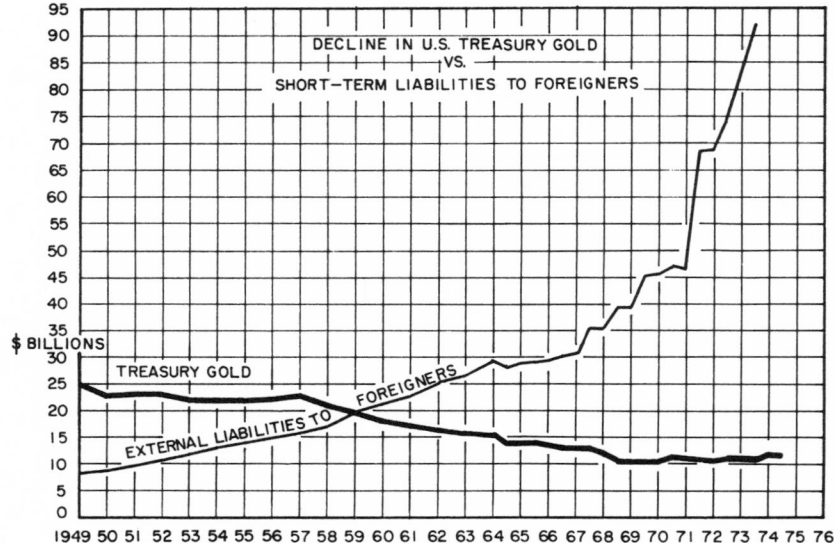

tably to the so-called "gunslingers of Wall Street" in 1968. Financial conservatives—including the publisher of *The Dines Letter*—were increasingly scoffed at or patronizingly tolerated.

Despite currency "swops," all kinds of controls, a so-called Interest Equalization Tax, "voluntary" programs, a gold pool, devaluations, import-export controls, subsidies, and an International Monetary Fund, nothing helped because the basic underlying problem of gold had not been solved. Distorted currency relationships created weird effects, and gave American tourists a terrific advantage. All the major cities such as Paris, London, and New York should cost around the same, barring unusual political phases.

In the 1960s, foreign automobiles made deep inroads into the U. S. automobile industry primarily because foreign currencies were undervalued in relationship to the U. S. dollar, which was not devalued or linked to gold so that it could fluctuate sensibly in relationship to other currencies. By 1974, two dollar devaluations plus heavy overseas inflation pushed the prices of many imported cars up 30% or more, wiping out their price edge and leaving them with a sizeable price disadvantage. The heavy layoffs in 1974 at Volkswagen and Italy's Fiat were casualties of gold, but I doubt the car dealers involved understood the reasons behind what was happening to them.

The Coda

By the time Nixon took office in 1969, elected as a conservative Republican to balance the budget and restore solvency to our government, he was facing the economic fact of life that inflation is cumulative and exponential. Like addictive drugs or pregnancy, once begun, it proceeds toward its logical conclusion. More and more money had to be printed. Nixon's first attempt to check inflation led to a credit "crunch" in 1970 which almost resulted in a stock market crash. The consequences of all the reckless borrowing, lending, and overspending of previous decades were beginning to come home to roost. The sixth-largest business in the country, the Penn Central Railroad, declared bankruptcy, and other companies, such as Lockheed, Chrysler, TWA, and Pan American Airways almost closed their doors. Brokerage houses began to fail, and interest rates skyrocketed to the shocking rate of 9% for a home loan. Looking back, it seems clear that the United States was on the brink of financial disaster in the summer of 1970.

A fateful decision was made on June 22, 1970, by an alarmed Federal Reserve Board. Just as the Fed had done in 1929 when it refused to cut back for fear of destroying the dollar, it decided to pump fresh fuel onto the fire even though inflation would return with greater force. Combined with large budget deficits over the next few years, ample funds were pumped into the system and the man in the street had no idea what had almost happened to him. Instead, a declining stock market and higher interest rates were simply thought to be a "Nixon recession." Charles Black writes in *Inflation: The Tragic Economic and Investment Consequences*, "Few people realized that they had witnessed the failure of the last gallant attempt to turn the tide of history away from our now inescapable meeting with runaway inflation and inevitable depression surpassing the one that might have taken place in 1970."

The mass printing of dollars inundating the world increased, so that by 1971 we owed $60 billion to foreigners, and this began to create tremendous pressures on our gold supply.

On August 13, 1971, Nixon received a note from the Bank of England requesting that its $3 billion in dollar holdings be guaranteed against devaluation, and two days later he slammed shut the gold window and ended all conversion of dollars to gold. This action meant that the innocent foreigners who had accepted our dollars over the years were now stuck with paper. It was an immoral action, but Nixon had little choice. August 15, 1971 ended the reign of the dollar as the strongest currency on earth. World trade was based on the guarantee of the dollar's convertibility, but now we were moving into the final phases of destructive inflation toward a depression. The United States was a great supporter of gold when it had most of it, and like a sore loser, denigrated it when its gold was about to run out. The United States began to lose the moral, financial and, therefore, political and military leadership it once had.

Foreigners tried to get rid of their dollars, but it was like musical chairs. Those who would still accept them became inundated by them, and finally they too refused to accept them at the old fixed exchange rates which were based on a fixed relationship to gold. Thus it was in 1973 that all currencies began to float freely, making long-term contracts and trade relationships difficult in the absence of some kind of currency futures market. So international trade declined, an early warning of the coming world depression. This floating at last cut us off from further European credit.

Europeans had been suckered into taking all the paper dollars they could stomach, marking the beginning of the end for the U.S. economy. Free-floating currencies led to a race to see which country could print money faster because the faster it is run off the presses the more it is in a sense devalued, which helps that country's exports and discourages imports; sort of a grand prix of competing immoralities. Instead of race horses, they used printing presses. Around this time, price inflation became increasingly obvious to the average citizen. The 1970 credit crunch restrained inflation temporarily, but resorting to the printing press in 1971 and 1972 resulted in a price explosion in early 1973.

Government refusal to treat the public like mature adults in the gold area, educating people to the necessity of facing up to recessions and a necessary weeding out of the inefficient, shows itself in other areas. In 1971, people began to demand some form of controls on prices, and instead of facing up to our responsibilities at the time, Nixon pandered to the masses, and instituted wage-price controls for the first time in our peacetime history. Nixon had often come out vehemently against wage and price controls as only leading to shortages, black markets, and even higher prices, but he nonetheless treated the American people as if they were incapable of rational choice. Perhaps he was right, and that is why we get the government we deserve. But perhaps not. No one will ever know because the opportunity passed. Shortages began to appear in the American economy after this fateful decision. Distortions of a free marketplace lead to dislocations. This is not capitalism. While some went hungry in 1972, farmers killed baby chicks rather than going to the expense of feeding and raising them, reminiscent of the massive potato burning by farmers during the first Great

Depression. Ironically, these unwanted surpluses occur at the same time that there is a scarcity due to consuming the world's resources much faster than could possibly have been done under a stable currency. We are squandering the legacy we should be leaving to future generations.

The United States is a very spoiled nation, accustomed to very cheap transportation, energy, and raw materials. Because of the Great Depression, wages have been considered sacrosanct and kept unreasonably high. Manufacturers have adjusted to the situation because the materials they used were so cheap in relation to labor. However, we are finally running out of cheap domestic natural resources and in future decades will compete on the world's open market for them. This will give us a chronic inflationary bias because we will have to pay increasingly higher prices for the depleting raw material reserves of the world. This puts long-term downward pressure on both wages and profits, out of whose hides this new source of inflation will have to come. Concurrently, newly efficient countries in Asia and Europe are beginning to compete with us in terms of labor efficiency. They, too, must compete for raw materials but their labor costs are much lower than ours. Yet to be heard from are emerging giants like Brazil, Canada, South Africa, and who knows who else.

The immorality of inflation begins to corrupt other acts of government. Just as Nixon went back on our word by refusing to honor our commitment to exchange gold for dollars to overseas holders (and just as the government refused to honor the gold contracts to our own citizens in 1933), thus it was that Nixon imposed export controls in 1973 on some farm commodities and steel scrap to hold down skyrocketing U. S. prices. This startled the world and made them realize once again that the United States would honor its word only when it was convenient. Japan was completely shocked, just as it was when the United States announced diplomatic relations with China without consulting Japan first. Such disgusting immorality of President after President in recent decades, of all political parties (and I spare no one) is predicated on an unsound currency dating back for decades. If it is truly inevitable that this system will collapse—then the faster the better. It will at least bring us closer to the day when a new and more sound economic system can be constructed on its ashes. Meanwhile a stock market and real estate crash appears inevitable.

The game lasted for a few decades, until Europe finally noticed the dollars piling up, first at their knees, then their navels, and finally their nostrils, and demanded that we slow down the printing presses. Charles de Gaulle tried to slow us down in the late 1960s, and the abuse he received was international. De Gaulle knew the only way to stop the U. S. Treasury was to demand gold for these paper dollars, a process which finally worked. Gold demands increased relentlessly despite U. S. political "pressures." On August 15, 1971, Nixon's refusal to exchange gold for paper dollars overseas exposed our bankruptcy for everyone to see. Nixon would not allow our gold stockpile to fall below the round number of $10 billion. This humiliating experience was followed by a second devaluation of the dollar on February 12, 1973, which brought the "official" price of gold from $38 to $42.22 an ounce. At this time, the free market price was much higher, which shows how far from reality Washington was from the free market. The puny devaluations of the dollar, by raising the price of gold first from $35 to $38, and then up to $42.22, can best be described as applying a band-aid to a severed jugular vein. The response was grotesquely inadequate, since the free market price of gold at the time was already around $100 per ounce, highlighting how degenerate the monetary system had become.

In the "Odyssey" chapter, more details are shown of these processes, and you can notice my language at the time—including my doubts and occasional confusions at the various smokescreens used by Washington to forestall the awful day of national bankruptcy. This includes Special Depository Receipts which, I think, are headed for the trash can of history, there to find the *assignat*, the American Continental, and the 1923 German mark.

In 1974, major Wall Street brokerage houses actually began to notice gold (miracle of miracles). It was high time, since gold shares were the only game in town, and *The Dines Letter* was becoming a bit too respectable for comfort.

Economist Paul A. Samuelson wrote in the July 8, 1974, *Newsweek*, ". . . for every two who in these last couple of years have profits to show in (gold) coins, bullion, or in South African mines shares, there are 10 who have yet recouped the losses in yield and price taken over the last decade in gold speculations . . . From the standpoint of economics—jobs, income, interest rates, inflation, lifetime savings—gold has not the slightest importance." With gold having gone from $35 to $190, I cannot imagine who has taken all those losses. Furthermore, England abandoned the gold standard in 1931 and the United States abandoned it in 1933, and since that time there has been nothing but accelerating world inflation. As *Barron's* so aptly put it on July 29, 1974, "To the extent that gold is 'phased out' of a currency system, bureaucratic arbitrariness and caprice—i.e. inflation—are phased in." Despite his popularity, Samuelson's comments are shockingly uninformed even at this late stage. In the "Odyssey" chapter you will read the chronological history of *The Dines Letters* and get a "feel" for how best to ignore those who are in error, even when that error extends to the highest government levels.

Will Our Banking System Collapse?

Banking is such a dull topic—until the place where you have left your life's savings closes its doors.

Banking did not react greatly to the inflationary crisis until the early 1960s. With our currency advantages, American business surged overseas and banks scurried in hot pursuit. According to former Federal Reserve Board Governor Andrew F. Brimmer, as recently as 1964 there were only 11 U. S. banks with foreign branches, and their combined assets were just $7 billion. Eight years later there were 107 such banks and their assets were $90 billion, a 1,300% increase. By mid-1974 the figure was even higher. This business is beyond the reach of Federal Reserve requirements.

When foreign central banks acquire dollars, they invest them in U. S. Treasury bills, which supplies credit to the U. S. economy and expands American banking liquidity. Unfortunately, when foreign currencies are under selling pressure, they sell U. S. Treasury bills to pay the seller. This reduces liquidity in both the American and the foreign banking systems. William Tehan points out that the 1970 dollar crisis saw massive purchases of U. S. Treasury bills by foreign central banks to the extent that 70% of the U. S. budget deficit had been financed by foreign central banks.

All that money slushing around the Eurodollar market is the Frankenstein monster getting bigger every year. It is now, in my opinion, out of control and will emerge at some unpredictable moment to wreak havoc on us all.

In the old days, banks primarily lent 60% of their depositors' money in demand deposits. Reserve requirements were so high against those demand deposits that expansion was limited. But "creative accounting" and other modern bookkeeping tricks enabled demand deposits to drop to less than 40% of total deposits today. So banks turned to Certificates of Deposit for money. In the last ten years CDs came from being a rarity to the point where there are now approximately $74 billion of them outstanding, and subject to only a fraction of the Reserve requirements that demand deposits must carry. Thus, any given quantity of capital now supports much greater loans. Furthermore, retail credit has ballooned and companies like Sears Roebuck extend more credit than many banks do, and are not subject to Federal Reserve requirements. Nothing is subject to the discipline of gold in the first place. According to a study by *Forbes* magazine, July 1, 1974,

"There have been nearly 300 banks formed in the last 30 months; but Fed membership has increased by only 16. At the moment, Fed membership is 5,700—about 41% of all the banks and representing about 70% of bank assets. That means $156 billion in bank assets are *outside* the Fed. That's huge money."

Lo and behold on May 21, 1974, Wall Street found its voice in the form of a fascinating study by First Albany Corporation* warning their customers who owned bank holding company stocks that there was great danger. Those interested in this facet of the problem should obtain that report.

First Albany ran statistical tests on fifty major banking companies, and came to these conclusions:

1. On average, over the 1964–1973 period, these fifty banks increased their deposits by 186%; they increased their loans by 241%; but their net worth went up only 116%.
2. The result was a deterioration in the traditional measures of bank financial strength and liquidity: the ratio of loans to deposits rose from 61% to 72%; the ratio of capital to deposits declined from 8.9% to 7.3%; and the ratio of U.S. Governments to all assets declined from 13.5% to 4.5%.
3. At the same time, although earnings were rising because of the continuing rise in aggregates, profit margins were eroding. In 1964, the fifty banks carried down 17.3% of gross income to operating earnings. In 1973, the margin was 8.9%.
4. Furthermore, especially in the latter part of the decade, there has been an explosive increase in short-term borrowings by banks, whether through commercial paper, federal funds purchased, acceptances, or interbank borrowings. We measured this by calculating the relationship of "other liabilities" (i.e., all liabilities except for deposits, loan loss reserve, and long-term-debt) to net worth. In 1971, this relationship was 198%. In 1972, it rose to 239%. And last year, it rose again to 338%.

For the top ten U. S. banks (measured by capital) the ten-year change has been even more striking. For them, on average, the ratio of capital to deposits declined from 9.2% to 5.5%; profit margins fell from 20.0% to 8.0%; and the ratio of "other liabilities" to net worth rose from 182% in 1971 to 251% in 1972 and to 372% in 1973.

The plain conclusion from this is that the bank stock owner

---

* First Albany Corporation, 90 State Street, Albany, N.Y. 12207.

today is accepting a lot more risk than was true ten years ago. Consider, for example, the impact on earnings and dividend paying ability of a bad loan experience; *first*, loans now make up a greater proportion of total assets, so even given loans of the same quality as ten years ago, write-offs have to constitute a larger fraction of gross income; *second*, as capital has declined in relation to deposits or loans, the impact of write-offs on net worth has necessarily increased; *third*, as profit margins have declined, write-offs constitute an increasing proportion of earnings. In fact, there are some banks whose gross write-offs in 1973 amounted to nearly half of pretax earnings.

It appeared to us that an increasing number of banks, over the last decade, have pursued the conglomerate objective of constantly increasing their bottom line EPS, which they sought to attain by diversification into "bank related" businesses, and in their banking operations, by lending money first and then buying money to support the loans.

3

Thou Shalt Not Inflate:
The Eleventh Commandment

Beggars mounted run their horses to death.
Shakespeare, King Henry VI

The Invisible Frankenstein

Who is Frankenstein? Frankenstein is the inflation created by man, destined to destroy his creators. Inflation inevitably spawns its own deflationary end.

Frankenstein is invisible. Even the inevitable crashes are invisible to inflation's creators. In the "Odyssey" chapter, many examples will be given of this fact, as you watch the posturing asses who got us into this mess declare that there is no mess, even at this late stage.

Whenever man created inflation by tampering with free gold, it ended by destroying its creators, as we have seen: the Greek and Roman Empires, France in 1719, the United States in 1775, and Germany in 1920. All inflations have ended the same way, and it is amazing that man has not learned it yet.

Our troubles derive from the fact that our economic system is based on an economic immorality of getting another's goods or services without fully paying for them. Inflation is evil, and when that side of inflation eventually shows, as it always has, its wrath is fearsome indeed.

Men walk around even today emotionally scarred by the Great Depression which was, after all, one of inflation's greatest triumphs.

It seems as if every politician is against inflation, yet it continues year after year. Perhaps inflation can be defined as something that all politicians are against and yet never conquer. The question cynically leaps to mind: Are they really against it? Let's face it, taxes are unpopular. Since politicians' eyes are bigger than their collective stomachs, how can they spend more without raising taxes and thereby arousing the voters' wrath? By inflation. Only governments create inflation. Inflation is addictive, and the U. S. Treasury has been the world's biggest junkie. The United States has had quite a joy ride these last forty years, but is rapidly approaching the end of the line. Inflation is a one-way ticket, and it always ends badly. Inflation is a loan to an armed robber. The inflation-deflation cycle will continue generating booms and depressions until the public figuratively hangs a U. S. Treasury official.

Now you see why politicians hate gold so much. Only gold can stop inflation, because the yellow metal is absolutely incorruptible. Gold cannot be multiplied, created, or corrupted by ambitious politicians, lobbyists, or other special interest groups.

How Did the World Get into This Mess?

Inflation, like democracy, is a vague word which can be defined in many ways. Here it means any increase in the money supply, whether from small quantities of new gold entering the system due to discoveries and mining, or any artificial creation of the money supply.

Wage and price controls are a logical solution to those who believe that higher inflation merely means "higher prices," which are really only one result of an increased money supply. Such controls failed in the United States in 1973–74, as they have always failed before. Yet, they will be tried again someday.

Money is modern society's most important commodity: the fulcrum of every financial transaction. Those who corrupt money are corrupting society right down to its most basic level, as cancer corrupts the cells of the body at the genetic level.

I grew up in an era when it was somehow chic to welcome a little bit of inflation. When the quantity of paper is increased, the value of each unit drops. At first, inflation is stimulative, like adding a small amount of seasoning to food. Inflation has been favored for centuries because it enables governments to secretly extract an arbitrarily equal amount of wealth from its citizens. It is an invisible tax. Some will benefit by it, while others will be harmed. Within inflation are invisible forces of destruction for our society. How sad that so few understand inflation and how it can do more damage than foreign armies.

Inflation really does not benefit society, but only those who get the new money first, and at the expense of those who get it next. Inflation never creates new wealth; it merely redistributes old wealth. It is an attempt to fool the people and it can only work for as long as the innocent public has confidence in that money. Loss of that confidence eventually expresses itself in a runaway flight to real value, which always ends the inflation. French President Georges Pompidou wrote, "Socialists know all about how to redistribute wealth, but nothing at all about how to create it."

"Credit expansion" adds to inflation because it increases the supply of goods and services. It occurs when a loan is made which is not offset by someone else's savings (one consumer postpones his desires and another uses this money, presumably for a solid business purpose, which will create jobs and profits). Purchasing power should be the same, but one of the curses of modern times arises when a bank creates credit by making a loan against which no one has saved. The bank lends money which is not offset by the postponed desires of another, so that suddenly there is more paper chasing the same amount of goods. Therefore, there must be more pieces of paper per unit of goods, since price is the device by which scarce resources are rationed.

Credit expansion also results in overconsumption. People buy more than they really need when they have excess paper money. This leads to ill-advised investments in industries for those with new money. What America really needs is to create new wealth, and thereby production, not more money. More investment to create real wealth eventually cuts prices and helps the workers. Inflation always ultimately hurts the workers.

Inflation comes in easy stages. Initially prices rise slightly, since the initial stages of inflation are usually small, and many postpone purchases while waiting for prices to come down. Credit expansion works without higher prices because someone else is postponing his desires so that another can use them. This can go on for many years, and most recently occurred from 1933 until the 1950s. The middle stage of inflation is when people begin to understand inflation, and anticipate it by spending money faster than the money supply is increasing. This occurs, for example, when a housewife buys a replacement vacuum cleaner now when that purchase really could wait until next year, which leads to bad investing, accelerated use of resources, pollution problems, youth rebelling against the rape of the countryside and a retreat from the "materialism" which some perceive to be infecting their elders. The third and final stage is like the German inflation of the early 1920s, when anything was better than paper money, and the headlong flight away from fantasy toward the ultimate reality begins. As the old money approaches worthlessness, either social upheaval or depression strikes. Gold is often reenthroned at this stage. Gold is the ultimate money and, therefore, the ultimate freedom for those who have it at the right time.

The Cost of Living Index since 1955 starkly reveals the nation's inflationary problems. The middle phase, the public's growing awareness of inflation, began in 1966—probably due to inadequate financing of the Vietnam war, when Johnson ran the printing presses instead of raising taxes to pay for the war. The annual rates of increase in the cost of living are as follows:

Year	% Rise in Cost of Living	Year	% Rise in Cost of Living
1955	0.4	1965	1.9
1956	2.9	1966	3.4
1957	3.0	1967	3.0
1958	1.8	1968	4.7
1959	1.5	1969	6.1
1960	1.5	1970	5.5
1961	0.7	1971	3.4
1962	1.2	1972	3.4
1963	1.6	1973	8.8
1964	1.2	1974	12.2
		(1st qtr.) (annual rate)	

U. S. inflation moved from the "creeping" stage of between 3%–5% to more than double that rate. This is nothing like the German experience in 1920–23, to be sure, but inflation is beginning to approach the 20%-per-year level which smug Americans had thought for decades belonged exclusively to "banana republics." In many countries, inflation is accompanied by unemployment and stagnant growth, hence Nixon's term "stagflation," from a combined stagnant economy and inflation. In other words, prices and wages are rising faster than jobs. I prefer the phrase "infression," or an inflationary depression, when workers get more dollars but are able to buy fewer goods and services with them.

The large U. S. balance-of-payments deficits in the 1960s generated a need to resort to the printing presses and expand the money supply. This, of course, prompted labor to demand bigger-than-usual wage increases to offset such inflation. Some people think higher prices are related to cartels, capitalism, droughts, floods, strikes, or even a rise in the price of gold, but all of these suppositions are false. The culprits are the U. S. Treasury, Lord Keynes, and the immorality of the idea of getting something for nothing.

It Wasn't My Fault!

After the coming crash, every Pontius Pilate will attempt to appear blameless in order to keep his job and begin constructing the next Frankenstein. The people who flocked to Lord Keynes in the '30s, who supported the Employment Act of 1946, who invented "pump" priming and deficit spending, who advocated removal of gold backing from our currency, who condoned deficit spending, and skyrocketing sums for "social" and "environmental" concerns which do not actually contribute to productivity, are the real culprits. They took your money from your bank accounts, bonds, and insurance like thieves in the night. The Arabs, who were content for many years to accept a low price per barrel of oil which was quietly being eaten away by inflation, suddenly threw the price on the open market and were probably shocked when the oil price skyrocketed. The consequent oil crisis probably helped bankrupt whatever was left to bankrupt in Western civilization's finances. The guilty also include those who called gold a "barbarous relic," and who advocated full employment at any price. *Any rate of inflation which is tolerated will soon be exceeded.* Once the public psychology is trained to expect more inflation, it becomes virtually impossible to stop it without an economic downturn, which politicians claim it will no longer accept. So, they will have to suffer worse.

The function of price is to ration what is scarce. The higher the price, the less people buy. But price is not the only factor. It depends on the values people have; some will pay a higher price for a certain item than others. The role of government should be as an impartial referee, deciding what the medium of exchange will be, and leaving the determination of price and value to the individual. When people lose their freedom, these decisions are made by their totalitarian rulers.

Keynes' policy of pumping money into the system to force interest rates to zero, and thereby stimulating demand with these funds which were theoretically limitless, was an abysmal failure. And it is a root cause of inflation, with the government in cahoots with the banking system in a clear case of embezzlement. America's incredible debt must be wiped out before the next sustained boom can begin. Debts now in the trillions which can never be paid off overhang our society and will have to be liquidated. Debts cannot be considered sound until the depositor postpones his material desires, and another, who needs the money, borrows it. With the highest interest rates in history and inflation approaching the runaway and ruinous stage, the bankruptcy of Keynes' "thinking" is exposed for all to see.

Inflation rates have been forced ever higher because there is no other way to service the government's enormous debt. So, inflation will wipe it out, and all the innocent government bond-holders will lose their postponed pleasures forever. This money was wasted and squandered on foreign projects, used in suppressing the price of gold, and other unwise and unsound policies. The long-term, planned embezzlement of lenders will eventually wipe out the lending class. It was planned that way. Who will the government gut next? Collapsing stock and bond markets, ruinously high long-term lending rates, and bankruptcy of cash-starved corporations, derive from that disastrous man "Lord" Keynes.

The Causal Relationship Between Budget Deficits and Inflation

> *Though it [paper money] has no intrinsic value,*
> *yet by limiting the quantity, its value in exchange*
> *is as great as an equal denomination of coins, or*
> *of bullion in that coin . . . Experience, however,*
> *shows that neither a state nor a bank ever have*
> *had the unrestricted power of issuing paper money*
> *without abusing that power; in all states, there-*
> *fore, the issue of paper money ought to be under*
> *some check and control and none seems so proper*
> *for that purpose as that of subjecting the issuers of*
> *paper money to the obligation of paying their*
> *notes either in gold coins or bullion.*
>
> *David Ricardo*
> *(written in 1817)*

Jesse Levin, in an article entitled "Budget Deficits and Inflation" (*Financial Analysts Journal*, July–August 1974), wrote that budget deficits are primarily to blame for inflation, although this is seemingly unknown to Congress and the majority of the voting public. This is logical, since deficits increase the amount of money in the hands of the public. To test this theory, Levin went back over a quarter of a century to see the connection, if any, between the years when the smallest deficits occurred, and the ensuing rises in the rate of inflation, as measured by the Consumer Price Index, either in the same or following years.

World War II created a $39.5 billion deficit in 1945, and the following year the inflation rate grew 4.6%. So, it took one year before the budget rise was translated into an inflation rise, and it reached its maximum effect two years later when the CPI rose 4.6 points.

In 1947, the budget surplus of $13.4 billion prompted a decline in the rate of inflation from 8.4% to 5.2% in 1948, and all the way down to −0.7% in 1949. Again, the effect was first felt after one year and the maximum after two years. The direction of the change in the rate of inflation followed changes in the federal budget. The 1950 budget contained a $9.1 billion surplus, and the inflation rate dropped from 5.7% in 1951 to 1.7% in 1952, again a lag of one year. But the drop continued until 1955, five years after the surplus was registered. The budget was again out of balance in 1952, but this time the impact on consumer prices took four years to culminate. Thus the pattern appears to be inflation rising or falling about a year after a change in the federal budget, and reaching its maximum two to five years later.

In 1955, the budget went from a $5.9 billion deficit to $4 billion surplus, a swing of almost $10 billion, and consumer prices reacted the same year. So, in some cases, the lag may be less than a year.

The budget shortage went to $10.2 billion in 1958, and inflation rates began to rise two years later. The 1960 surplus of $3.5 billion reduced the rate of inflation in 1961. Consumer prices did not change significantly between 1962 and 1966, again following the pattern set by minor changes in the budget.

The 1967 deficit of $12.3 billion brought the CPI up the next year and in 1970 the peak rise was 6.5%—three years later. Another surplus, this time $8.1 billion in 1969, was followed by slow inflation in 1970 after a two-year lag, with peak acceleration after three years.

From 1970–72, Congress went on a spending spree with deficits totalling $48.3 billion, so it is not surprising that inflation went wild beginning in 1973. Even without the world food shortage, oil crisis, the severing of the gold link to the dollar, or dollar devaluations, budget deficits alone would have brought on an increased rate of inflation.

The cause and effect were present in every comparison without exception, and it can be seen that the impact is usually felt within the first year and a half, on the average, and the maximum effect is felt in about two and a half years. As Levin puts it, "Under normal circumstances, increased (decreased) inflation usually follows greater (lesser) Federal budgets."

Symptoms of Distress Finally Surface in 1974

I have shown how, by short-circuiting gold, the United States consumed more than it produced. Foreigners gave us goods for paper dollars, and when they could no longer trade those dollars for gold, they took our food, common stocks and real estate. Thus, shortages began to appear in the American economy as early as 1973, representing a flight from paper money generally unrecognized by the press or the public. Americans could not retaliate by buying these same goods in different countries because this would create unemployment and loss of jobs at home. Also foreigners already had their fill of paper dollars. Removal of the gold mechanism for repayment was leading to a catastrophic breakdown in international trade.

Interest rates have begun to chip in their two cents. The rate of inflation dictates the level of interest rates. While not a close mathematical relationship, obviously the higher rate of inflation, the more lenders demand as protection. Inflation in the double-digit phase has spurred interest rates to spectacularly high levels already, which is choking off business. High interest rates are notorious boom-killers, which must lead to a deflationary crash followed by plunging interest rates. Excessively high interest rates to compensate lenders for inflation are the gut cause of depressions. The inflationary boosting of interest rates aims a dagger at the heart of capitalism. This is the self-destruct feature of inflation, and is why it always leads to deflation.

In 1974, the pincers of skyrocketing interest rates and soaring fuel prices left governments speechless. This time they could not print their way out of trouble. They had printed too much paper and bought the workers off too many times, and the inbred inflationary psychological bias was adding to its own problems.

As interest rates continued to skyrocket in 1974, those holding stocks on margin became increasingly squeezed because they now paid higher rates for the money borrowed to buy stocks. With stocks plunging because the inflationary forces were beginning to self-destruct, investors were caught between a shrinkage of assets and a colossal debt structure. Unsound debts, no longer hidden by inflation, surfaced. A stock market crash became a foregone conclusion and throughout early 1974 *The Dines Letter* ran huge ads declaring *"The Dines Letter* has never been more bearish." It took a lot of nerve placing ads in publications months in advance with that definite a position. By mid-1974, housing costs showed unbelievable increases as evidenced below.

INCREASING HOUSING COST*

	1969	% of Total Cost	1974	% of Total Cost	% Change
Land	$5,630	22%	$8,950	25%	+59%
Financing	1,790	7%	3,580	10%	+100%
Overhead & Profit	3,330	13%	4,650	13%	+40%
Labor	4,430	17%	5,380	15%	+21%
Materials	9,400	37%	11,450	32%	+22%
Other	1,020	4%	1,790	5%	+75%
Total	$25,600	100%	$35,800	100%	+40%

Courtesy of *The New York Times*

At the time the above figures were being registered, Mr. Nathaniel Rogg, of the National Association of Home Builders, offered his views:

I'm all for efforts to control inflation. You can hardly argue against motherhood. But you argue that this inflation is not responsive to the classic remedies. This inflation is geo-political, caused 55 per cent by the energy crisis and 30 per cent by food. How in hell can tight money help that?*

In mid-1974, newspaper and magazine articles on inflation began to appear everywhere. The problems were truly beginning to surface. Influential *Money* magazine, in an article entitled "The Superinflation Squeeze" (August 1974), quoted Federal Reserve Board Chairman Arthur F. Burns: "If long continued, inflation at anything like the present rate would threaten the very foundations of our society." Remember this phrase when reading the "Odyssey" chapter and that Jacques Rueff said it long before Dr. Burns thought of it. The *Money* article pointed out that in current dollars the typical U. S. household had more savings and income than ever, but after inflation the real value had declined after a steady twenty-year rise. Bank accounts, bonds, insurance cash value, and other such "safe" savings lost value, while debt payments soaked up an increasing share of the household dollar. Families increased their credit buying at ever-higher interest rates, a sure-fire recipe for financial disaster.

Borrowing became a fantastic buy when the government "jawboned" interest rates below what they should have been (after deducting inflation). This was done at the expense of the moneylenders. This immoral contract overstimulated borrowing and got people in debt way over their heads, setting the stage for a credit collapse. *Money* concluded the article by citing the woes of inflation and urging people to comparison shop for loans. This advice is always good, except when inflation gets so rapid that it corrupts all desire to save or to be frugal.

Money quoted Milton Friedman, the "conservative" University of Chicago economist, as suggesting "indexing" the U. S. economy so that wages, taxes, and interest rates would automatically rise or fall with the cost of living. Indexing must fail, for there is no substitute for sound money. Some 5.1 million U. S. workers already have cost of living escalator clauses in their union contracts but this is at the expense of those who do not have union contracts. We are not all members of a union. Hence, it is the same old game of getting something for nothing at the expense of someone else: just the Frankenstein monster with a new face.

The most interesting aspect of the *Money* article is proposed legislation, introduced by Senator James L. Buckley (C-N.Y.), to index the Federal Income Tax. If the taxpayer received extra income which did no more than keep him abreast of inflation, he would not be catapulted

* From *The New York Times,* May 22, 1974.

by that increase into a higher tax bracket. Also, the current personal exemption of $750 would rise proportionally with an increase in the cost of living. Buckley is turning the inflationary dagger right back at those who are wielding it. But two wrongs do not make a right. The cure is to get rid of so-called "progressive" tax rates and the true source of inflation.

What troubles me most now are the preconditions of social breakdown cropping up everywhere. Pan American World Airways, in full-page newspaper ads, urges, "Live today. Tomorrow it will cost more." In San Diego, the CNA Financial Corporation's home building unit advertised, "Buy now. Or pay later." Indeed, all is not well in America. The instability built into this system, when associated with pessimism, could yield anti-intellectual and anti-scientific attitudes. The threat here is that society will seek irrational ways to solve problems, as Germany did in the 1930s. The possibilities of increased riots and strikes, an atmosphere of living more "for the moment," an increasing lack of real values, and growing pride in selfishness all add to the stress on individuals as they lose their sense of predictability and order, which a sound currency could have provided. Some react by living for today, others become more careful, still others become depressed and overwhelmed, while others again will be seeking scapegoats.

Underneath all this is a rising tide of resentment, leading to suppressed anger and frustration—socially dangerous ingredients. People on fixed incomes who have worked for pensions for decades are having this security knocked out from under them by a stock market crash, and this must be creating extraordinary social pressures. Frankenstein's insidious and cancerous tentacles will eventually reach to every last cell. According to Dr. Abraham Jackman of Chicago's Ridgeway Hospital, "Last year it cost $22,000 to maintain the average drug habit and today it's $29,000," which means more crime to raise the money for drugs. Even vice is taxed by inflation.

In the last half-century, world attitudes have changed. Equating themselves with gods, no longer needing to be concerned about their daily bread, electorates insist they no longer "tolerate" high unemployment, depressions, and stagnations. In fact, the public demands high growth consistently, not understanding the sine wave, the Yin Yang, the inner harmony of the universe. And politicians spring up to pander to the masses. Because of this new rigidity, downward movements of wages have become "politically impossible." Even if the workers were willing to accept lower wages, skyrocketing prices would not permit it. Due to government assistance, the unemployed have put little effective downward pressure on general wages, and so, cost levels. Therefore, price levels have a built-in upward bias, which is uncontrollable under the present system. Society is so riddled with inefficiencies, which should have been rooted out long ago, that it is difficult to know where to begin. The golden goose is being killed, and there is no hope of improving the system peacefully or voluntarily. To remove excesses like union featherbedding which were forty years in the making, perhaps what is required is an economic smash, which, although it would be regrettable, would at least purge the system all at once. It cannot be effected gradually. In other words, we are suffering from a cumulation of all the mistakes since the New Deal. There is too much government, too much paper money, unions are too strong, and wages are too inflexible, and therefore there is not enough left for profit and reinvestment. All these failings will have to be corrected before the United States can once again move on to great new heights. I am confident it will do so in the 1980s—unless our next political system aborts it.

The Painful Cure

> *But some worried observers fear that much worse lies ahead. Ashby Bladen, senior vice president for investments of the Guardian Life Insurance Company of America, says:*
>
> *"We have now reached a point at which our economy could not support the existing debt structure if its real burden were not being steadily reduced by accelerating inflation. In hard, practical terms, this means that any significant reduction in the availability of credit is likely to produce massive bankruptcies of overextended people and businesses, leading to a deflationary collapse."*
>
> *Mr. Bladen plainly says he thinks that crash will come. Given the current rate of inflation and the enormous accumulation of debt to finance it, he says, "A return to either price stability or financial stability without an intervening crash appears to me to be practically impossible. And the longer the crash is postponed by continuing the inflationary process of excessive credit expansion, the worse it will be when it does come."*
>
> The New York Times
> *May 22, 1974*

There is little chance any U. S. government will voluntarily adopt the economic policies necessary to remove the large amounts of inflationary purchasing power pumped into the economy in recent decades. This would involve gradual deflating. Unfortunately, further inflation will aggravate present economic trends, with or without price controls. This inflation will result in further severe losses to holders of savings and life insurance. The enormous amount of debt, a lack of adequate business capital, excessive borrowing on land and securities, and installment borrowing by individuals on an unprecedented scale make the currently unstable situation intensely vulnerable to a depression. In my opinion, an international crisis justifies the immediate placement of substantial funds abroad before exchange controls, which could come at any time, are imposed.

All paper currencies have been, and are being, steadily degraded by inflation. All leading currencies have lost at least two-thirds of their pre-World War II buying power, and the trend toward worthlessness is clear. The world needs a monetary system which will elicit confidence. Instead it is being inundated by a continuous flood of paper "money," which is not a reliable medium of exchange for daily transactions; nor does it offer assurance that it will retain future purchasing power when used as a store of value. Lack of confidence is increasing at a frightening rate. Eventually politicians will have to acknowledge that paper is not a satisfactory basic unit of a monetary system. With central bankers as accomplices, politicians have been buying their reelection with promises and easy credit, and express their gratitude by embezzling the savings and life insurance of the people who elect them. Sound money will return someday, but bitter lessons will have to be relearned. People will turn increasingly to gold, which through the centuries has proven to be unsurpassed both as a standard of measurement and as a store of value. The more desperate the situation becomes, the more politicians will try to tax whatever they can get their hands on (which is bad news for real estate because it represents the ultimate in visibility), and the more citizens will rush to invest in less-visible gold as a means of retaining their wealth and avoiding the grasp of the tax collector.

This is the time to disperse your assets, spread your political risks and aggressively reduce your property holdings. Spread out into different currencies with heavy gold positions, with the majority in Switzerland. How tragic that everything in this world is increasingly measured by money, which is currently the most unstable standard in the world.

The coming economic stress will be profound, as the fantasies of several generations are smashed. Politicians are accustomed to an unlimited amount of money, so it can be expected that they will try various methods of raising money to continue their spending proclivities which, of course, will be the last thing to go. Authorities might even recall the good old copper penny and attempt to exchange that for paper pennies.

Politicians are not yet aware that their "secret tax" of inflation is suddenly disappearing. They will even try slowing inflation down, but governments are too weak for the Draconian measures which would be necessary at this late stage. Furthermore, there will be a painful reexamination of depreciation allowances which, in view of inflation, are currently inadequate to replace capital. This source of embezzlement for the inflationary process will definitely be closed off, but the accounting profession will go through considerable agony before it figures out how to do it. Finally, governments will no longer be able to float loans at yields that do not allow for current rates of inflation. They will find it increasingly difficult to borrow more money, so if they decide to reinflate, look for incredibly large federal deficits at first, which could unleash the final and most explosive inflationary phase.

It will be hard for people to believe this but, via inflation, their own government did them in. Practically on a daily basis in 1974 people saw rising prices in grocery stores, as they received fewer goods for their dollars. A full-fledged panic away from paper money could start.

This generation is finally learning that inflation is an enemy, the great destroyer of freedoms and societies, just as all the other generations did in previous inflationary cycles. Now Frankenstein is at such an advanced stage that a collapse and depression will probably be mandatory to absolve us of our past economic sins. Business is going to realize what is meant in this following passage from the *American Institute for Economic Research*,* one of the earliest and soundest observers of the gold situation. Colonel E. C. Harwood writes:

> A modern industrial society cannot be maintained without a sound accounting unit with which to calculate long-term depreciation schedules for business and long-term contracts of all kinds, and to record savings, life insurance, and pension funds in real rather than fictitious values. The tremendous capital investments required for a modern industrial society cannot be maintained, replaced, and enlarged without correct instead of erroneous or false accounting. Accounting in terms of paper units has become a fiction incapable of providing the information essential to the operation of large-scale and long-lasting enterprises . . . Of all the possible accounting units that men have used, gold has proven to be by far the best.

To me, it is obvious that we are at a great historical turning point, like 1776 or 1929, and current heroes will be future goats. The New

---

* Great Barrington, Massachusetts 02130.

Deal will probably be repudiated, and John Maynard Keynes will be relegated to the trash can of history. Nearly all of the world's leaders now believe in currencies unbacked by gold. This will change.

When people see gold and silver standing alone amidst the economic ruins, they will realize that we gloom-and-doomers were actually right. Hopefully, eternal optimists will pay more heed to warnings the next time around. This generation will be scarred by the memory that Wrigley abruptly raised its chewing gum prices from 10¢ to 15¢, and *The New York Times* from 15¢ to 20¢. Figure out those percentage increases.

It will be gradually realized that the roots of inflation and the ills derived are political rather than economic. Ideas such as "indexation," which are nothing more than price-escalator clauses in contracts, will be attempted. Milton Friedman likes this idea because, he says, it reduces the money governments get via inflation and therefore reduces their incentive to inflate. Not at all. First of all, not everyone will have such clauses, so non-union workers will bear a greater burden of this invisible tax. Furthermore, indexation institutionalizes inflation and makes it a fact of life, and there is simply not that much time left. The United States would speed up inflation to make up for lost revenues, then twist statistics around to prevent escalator clauses from protecting the workers while they printed paper money. If escalator clauses rose adequately there would be no point in indexing. It would be just one more sophisticated device to cheat the people who are already being skinned by higher prices and higher tax brackets.

This is the end of a way of life, the end of an era, a watershed. Automatic wage increases followed by automatic inflation to wipe out these increases are an idea gone with the wind. To pay for the higher prices of raw materials, power will shift from the industrialized nations toward those with real in-the-ground wealth such as the Arabs, so the price of gold will have to be raised to pay for oil. When fuel was cheap we fouled our country. Using cheap transportation, people spread out from the cities to the suburbs, which led to urban decay. Steel needed for shelter was squandered to produce snowmobiles which belched more carbon monoxide into our already foul air. A great deal will come out of the restructuring, finally, of the proper price relationships of everything. The United States of course will survive. We will be punished, not destroyed.

The crash *The Dines Letter* has long predicted appears to be starting. Wall Street is so upset at how low Price/Earnings Ratios are that it refuses to understand that those low P/E Ratios are a terrifying warning that earnings simply are not going to hold up. After all, Wall Street does not give away its wares. This suggests that inflation is apparently near its climax. This could lead to food riots. The United States might turn in its green currency for a purple one, the government could force bank vaults open or otherwise make your assets inaccessible. Or worst of all, like Germany in 1920 when the middle classes were wiped out, the disillusioned masses could turn to another "man on horseback."

The American people veered sharply toward more government after the 1929 Crash. It remains to be seen whether there will be an additional turn in that direction now, or whether we will turn the other way. Perhaps an insecure people will feel that only a more authoritarian type of government can properly allocate income and resources. This concept would be alien to Western democracy, but the process of radicalizing society has already started. Hopefully, if there is a future world monetary order built soundly perhaps on the "Dines Plan," which involves permitting everything to float, letting money manage itself, and keeping everybody's hands off the gold lever, there might actually be a better balance than now.

Every depression offers opportunities. Few will see the opportunities and will, instead, negatively cling to the gloom through which I have been suffering for over a decade. My gloom is now ending, and I am looking forward to a great boom on the other side of the valley.

There will be a massive redistribution of wealth. The rich are usually able to protect themselves against inflation (a form of economic imperialism), using fiat money as their tool. But I do not think this crash can be ridden out by the rich. The next bull market might involve new leaders such as coal, rails, and cruise ships. On the other hand, declines could be seen for the automobile, pollution equipment, suburban real estate, oilwell drilling equipment and snowmobiles. No one knows for certain.

> *There is a tide in the affairs of men, which, taken*
> *at the flood, leads on to fortune; omitted, all the*
> *voyage of their life is bound in shallows and in*
> *miseries.*
>
> *Shakespeare, Julius Caesar,*
> *Act IV, Scene III*

I have shepherded my subscribers into gold- and silver-related investments on the grounds that they are the best stores of value, and that capital gains are only incidental for this hedge against both inflation and deflation. The goal here is to survive, to have buying power at the end of a major market crash. The people who will ride up the crest of the next wave will be those who survive. I hope I will be able to lead my subscribers out of the golds and silver-mining shares near their peaks, and back into the stock market within the next few years.

There will be repeated monetary crises until the world turns back to gold. Until then, I am bearish on all stock markets. There will be periods of rising stock prices, to be sure, but *the main trend* will be down until gold is reenthroned. This is not a good time to borrow or lend. It is not too late to liquidate your assets and move into the precious metals, spreading out geographically. Sooner or later, the world will stumble onto the solution of a much higher gold price to wipe out the mountain of paper printed in the last four decades. Hopefully, there will be a return to some form of gold standard, and the dollar will once again become convertible into gold. But no one knows what the next economic system will be. It certainly will not be the one which is about to be discredited. I hope people will realize that what we need now is to turn the clock back, to what made this country great, not forward to what has been dragging it down. All paper will weaken in relation to gold, and those who relied on paper such as pension funds will be wiped out—a tribute to the folly of building houses on quicksand. As the price of gold skyrockets, everything is plunging against gold—art, land, antiques, stocks, bonds—while the number of pieces of paper currency to buy an ounce of gold is soaring. The price of gold has sextupled from $35 to nearly $200 in the last few years, handsomely rewarding those who saw the light I unflinchingly held aloft.

There could be political chaos. Governments will change as they attempt the same, tired, old Keynesian remedies, all of which, I must inform you now, are inevitably doomed to failure. Debts will be liquidated, land prices will come back down to reality, and if gold is made the core of the new monetary system, I will at last begin to feel exhilaration, after my many years of personal agonizing over the future.

Perhaps the inherent immorality behind planned, conscious inflation is at last being recognized by the American people, if only on a subliminal level. Perhaps Watergate was the beginning of mass revulsion against general immorality and apathy. One would hope so.

During this next crisis and to avoid another depression in the future, the American people will have to realize that the sole responsibility for inflation rests on government. Not only is money printed too quickly because there is no gold discipline, but costs are held artificially high by government in transportation, construction, farming, oil, power, and dozens of other fields. Some interstate truckers are prevented by the Agriculture Department from shipping some foods from one part of the country to another. The Jones Act restricts coastal shipping to American vessels. The government buffers sharp shifts in livestock prices, Presidents demand "voluntary" limits on what other nations can send us, and Congress has set compulsory import quotas on food and other items. The Davis-Bacon Act fixes wages paid under government contracts. All of this represents hardening of the arteries because it eliminates the flexibility, the give and take, the natural competitive play so that some can fail. The moment governments prevent failure, they guarantee failure of the whole system; it's just like thinking that if the human body could repair a single vein instead of letting it get hard, the entire body would not decline.

4

The Function of Gold:
Capitalism's Pituitary Gland

It is an interesting historical footnote that the private ownership of gold was banned by Lenin, Hitler, Mussolini, Mao Tse-tung, and Franklin Delano Roosevelt. The moment citizens are no longer allowed to own gold, let the future note how quickly their freedoms will be lost. The loss of freedom is indivisible. It is a chain reaction—the loss of the first freedom guarantees the loss of the last. Or, as was stated with deceptive simplicity by Garrett Hardin, "You can never do merely one thing." If this makes some accuse me of an anarchism harking back to Walden's primeval pond, so be it, but economic misery will increase for as long as my warnings are ignored. Too much government is the root of all of our evils and only the merciless discipline of gold is our primary salvation.

Collectivist Lenin sneered at gold and said, "When we conquer on a world scale, I think we shall use gold for the purpose of building public lavatories in the streets." However, he did try to sell Soviet gold at the highest possible price on the theory "When living among wolves, howl like wolves." Yet, hostility to gold has deep roots, much of which probably derives from socialist origins. Perhaps only socialists understand that gold is the policeman, the superego of our collective economic minds. Gold is the governor of governments, the conscience of Treasury officials, the common denominator of all financial transactions. Considering the fact that gold represents the ultimate economic discipline, it can be expected to receive a rather negative reaction whenever it confronts any politician who favors loose or irresponsible fiscal policies. Few people truly understand gold because the powers that be do not teach it. Ignorance aids their name-calling campaign against gold, and they invent ingenious arguments against it. Some have said gold is a dead item, that it is dug up from a hole in South Africa and reburied in a hole in Fort

Knox. They assert it costs money to store, insure and protect, earns no interest, and if the price is fixed, an investment in gold is vulnerable to inflation. Politicians say it can only enrich Russians, South Africans, and European "speculators." They claim devaluation would hurt other nations holding dollars overseas, damage their reserves, and aid our enemies. There is no limit to the ingenious ways they find to put gold down. Politicians must fear it indeed, and the "Odyssey" chapter gives many examples of this.

Yet, as our monetary problems snowballed, the public did not know whom to blame for the mishandling of gold. They were left with a deep sense of frustration, and did not remember that Lenin, by subtle debasement of Western currencies, urged the downfall of capitalism.

Repeated monetary crises is the only weapon gold can use to teach its lessons, just as a malfunctioning pituitary gland would issue its own signals. All the anti-gold propaganda in the world has not stopped gold.

In view of this, it should not be surprising that C. L. Sulzberger wrote in the July 14, 1974, *New York Times*, "Monolithic Marxist systems are free of today's ominous curses . . . famine among the poor, inflation among the rich, diminished energy supplies for everyone." What frightens me is that people could conceivably believe such rubbish. Alice Widener, publisher of *U.S.A.* magazine, aptly pointed out that Russia and Red China do not have a mass fuel consumption problem because neither one has entered the automobile age, there is no national network of roads and filling stations, and no large-scale commercial air travel. Nor is there free consumer choice. People accept instructions about what they are permitted to own, use, wear and eat, where they work, and for what pay. The ruble has been devalued repeatedly but there is no free market to measure inflation. Furthermore, only massive wheat purchases from capitalists have staved off famine in Russia and China. One of my theses is that the world is about to make a major change in its political systems because of an economic crisis. I am not certain yet in which direction it will be, but I have a mixture of both fear and hope. I hope it is not Sulzberger's path.

> *A gold country is like a bank. Its first responsibility is to itself, for the integrity of its money, its credit and its assets, and if it suffers this imperative to be overcome by a sense of responsibility to others, no matter with what intention, it will fail in its responsibility to others because it has forgotten that first responsibility to itself.*
>
> *. . . The value of gold is arbitrary; so is the length of a yardstick. But just as it is necessary to*

*sell cloth by the yard or coal by the ton, so it is
necessary to have some arbitrary unit of value in
which to price the yard of cloth and the ton of
coal. It would be ideal to have something of absolutely invariable value in which to price them.
But, there is no absolutely invariable thing in the
world. The relative constancy of gold supply, the
durability of the metal, the fact that over the centuries the amount of human exertion necessary to
get it out of the rocks changes very slowly—for
these and other reasons gold is the least unstable
thing man has found for purposes of money, hence
his preference for it . . . We know about ourselves
that we have seizures of ecstasy and mass delusion;
that again a time may come when the temptation
to throw the monetary machine into wild motion
so that everybody may become infinitely rich by
means of infinite debt will rise to the pitch of
mania, as it did for example in 1928 and 1929.
With this intelligent knowledge of ourselves we
make bargains beforehand with reason; we agree
that money, credit and debt shall not be inflated
beyond a certain ratio to gold, under certain penalties such as we shall be very loath to pay and yet
such as we cannot refuse to pay under worse
penalties still.*

Garet Garrett,
The Bubble That Broke The World, *1932*

Time and again politicians swore that gold would be demonetized,
that it would be made just an ordinary commodity, that the king would
be dethroned. It has not happened. Presidents Kennedy and Johnson said
gold was not necessary for backing our currency and should be demonetized because they understood that gold is a brake on the printing press.

*Gold is the only way to stop selfish politicians from buying their
reelections with your money via inflation.*

Gold has been used as money since the beginning of recorded time,
and probably before that. Virtually everything has been used as a medium of exchange, but gold is the only material which has not been ultimately rejected. Credit is used, but that only works when it is backed
by gold. Even Communist countries, knowing their paper is worthless
outside their borders, make payment in gold for overseas purchases.
Gold satisfies the demand for the perfect: that it be indestructible, an

element, so that it is homogenous, easily divisible, portable, clearly recognizable, and stable in value. Gold fulfills all these requirements. All the gold mined between the discovery of America and the start of World War II only fills a cube as tall and wide as a four-story building and is worth about $200 billion at current prices. With the U. S. Gross National Product at around the $1 trillion mark, gold is obviously valuable not only in its quantity and its price, but as protection against man's own follies. It is the regulator of the economic body, like the pituitary gland in the human body.

Money is a medium of exchange, a store of value, and a unit of measure. Every contract is based on money, so if the money is corrupt, commerce winds up with terminal cancer. Money is the foundation of nearly all of our economic decisions. For a stable society we must have a stable currency. Without it, capital formation is inadequate and it is difficult to engage in the long-term contracts which enhance the values to which we aspire.

Gold fits all our requirements for economic stability. It is as if this element were given to us for that express purpose. Yet, the anti-gold fanatics have been so successful that gold has been overlooked by many nations for decades.

Politicians warn against "enthroning gold" and instead appear to enthrone the dollar. Yet, gold is a better king than the dollar precisely because it is impartial, indestructible, universally acceptable, and no politician can erode it by printing more gold, though they have tried since the days of the alchemists. Worst of all for greedy politicians, gold affords the people a chance to avoid the evils of inflation. It frustrates the immoral government leaders who ostensibly dedicate their lives to gutting moneylenders on the mistaken "Robin Hood delusion" that those who lend money have too much of it, and therefore it should be stolen from them and given to some mystical concept like the poor. Their goal is not really to help the poor because there are better ways of doing it. Some demand that the people trust their government. If so, the people should be trusted to own gold, rather than being scolded like naughty children for wanting protection from unscrupulous politicians.

On a gold standard, the money supply would grow gradually and steadily as new gold is found, and be insulated from politicians' "needs" for more money via the printing press. The same policies which led to the Great Depression have been followed in recent years. I have revealed the shocking parallels between the early part of this century and recent years. When England fixed its gold parity in 1925 at the pre-1914 level, it was a major factor in disrupting the international monetary system, leading to the Great Depression. When the price of gold is fixed too low, monetary upheavals are inevitable, and economic chaos must ensue.

Those who believe in capitalism must learn to understand their own system and not be defensive in the face of communism's relentless criticisms. Gold is the heart of capitalism and, until it is understood, capitalism, and therefore freedom, is in grave danger.

As laughable as this will sound to some, a gold standard would resurrect traditional values of frugality and thrift, qualities which helped make our country great. A gold standard would restore respectability to bonds, life insurance, pensions and long-term contracts. A stable currency would rid our economy of its incessant drive for higher wages and prices and its self-defeating, dog-chasing-its-tail syndrome. There would be more tranquillity in the land, more honesty, and less government. If these somewhat idyllic conditions prevailed, the moment the government began printing too much paper, an informed citizenry would cash their paper in for gold and, as the gold gushed out of the banking system, the government would be forced to call a halt to such policies. Thus, my book's point is: *I want power given back to the people.*

In 1933, Roosevelt tried to correct the previous inflationary mania by starting yet another one. A gold standard would have stopped him. However, he was allowed to spay the gold standard, and forty years later the capons are coming home to roost. Consider the horrendous confusion of two-tier currencies, the artificial ruble block, "hard" currencies, SDRs, Eurodollars, swops, and a complete hodgepodge of barter and gold. Runaway inflation is a symptom of these policies because no one any longer knows what money is worth.

Our cockeyed monetary system has been vulnerable to every nut with his own plan. Yet, all of them refuse to recognize the only solution which could possibly work: a completely untrammeled and free gold price. One of the better descriptions I have read of the function of the gold standard was published on September 21, 1964, by the aforementioned American Institute for Economic Research. It is a description of Roosevelt's efforts to raise commodity prices in the 1930s by raising the price of gold.

> The attempt to raise prices by Government authorities indicated that they did not understand the triangular relation that has long existed among gold production, purchasing media, and commodity prices. To describe this relation briefly, the quantity of gold produced varies inversely with the cost of producing it. Rising prices of commodities other than gold including wages and equipment used for gold production reduce such production, relatively if not absolutely. Inasmuch as gold not used in industry nor hoarded becomes purchasing media, the total amount of purchasing media in use or its rate of increase is reduced when gold production is re-

duced or curtailed by rising prices. Thus, the attempt to raise prices during the mid-1930s had the undesirable effect of relatively reducing gold production and the amount of purchasing media derived from it. The low level to which commodity prices had fallen early in 1933 had stimulated gold production, which was increasing the amount of purchasing media in use and which tended to raise commodity prices, albeit slowly. Attempting to raise prices rapidly by inflating or creating excess purchasing media tended to discourage gold production, especially when the rise in prices became great during the 1940s. In short, the notions for alleviating the depression that were tried were unsound, because they did not take into account the function of gold in a monetary system.

In the early 1930s U. S. Government authorities lost sight of whatever understanding they may have had of the beneficial functioning of a full gold standard and took steps that hampered its functioning. Acting on the mistaken notion that raising prices would alleviate the Great Depression (which was the aftermath of a speculative mania that not even the gold standard could prevent), the Government rescinded the right of citizens to redeem currency in gold in 1933, devalued the dollar in 1934, and began deliberately to incur budget deficits. The banking system was virtually forced to create excess purchasing media with which to finance such deficits. Never before had the U. S. Government attempted to correct the aftermath of the issue of excess purchasing media by augmenting such an excess. The experiment was a costly failure; average unemployment remained above 14 percent for the unprecedented period of 6 years, and production remained below its long-term trend for that unusually long period. Had the gold standard been permitted to operate, the distortions that had been introduced into the economy by the speculative excesses of the 'twenties probably would have been removed and recovery could have proceeded much more quickly. Such, at least, had been the Nation's experience in the past.

In other words, when gold flows out of a country, its economy deflates. As prices go down, its exports become more competitive in world markets, and soon the balance-of-payments is equalized. Conversely, a gold inflow inflates a country's economy and its exports drop, while imports rise. Theoretically, the process is beautifully self-adjusting, and involves no more than minor ups and downs in domestic economies. *This is the way to get rid of big binges and big depressions, in my opinion.* Yet, despite how badly politicians mismanage the economy, they refuse to relinquish control. Thus, it can be seen that depressions are a result of selfish political power struggles.

It seems inconceivable that an Administration which cannot even handle the complexities of managing the U. S. postal system should express confidence in its ability to "manage" the vast complex of international money markets without a tie to gold or free markets.

The unbelievable decline in the value of the English pound in recent decades would have been thought impossible by Englishmen only fifty years ago. The dollar is also on its way downhill. The destiny of the United States, and therefore of the world, is tied to the destiny of the dollar. Our currency has been willfully and knowingly debased. Americans have not only voted for it, they have insisted on it, and the press and politicians have endorsed it. The net result will be horrible. Yet, since the United States would be a prime beneficiary of a gold price increase (since the United States has more gold than any other nation, including South Africa and Russia), I can only conclude that politicians are more concerned with reelection than the welfare of their country. That is why the end will be grotesque.

Some feel that higher gold prices will help South Africa and Russia. This argument is weak. First of all, the free market is already giving them higher prices. Secondly, greater wealth for those two countries will in no way make them more tolerable to us. In fact, their prosperity might actually help strengthen the hand of the doves in Russia and aid the progress of blacks in South Africa. At any rate, it is ridiculous to compare their benefitting by a higher gold price against the collapse of the world's economy because of a stubborn refusal to raise the price of gold. All paper currencies end alike—in a loss of purchasing power, devaluations, and devastating depressions.

People might, only now, be opening their eyes to the enormous importance of gold, to the fact that it is the leash, the discipline, on a stable international monetary system. It leads to stability of world financial sysems, since there can be no arbitrary creation of money. All nations accept gold as final, definite and simple. When you toss a gold coin on the counter to pay for something, its solid ring sends credit scurrying to its vermin-infested quarters and inspires confidence all around. Disraeli once said, "Confidence is suspicion asleep," to which I add, "Free gold is confidence." Tampering with gold is insidious because its negative results can be so long postponed.

The currency is rotten, so we must return to gold, the ultimate currency. There can be no end to monetary crisis until the United States corrects its balance-of-payments deficit and balances its national budget. These must be made targets of urgent national priority, enabling gold to enforce it with dictatorial and incorruptible finality.

So it can be seen that the true function of gold is as an abstract device facilitating exchanges between individuals or nations. There can be no fixing of parities between currencies, and no International Monetary Fund is needed, contrary to what the public has been led to believe. Our motto must be "In Gold We Trust."

In my opinion, climactic events could lead, in the next few years, to a massive depression which could wipe out the debt accumulated by short-circuiting gold, and the resultant international strains could possibly lead to war (economic or shooting). All currencies will float more or less freely against one another.

Those who think gold will be demonetized and replaced by paper (to free man from gold's "arbitrariness" and enthrone the reason of human judgment) will learn there are no utopias here on earth. Gold will emerge as the next currency, as the only standard of reference indisputably accepted by Russians, Chinese, Americans, Europeans, Arabs, and so on. Only a police state can force a people to accept a paper currency instead of gold, because gold is the ultimate freedom. People will be driven to gold. They will have learned that without it the consequent inflation is only curable by painful depression. Unfortunately, deflation often yields totalitarian governments, so the future must be looked at with some trepidation. A big showdown over gold is already on the horizon because governments will realize that a higher gold price is needed to write up their assets to pay for higher oil prices. In time, due to basic capitalistic laws of supply and demand, oil prices will also come down and glut the world for years to come, hard as that is to believe now.

Since 1961 I have unflinchingly predicted a massive devaluation of the dollar. Despite two trivial devaluations, I do not feel my prediction has been fulfilled. The dollar is worth a fraction of what it used to be in terms of gold, and the price of gold will have to rise much more to reflect this. Since no nation would tolerate America's unilateral devaluation, there will have to be an international devaluation, or a radically higher "official" gold price. Thus, in terms of exchanging dollars for other currencies, people might not notice the difference. But when the time comes when they want to get out of paper and into gold, they will see that they are getting much less gold for a handful of paper. The missing difference will have gone to pay for spiraling government debts of recent decades. What will be missing will be your share.

Some will say that the higher gold price "caused" inflation. Laugh at them and give them this book. It is just one more distortion of the truth. Those who held dollars abroad also will get less gold for those dollars. Dental work and jewelry will go up in price. Otherwise, though, prices would not double, or be cut in half, no matter what the price of gold, because there is now no connection between paper and gold.

Internationally, devaluation will hurt those nations that used dollars as backing for their own currencies. They will not make that mistake again for a long time. On the other hand, gold-producing nations such as South Africa, Canada, the United States, and Russia would be helped inasmuch as they will be able to sell their domestically mined gold internationally. Look for a big boom in gold mining towns and all the attendant industries, including the chemicals in the processing of gold ore, which will be a great opportunity for the alert.

There will be considerable misinformation published when the gold crisis enters its final stages. Regard all comments cautiously, because few Americans are accustomed to thinking in terms of gold. The national gold stock will be worth more dollars, and the U. S. government will consider this a "profit." That profit belongs to the American people, so complain loudly and clearly if they give it away or spend it. That is the national patrimony backing our currency.

Some distortions in our society will disappear when currencies are finally in proper alignment. For example, aluminum and cement require a great deal more energy to prepare for construction than wood. Transforming bauxite into aluminum especially requires a lot of electricity, and higher prices there could lead the construction industry back to more abundant use of wood. Those who make their living from the wood industry probably never knew they were victims of the gold crisis, but they will later find great relief. Our tragic misuse of natural resources will subside. Many of our present ecological problems will be corrected when costs become more realistic.

Make no mistake. The road to a much higher gold price is loaded with booby traps, and will be fought tooth and nail. It will be said that a large dollar devaluation would be immoral, a repudiation of so many promises made by Americans. Yet, these same people are enthusiastically in favor of government bonds which goldbug John Exter of the American Institute for Economic Research describes so elegantly as "certificates of guaranteed confiscation practiced on innocent individuals living on fixed incomes." We truly get the government we deserve. Moneylenders responded to this immoral inflationary confiscation of their loans by demanding higher interest rates to compensate for what they were to lose through inflation. When this process proved inadequate, they tried to escape inflation by investing in land, art, and philately, by moving money to Switzerland, or by spending their money, since there was no way to store it. (This is much like someone who eats ice cream rather than keeping it for tomorrow because he does not own a refrigerator.) The overconsumption of our society will moderate and, who knows, excess weight might become less of a problem in the United States.

Some will believe that if we pitch in and cut our prices in half, foreigners would spend those excess dollars here and the gold problem would disappear. Sorry, but this will not work either. If we did that, foreigners would take everything and there would be starvation in the streets. Too much paper has been printed in the past, and will have to be wiped out no matter what.

We will have to earn our way in the world to eliminate our trade deficits. Then any surplus can be treated as foreign aid or charity, or whatever else the American people wish, including saving or investing it. In no event should this lead to a balance-of-payments deficit more than temporarily.

In the "Odyssey" chapter is the story of Italy's economic plight. On June 12, 1974, Italy was allowed to ignore the U. S. "official" price of gold of $42.22 and revalue its $3.4 billion gold reserves on the basis of $150 per ounce. Thus, Italy's reserves leaped overnight from around $3 billion to $14 billion. This lesson surely was not lost on other nations. I predict that as more countries threaten to go broke, they will pull the Italy caper and score an end run around the U. S. Treasury.

On August 14, 1974, President Ford signed Senate bill No. S2665 permitting Americans to own and trade gold bullion on December 31, 1974, for the first time in forty-one years. This profoundly important event sounds the death knell for the Keynesian inflationary cycle. It was inevitable, and now that Americans have an alternative to paper dollars they will flee by this escape hatch. This new freedom is one of those little-noticed events which will spawn a great deal of history.

We are indebted to *Barron's* for its sensational August 26, 1974, article on the true story of how narrowly the right to own gold was won. Apparently, the Senate had been favorably inclined toward this gold legislation for several years. But the bill had been bogged down in the House, due to the opposition of Representative Wright Patman (D-Tex.) and other supporters of the Treasury Department's anti-gold ownership position. Since 1961 the United States has been giving aid to under-developed nations through the International Development Association (affiliated with the World Bank). New legislation providing funding was proposed by the House but defeated by the conservative bloc. However, in the spring of 1974, Representative Phillip Crane (R-Ill.), one of the few people in government to fight for gold, did some horsetrading. He offered to support the IDA bill and bring along some eighteen votes if Patman would tack on a clause legalizing American ownership of gold, no longer leaving it to the discretion of the President, as previously proposed. House liberals accepted the deal, and the bill sailed through the House by an 85-vote margin, far more than the liberals needed. Even at that late date, if it had not been for liberal miscalculations on that vote, the gold clause might not have been accepted.

A substantial upvaluing of gold in terms of all currencies is an essential first step in the restoration of monetary order. At the same time, once the value of gold is high enough, fixed exchange rates will be just a memory. All currencies will float, with supply and demand ruling. International business should get used to it.

The Dines Letter has favored floating exchange rates for over a decade. When the currency is sound, it will not matter whether exchange rates are fixed or not because fluctuations will be trivially small. I predict that the exchange rates abandoned in February 1973 will be described as the "cause" of wild inflation. As a matter of fact, the opening salvo of that argument may be found in *The Wall Street Journal* editorial of June 14, 1974, which shows how, even then, the leading publication of the capitalist world failed to understand what gold was all about. Apparently the editors never listened to their sister publication *Barron's*, which clearly saw the gold crisis way ahead of the pack. If I trusted governments, I would not object to fixed exchange rates. Fixed rates could work only if the price of gold were set artificially high, perhaps close to $1,000 per ounce. At such prices, there would be massive dumping of gold onto the open market by dishoarding, and the flow of gold into the economy would probably fuel a great boom. Hopefully, the government will understand this position in the near future. Instead, the Washington Economic Establishment will probably try to set the price of gold artificially low again, and fixed exchange rates will break down once more. Washington does not grasp the magnitude of the problem. This can be seen by the fact that the last two devaluations of the dollar were so inadequately small.

In the absence of government leaders who understand gold, I must favor floating exchange rates because this method will sound the alarm immediately when the Treasury printophiles begin doing their dirty work in the dark of night again. Over the centuries no government could be trusted with sovereign power over money without some kind of restraint. Anchoring paper to gold was an effective rein on political squanderings and on the voters themselves who condoned such spending.

5

Who Gets the Booby Prize?

Decades of anti-gold propaganda have prejudiced the public and institutions and therefore kept them out of gold shares. The fact that it was also illegal to own gold degraded the yellow metal in the public's eyes. Thus, people clung to the pathetic belief that paper dollars, which ostensibly represented gold, were better than the real value behind them.

Since 1961 *The Dines Letter* has withstood a great deal of punishment. Now it is time to print some of the opposition for the permanent record. This is not done out of a sense of bitterness or retaliation, although of course there is some satisfaction in having been right. But the real reason is to point out who the false prophets were so that when the next monetary system is constructed we may, at the very least, get a different collection of bunglers. And perhaps, years from now, some isolated and ignored individual, finding himself in the same position I found myself in, will have some comfort in knowing that all this happened before. Incidentally, some of the following quotes are from men who *do* deserve to be listened to next time.

> If a policy of active or permissive inflation is to be a fact, then we can secure the shreds of our self-respect only by announcing the policy. This is the least of the canons of decency that should prevail. We should have the decency to say to the money saver, "Hold still, Little Fish! All we intend is to gut you!"
>
> Malcolm Bryan, president,
> Federal Reserve Bank of Atlanta,
> October 11, 1957

●

A major reason why foreigners and others wish to hold gold is because it is convertible into dollars at a fixed price. If we abandoned support of the price of gold, but, let us say, retained our present stock, the demand for gold would be altered since gold would no longer have the property of conferring command over a fixed number of dollars. The result might be a decline rather than a rise in the world price of gold.

Milton Friedman
A Program for Monetary Stability,
Fordham University Press, 1960

●

... The United States official dollar price of gold can and will be maintained at $35 an ounce. Exchange controls over trade and investment will not be invoked. Our national security and economic assistance programs will be carried forward. Those who fear weakness in the dollar will find their fears unfounded. Those who hope for speculative reasons for an increase in the price of gold will find their hopes in vain.

White House press release
February 6, 1961

●

My specific interest at this time is in maintaining a competitive world position that will not further stir the gold of Fort Knox ... We must avoid inflation, modernize American industry and improve our relative position in the world markets.

President John F. Kennedy
to U. S. Chamber of Commerce
May 1, 1962

●

EUROPE UNFAZED BY U.S. GOLD LOSS—Faith in American Economy Is Termed Undiminished Despite the Outflow—DOLLAR HOLDS RESPECT

The New York Times
May 21, 1962, headline

●

NEW U.S. STRATEGY ON CURRENCY STIRS WORLD MONETARY CIRCLES —Plan Viewed As Daring Solution to How U.S. Can Have a Payments Deficit and a Stable Dollar As Well—
... The plan, announced in Rome ... by Robert V. Roosa, Under Secretary of the Treasury, involved a new strategy for the holding and acquiring of foreign currencies in the United States ...

Mr. Roosa made plain that the foreign currencies held as reserves by the United States would be sold (swapped) for dollars at times when the United States international payments were in deficit. This would mean that the United States could run deficits without losing much gold, because foreign dollar holdings—the original claim on gold—would not be rising.

Edwin L. Dale, Jr.
The New York Times
May 21, 1962

•

Surely there is something ludicrous about our obsession with digging up a pretty but rather useless yellow metal in South Africa in order to rebury it in Fort Knox.

The Burnham View, Vol. 11, No. 8
Burnham & Company
Wall Street
July/August 1963

•

On July 16, the Federal Reserve Board, with President Kennedy's approval, had acted to increase interest rates as one gesture aimed hopefully at easing some of the dollar's problems.

Still more moves are under study for possible use if the measures now being tried fail to do the trick.

One idea is to remove entirely the present legal requirement that U. S. currency and reserve deposits be backed with gold. Another is to put an embargo on sales of gold abroad. A third is to create a world superbank to help finance international payments.

Not under study, and flatly rejected by the U. S. government, is a device that comes to the public mind whenever money is discussed.

This is the device of devaluing the dollar—that is, raising the official price of gold in terms of the dollar.

. . . Of one thing you can be sure. Devaluation will not occur suddenly, by surprise. An act of Congress would be required to raise the price of gold . . .

The devaluation of 1934 had little effect on prices then or for some years afterward. At that time, the White House announced almost daily changes in the price of gold. The idea was that prices of farm products and other things would rise along with gold prices. Nothing of the sort happened, much to the surprise of President Roosevelt and the authors of the devaluation program . . .

This Government is jealous of the dollar's standing in the world, and now shies away from any step that might suggest the dollar is no longer "as good as gold."

This means that devaluation would be undertaken only as a last resort. Mr. Kennedy and his Secretary of the Treasury, Douglas Dillon, will not even consider it unless all other measures fail.

President Kennedy told Congress in his message of July 18: "I want to make it . . . clear that this nation will maintain the dollar as good as gold, freely interchangeable with gold at $35 an ounce, the foundation-stone of the free world's trade and payments system."

> *U.S. News & World Report*
> July 29, 1963

●

In the autumn of 1962 the [Kennedy] Administration had quietly committed itself to a radical principle: the deliberate creation of budgetary deficits at a time when there was no economic emergency —when, indeed, the budget was already in deficit and the economy was actually moving upward. This idea was the wildest heresy to those like George Humphrey [Treasury Secretary in the Eisenhower Administration] who used to predict a depression to curl a man's hair if the government did not balance its books.

> Arthur M. Schlesinger, Jr.
> *A Thousand Days*, Houghton Mifflin,
> 1965

●

Our balance of payments deficit has declined and the soundness of our dollar is unquestioned. I pledge to keep it that way.

> President Lyndon B. Johnson
> State of the Union Message
> January 4, 1965

●

If we continue to operate under the same system, we shall some day arrive at the end of the means of external payments by the United States. This will mean that, whether it wants to or not, despite the agreements in the IMF and GATT, it will have to establish an embargo on gold, establish quotas on imports and impose restrictions.

> Jacques Rueff, French economist
> *The Economist*, London
> February 1965

●

The United States is not a nation in real deficit, but, on the contrary, of global excess.

> Roger Auboin, ex-director,
> Bank for International Settlements
> *Le Figaro*, Paris
> May 1965

●

GOLD BUYERS SURE TO GET BURNED

Gold is on the way out. Paper money is in. And the greedy gold speculators are on the brink of taking one of the worst pastings in financial history.

The speculators still don't know what has hit them. They have yet to grasp the full meaning of the Free World decision to abandon the outmoded gold exchange standard.

The gold price in the tiny Paris and Zurich markets still is holding above the old $35 an ounce support price and that is a sign that the plungers still think they will win out.

But the happy conclusion in informed official quarters is that the speculators' days are numbered. The moment of reckoning is approaching and it couldn't happen to a more deserving bunch of highstake gamblers.

The "Great Gold Panic of 1968" was a close thing. The speculators grabbed several billions of dollars of gold and almost brought the Free World's financial system tumbling around its ears.

It is now clear, though, that the hoped-for profits will not materialize. The dollar is not going to be devalued. The price of gold will not be doubled to $70 an ounce.

On the contrary, it will fall once the speculators get the message and the more jittery plungers begin to unload.

The glorious irony of the gold panic has been the self-destructive behavior of French President Charles de Gaulle. The aging general loves gold and the archaic gold standard yet he did more than anyone to destroy it.

De Gaulle's bitter public attacks against the dollar and the pound and the rumors that he had French finance ministry officials float in "Le Monde" fueled the speculative attacks that ended the gold exchange standard as the post-World War II decades have known it.

But there is only $40 billion of monetary gold in the central bank vaults and the Free World countries will be printing up to $2 billion a year of paper SDRs to start. The paper gold gradually will dominate international dealings. The $40 billion of ingots will steadily decline in significance.

The world will have two kinds of gold, despite speculator and banker talk that the two-price system cannot last. There will be the commodity that the free markets will buy and sell for industrial and hoarding purposes and there will be the special $40 billion pile of monetary gold that the central banks never intend to make larger or smaller.

> Joseph R. Slevin, a columnist for the
> Philadelphia *Inquirer,*
> Miami *Herald*
> March 26, 1968

●

FALLACIES ABOUT GOLD AS MONEY

Gold has never functioned as a satisfactory money in a modern industrial economy, and in the last half century it has not functioned as a satisfactory monetary reserve. Increasing its price in 1934 did not end the Depression; there were still 10 million unemployed in 1939. Holding its price steady after World War II prevented neither expansion of money supplies nor inflation of prices.

Its value has not been stable. It has fluctuated with the dollar and is worth less than half as much in terms of goods and services as in the late 1930s. Now the gold interests want a price increase— not by market action, but by government edict.

The government need not be forced into a price increase when its stock of gold runs out. Nothing dreadful will happen if it just sits tight and announces that henceforth it will treat gold as a commodity like any other. The world has been on the dollar standard for two decades and may continue until an international money system is created. There will be a sudden realization that gold is not needed, and its price will fall.

Like coins with little silver, paper money that commands all kinds of goods and services will be acceptable after the pretense of gold backing is gone.

No other country can force us to raise the price of gold. In trying, it would cheapen its own currency and have to pay too much in real goods and services for an unproductive stockpile of gold. It is not a tolerable policy under present inflationary conditions, and we should give no country any alternative to cooperation in setting up an international monetary system.

To say that gold is the only universally acceptable medium of exchange is true only because we accept it. We can change the facts simply by changing our policy.

Once we go off gold, gold is finished. Eventually, it has to be displaced in any case. Right now, it threatens to work too well in restricting reserves. The world's monetary system cannot work with declining reserves. If its price were doubled, it might come back to us in a flood. But that only stresses its instability; the world economy cannot tolerate any such lack of control.

THE VALUE OF GOLD AS A COMMODITY

Official stocks held as monetary reserves total $42 billion. Private stocks, excluding jewelry, amount to over half again as much. The total inventory is over 100 years' supply at the current rate of use. Hence, the easy way to solve the problem would be to let all those who may feel secure holding and guarding unproductive stocks of uncertain value take all the gold they are willing to purchase. Unfortunately, in the international political muddle we have created it may be difficult to get agreement on the easy way.

V. Lewis Bassie
Illinois Business Review
March 1968

•

The fact is Joe [Henry H.] Fowler will leave a real score in the history books. With the dogged persistence and excruciating attention to homework that marks all his efforts, Fowler forced the biggest international monetary reform since the Bretton Woods agreement in 1944 with the creation of "paper gold"—the special drawing rights system of the International Monetary Fund.

Still to be made effective, the "paper gold" scheme is a solid achievement, especially measured against the selfishness of M. de Gaulle, and the wily planning of the European central bankers.

When he announced that the price of gold "will not be raised in my lifetime as secretary," the gnomes of Zurich got and believed the message.

Hobart Rowan
The Washington *Post*
November 13, 1968

•

People buying gold at skyrocket prices are "a lot of suckers."

Henry S. Reuss, chairman,
Subcommittee of the Congressional
Joint Committee
The New York Times
March 1972

•

Paper gold must weaken the power of the gold metal in the sphere of international finance . . . It is the beginning of the end of the tyranny of gold as a monetary metal . . . It means that after centuries of absolute power, gold is being toppled as the prime yardstick to measure the value of currencies . . . It means that on the 25th anniversary the monetary system created at Bretton Woods is being refined and basically strengthened.

> Sylvia Porter
> *Myers' Finance & Energy*
> September 29, 1969

●

The dollar remains the pivot of the international monetary system, gold continues in its twilight, despite scare headlines . . . there has been no dangerous tearing apart of the Bretton Woods money machine.

Gold will continue in its twilight as an international monetary metal . . . SDRs are far more suitable as the basic assets in the new monetary system.

> Sylvia Porter
> *Myers' Finance & Energy*
> May 18, 1971

●

Gold's performance must be mortifying for investment advisors who urged clients to buy gold shares. It would serve no purpose to recite names here. Investors who have lost their shirts on gold stocks in recent months know well enough who advised them. The advice came from some of the most prestigious firms on Wall Street, among other places.

For many Americans, however, gold's 1969 price slide must be exceedingly gratifying—and not just because it marks something getting cheaper. If the price has soared this year, contemplate the major beneficiaries: Americans who break U. S. law and South Africa and Russia, the world's top gold producers, countries that many U. S. citizens not unreasonably hold in less than the highest esteem.

Can it be that virtue occasionally is allowed to triumph?

. . . Rather, gold's remarkable 1969 record reflects political developments. Private gold hoarders, who own nearly a third of the more than $70 billion worth of gold in the world, began to realize in 1969 that the U. S. really meant what it had been saying for years—that it would not increase the price of $35 per ounce at which it exchanges gold for dollars in dealing with other governments.

... Gold's diminished role in international monetary dealings has not escaped notice in recent Wall Street commentary. Typical is an advisory item distributed December 10 by Francis I. duPont & Co., under the headline "The King Is Dead." Proclaiming "King Gold is dead," the item goes on to warn investors that "the outlook for gold is one of substantial risk."

Alfred L. Malabre, Jr.
The Wall Street Journal
December 29, 1969

●

GOLD ENTERS A NOT SO GILDED AGE

In the future, gold will play an even smaller part in the growth of reserves, as the newly created Special Drawing Rights gain acceptance as "paper gold."

The smoothly functioning two-tier system and the introduction of SDRs support the conclusion that if gold is not dead, as some observers claim, at least its volatile role in the world's monetary affairs will gradually and steadily decline. Recent market developments bear this out. The price of gold, which reached a peak of close to $44 an ounce early last year on the private markets, probably would have sunk well below the $35-an-ounce floor this past winter without support from Swiss banks and the international agreement that allowed the IMF to buy some African gold. The precipitous price drop reflected a general consensus that all hope of a large increase in the price of gold, on which speculators had wagered something between $4 billion and $6 billion since 1958, was dead.

As Volcker observed, "In a decade or so, news of gold will be chiefly in the commodity tables of newspapers."

The propensity to salt away wealth in unproductive gold, whose value is eroded by inflation, is steadily waning as sophistication increases.

Eventually, as gold loses its special status and becomes more of an ordinary commodity, its price may ride along with inflation, as is the case with other metals.

Gold, clearly, is entering a new era. As Robert Roosa says: "We're just getting to know that there are two characters of gold— that of money and that of a commodity." In a few years' time gold, like silver, could become too valuable to use as money. Then,

Roosa remarks, "People will ask why we keep it in monetary reserves."

> Juan Cameron
> *Fortune* magazine
> April 1970

●

The run on the dollar in Europe represents an immediate crisis only to a handful of free world monetary officials, a few international bankers, and those American tourists abroad who have been shocked to find that their Yankee dollar is not as welcome as it once was . . .

As far as the average American wage-earner is concerned, it will mean very little, except perhaps a hike in the cost of imported goods.

This is basically the view the Nixon Administration takes of the dollar crisis in Europe that yesterday prompted the central banks . . . to suspend their support purchases of the dollar that maintained its value . . .

> James Wieghart
> The New York *Daily News*
> May 7, 1971

●

Secretary of the Treasury Connally said on a National Broadcasting Company interview program that devaluation of the dollar would be "very, very beneficial to the United States." He added, "I don't think the average American will even be conscious of it."

> *The New York Times*
> December 15, 1971

●

At a White House cocktail party recently, Paul Volcker, Under Secretary of the Treasury for Monetary Affairs, asked the president of a major company, "What would be the public reaction to a devaluation of the dollar?" Said the executive, "I don't know anyone who gives a damn."

> *The New York Times*
> December 15, 1971

●

THE DOLLAR AS KINGPIN IS DEAD

The accord in Washington to realign the world's currencies represents the beginning of a long road toward reconstruction of the

international monetary system . . . After the inevitable rush back
into the dollar that will follow the agreement . . . world money markets should calm down.

H. Erich Heinemann
The New York Times
December 20, 1971

●

. . . The French would have preferred more than the token raising
of the official price of gold . . . But de Gaulle was dead and Pompidou was willing to compromise . . .

. . . the odds are 99 to 1 there will be no doubling of the price
of official gold. Just because a rumor is denied in Washington
doesn't mean it's true. (The free-market gold price could go anywhere—to $80 or down to $34 depending on what some gnomes,
peasants and gangsters think other gnomes, peasants and gangsters
are thinking about the crisis.)

Paul A. Samuelson
Newsweek
March 1972

●

Representative Henry Reuss . . . proposed that the International
Monetary Fund should stop buying gold from South Africa as a
move to curb speculation in world gold markets.

In a speech critical of recent speculative activities that drove
up the price of gold in Western Europe, Mr. Reuss also told the
House of Representatives that he was convinced that "no further
devaluation of the dollar relative to other currencies is in order."

Reuters
Washington, D.C.
June 13, 1972

●

Dr. Arthur F. Burns, chairman of the Federal Reserve Board, said
today the dollar is "not in any danger at all."

UPI
Buenos Aires
June 29, 1972

●

History may prove that "closing of the gold window" was the most
fundamental of President Nixon's economic policy moves last Aug.
15, Treasury Secretary George Shultz said.

Until then, foreign governments could cash in surplus dollars for Treasury gold at the old fixed price of $35 an ounce, a practice that allowed them to exert anti-inflationary pressure on the U. S.

The President's suspension of the dollar's convertibility into gold "removed that pressure," Mr. Shultz said . . .

The Wall Street Journal
August 14, 1972

●

Paul A. Volcker, Under Secretary of the Treasury for Monetary Affairs, warned European nations here today not to try to impose their ideas of monetary order on the rest of the world.

At the same time, Mr. Volcker reaffirmed in the strongest possible terms the United States' determination not to increase the official gold price of $38 an ounce, and eventually, to eliminate gold as a monetary metal.

H. Erich Heinemann,
Austria, to
The New York Times
September 4, 1972

●

. . . That key indicator of a nation's economic health, the value of its money in world markets, is sending out an important and hopeful message about the United States: The dollar is making a comeback . . .

"The bears are being squeezed," says Charles R. Stahl, who writes a newsletter for specialists on international monetary affairs for the Economic News Agency of New York. He predicts the vise will get tighter and tighter and that "you will soon see the dollar at a premium" as debtor-speculators are forced to bid up its price . . .

The Wall Street Journal
September 26, 1972

●

The United States tonight put into effect a 10% devaluation of the dollar . . . The devaluation, second in 14 months, will have little impact on American consumers . . . Under the new 10% devaluation, the price of gold will rise from $38 an ounce to $42.22.

The New York *Daily News*
February 13, 1973

●

BURNS VOWS TO END DEVALUATIONS

Arthur F. Burns . . . told Congress . . . that "as far as I am concerned, this is the last devaluation of the dollar.". . . Dr. Burns said in response to a question, "I hope we can avoid any significant increase in long-term interest rates."

The New York Times
February 21, 1973

●

A Robert Abboud, Executive Vice President, Chicago's First National Bank, put the blame for the continuing dollar crisis squarely on Uncle Sam's shoulders. "We are doing it to ourselves deliberately by an almost fanatical determination not to defend the dollar's parity in world markets. We don't believe in gold; we don't believe in par values; we don't believe in balanced budgets; we don't believe in interest rates responsive to supply and demand; and yet we seek world-wide cooperation in the ordering of a new monetary system."

The New York *Daily News*
March 18, 1973

●

Confidence in the dollar has been thoroughly shaken among the foreign holders of that currency, not only by the erosion of its purchasing power and two successive devaluations, but even more so by the declarations made by members of the American government. These statements show very little concern about this development, if not openly welcoming it; moreover, they do not imply a desire to defend the parity of the dollar with other leading currencies.

It is evident that such an attitude must deeply disturb the foreign holders of dollars . . .

Confidence once lost is not easily retrieved. Therefore it is to be expected that dollars will be on offer whenever this currency shows a tendency to harden in the market, and this will bring the price steadily down. This development will necessarily increase the distrust of the American currency.

Fernand Collin, chairman,
Kredietbank, Brussels,
April 10, 1973

●

Only time will tell whether the world's gold bugs or its equally dogmatic anti-gold nuts will be proved right on monetary matters. But, judging by the way the free gold price has climbed above $100 an ounce, the gold bugs have more grounds for chortling than do advocates of managed cellulose pulp currencies. Today, gold is anything but dead as a store of wealth . . .

Instead of fighting gold as an evil metal to cut from the monetary system at all costs, America should be examining how to use gold to further its own interests . . . it does seem that American policy makers should be paying more attention to how gold could be utilized as an ally of American policy rather than to waste any more time fighting it as if it were a monetary cancer to remove.

> Ray Vicker,
> chief European correspondent,
> *The Wall Street Journal*
> May 16, 1973

•

The unfortunate aspect of this tortured effort to make the sagging dollar look like a backlash from the current rumble in Washington is that it diverts attention from an unpleasant fact of life:

As Switzerland's Foreign Commerce Bank said in its *Economic Viewpoint* newsletter a few weeks ago:

Unbeknown to the majority of the American electorate, but obvious to most foreign central bankers and political heads of state, the turning point from the collapse of confidence to a resort to force came with the August 1971 double bomb-shell released by Mr. Nixon: (1) the official gold embargo; (2) the sudden imposition of an across-the-board 10% surtax on imports. This was a plain declaration of monetary and trade war . . .

The political significance of the Common Market's decision (March 1973) to jointly withdraw dollar support is that Western Europe now believes it is strong enough to refuse U. S. demands for special monetary privilege . . .

The economic significance of this European reverse-embargo on the dollar is that the trade war is now out in the open—for good or bad. The monetary significance is that in 1973 the monetary officials of the Western nations are faced with a clear choice: (1) return to metal-based currency and a period of sound policy; or (2) an accelerated resort to inflation and force. More directly put, the choice is between sound money and chaos.

> Don G. Campbell
> The New York *Daily News*
> May 20, 1973

•

The danger of another world monetary crisis has grown more intense. Some foreign officials believe it is already here and could degenerate into a more severe economic wrench than the world has seen since the 1930s . . .

To deal with the . . . crisis . . . one strategy is to let the dollar continue sinking until markets come to their senses and the dollar starts rising to its true equilibrium . . . The alternative strategy is for the United States, with the cooperation of other governments, to try to arrest the fall of the dollar and if possible turn it around. This cannot be done in the present state of market psychology without a massive mustering of resources, possibly totaling scores of billions of dollars. The United States alone cannot provide these resources since its monetary reserves, primarily gold, amount to only $12 billion. The support would have to come chiefly through American borrowings of foreign currencies, possibly augmented by renewed foreign central bank purchases of dollars.

This course is fraught with danger since, if it failed, it would pile up still greater liquidity in the world and intensify inflationary pressures. But there is no middle course between continued floating or full-scale support of the dollar . . .

> *The New York Times*
> July 8, 1973

●

A pair of important conclusions are emerging at high levels of the Federal Reserve Board . . . The first is that . . . it takes a higher level of interest rates to achieve a given degree of monetary restraint than it did several years ago.

The second is that inflationary private supply-and-demand forces in the world economy—above all forces affecting farm and other new commodity prices—can swamp anything the Federal Reserve does to curb inflation . . . That word is an on-the-record one in a recent interview with Arthur F. Burns, the chairman of the Federal Reserve Board.

> Edwin L. Dale, Jr.
> *The New York Times*
> August 1, 1973

●

Obviously, with interest rates in the United States and other advanced industrial countries climbing to levels that would once have been thought high for underdeveloped countries, the Keynesian forecast of a zero "pure" interest rate is ludicrously wrong.

What is the explanation? In part it is the chronic inflation that haunts modern capitalist economies. And the inflation, ironically enough, results in a large measure from acceptance of the basic Keynesian fiscal and monetary policies by nations for achieving high employment and economic growth . . . Chronic inflation requires a positive rate of interest, even on risk-free loans, to compensate lenders for anticipated shrinkage in the value of a given amount of money . . .

Leonard Silk
The New York Times
August 8, 1973

•

Sometime before November 30, Congress will have to take another look at the temporary $465 billion ceiling on the Federal debt. The ceiling has been "temporarily" raised 19 times since 1961 and "permanently" raised four times.

Tax Foundation, Inc.
August 1973

•

. . . The net flow of savings to mutual savings banks and savings and loan associations is dropping . . .

The commercial banks are hard pressed for liquidity . . . Desperate for cash, and paying record rates to get it, the commercial banks must worry about the deteriorating quality of credit—the danger in times of economic instability that borrowers may over-extend themselves . . .

Some economists regard the major area of weakness in the business outlook as the American consumer, because of the weight of his debts relative to his income.

The international scene is adding its own uncertainties. The world is struggling to adapt to flexible exchange rates—or dirty floating, with nations still intervening in foreign exchange markets and resorting to export controls to keep desirable goods at home in this time of shortages and inflation.

Should present food and energy shortages and inflation continue to intensify, the result could be a severe disruption of world trade and a hard blow to many domestic economies.

International commodity and gold markets are swinging from manic to depressive stages overnight, and back again . . .

World market gyrations, aside from simply bringing the doom-sayers large and small to the forefront in a most unpredictable world, recently have turned the more uninhibited soothsayer from visions of cataclysmic inflation to ones of cataclysmic deflation.* Their advice, however, tends to remain on the side of objects of intrinsic value, whatever the particular danger seen . . .

Leonard Silk
The New York Times
August 29, 1973

●

Nevertheless, I do not believe that the price of gold will have a specific rise above the previous intraday high of $186 per ounce, despite the fact that the central bankers will not press their sales. Barring a major war, or small war near South Africa's border, I expect that for many years gold will be trading in a range between $125 on the downside and $185 on the upside, give or take $10 to $20 either way, with central bankers being purchasers when gold trades between $125 and $150, and sellers when it trades between $150 and $185.

Charles R. Stahl, publisher,
Green's Commodity Market Comments
1974

●

U.S. RE-EXAMINES THE ROLE OF GOLD

. . . The Government owns $15.5 billion in gold . . . Technically, $13 billion of the United States gold stock is tied up as backing for deposits at the 12 Federal Reserve Banks for the currency—one-dollar, five-dollar, and ten-dollar bills—issued by the reserve banks.

Federal law prescribes that these Federal Reserve obligations be backed at least 25 per cent by gold. Yet the Federal Reserve Board has authority to suspend the 25 per cent requirement from the Administration's point of view:

It would dramatically underscore that all of the gold stock is in fact available for purchase by foreign governments;

And, it would remove the possibility that as the Federal Reserve increases the money supply to accommodate the expansion of the American economy, it might bump against the 25 per cent requirement . . .

The New York Times
January 11, 1974

●

* He means me.

Many more individual quotes will be found in the "Odyssey" chapter. In *The Dines Letter* I quoted many who commented on gold over the years so that it could be seen later who was right and who was wrong. For once, let the American public choose the right people to fashion the next international monetary system.

●

6

The Face of the Future

*When the paper system collapses, the survivors
will dig in the rubble and they will find gold.*
                                    The London Times
                                    May 1974

Few have earned the right to predict what is coming next. Since I am
not a member of the soon-to-be discredited Washington Economic Establishment, I feel I qualify. The so-called "top economists" who have
tainted themselves by working for an inherently immoral inflationary
system are hopelessly entangled in their own theories. Asking them to
get us out of the mess they created would be like putting Dracula in
charge of a blood bank.

A Clarion Call for a New National Direction

In 1776 this nation knew where it was going. Nearly two hundred years
later we seem to have forgotten that purpose. The United States of
America is about to be tested as never before, hard as that is to believe
now. Hopefully, an intense period of national soul-searching will ensue,
featuring a reexamination of our goals and a reevaluation of the nature
of democracy and capitalism. Some will say capitalism failed. Did it?
The future of the republic hangs on that decision. I still have great faith
in the American people, provided they are not misled by the same false
prophets who have done such a catastrophically poor job so far this century, leading us through war after bloody war and a seemingly unshak-
able boom-depression cycle. For reasons I will outline below, I see
violent social changes ahead, including national bankruptcies, hyperin-

flation and then deflation. Detonate your imagination and avoid time-gulping bores.

The primary function of government is the protection of its citizens. Yet, it seems we are getting everything but that protection. Our parks are becoming terror-filled enclaves, harking back to the "highwaymen" of medieval England. Each year the government devises clever new ways to tax for its spending purposes, and it always overlooks the basic, original, and primary function of government: protection of its citizens. Ingenious ways are found to release criminals back into society. A situation like this is ripe for vigilantism. The fact that in 1974 a movie like *Death Wish* could be so popular, at the same time that vigilantism is resurgent in England, suggests that the time is not too far ahead. While the English groups were quickly written off as a "near-fascist ground-swell" by the Labor government's Ministry of Defense, there is a basic human need here which government is not satisfying. Vacuums make me nervous because one never knows in advance how they are going to be filled.

As you will see in the "Odyssey" chapter, my blueprint for the coming crash and depression is rapidly being fulfilled, and I am struck by the eerie paradox that those who claim to help our working class, albeit doing so for the short-term, wind up massacring them. It happened in 1932 and it will happen again.

Our political leaders have shirked their duty by not educating the public to realize that when they demand more services, they must pay for them. They are taxed by the insidiously invisible inflations I have described in this book. Is there a politician alive today with enough guts to stand up and ask the American people to lower their expectations? It takes a lot of patience to learn patience.

I am not, by profession, a political analyst. I am a security analyst. Therefore, I develop a deep sense of frustration when the topic of gold, of which I do have a command, begins to approach its political roots. It is my fervent hope that someone will trace back through history the basic political dichotomy which is manifesting itself in an economic fashion now, and who can also "reach" the public. By this I mean that the decision must be made defining who is more important in our society, the *individual* or the *government*.

Totalitarian regimes, such as Nazi Germany, Russia, and China, value the state more highly than the individual. On the other hand, democracies value the individual more highly than the state. In the United States this has become decreasingly true, and the trend is alarming. Computerized records of individuals, eavesdropping devices, and increasing government intervention in our affairs are parts of trends which stretch back perhaps for centuries. Maybe one needs to go back beyond Aristotle and Plato to find out where these two disparate main-springs of modern political thought originated.

Adam Smith wrote his famous *Wealth of Nations* in 1776, favoring free markets and minimizing the role of government. In 1936, Lord Keynes wrote *General Theory*, which strongly favored government in-tervention. Keynes asserted that national debt was irrelevant because it did not have to be repaid; we owed it "to ourselves." He completely ignored the fact that some Americans own more government bonds than others, and his declaration that we owe this money to ourselves assumes the United States is one big family, which we know is untrue. Keynes was confident he could improve on market solutions by intervention, especially by fiscal and monetary methods. This, theoretically, would eliminate depressions and give man control over his own economic des-tiny. So, to Keynes, governments were wiser and more important than individuals.

Perhaps control of our economic destiny will come one day, but the basic fallacy in this position is the assumption that individuals acting as government economists are much smarter than the very same individuals acting for themselves. Where men are ignorant, they are widely ignorant, and the effort to find a few men who are more intelligent than we mor-tals must fail. That is partially why the world is suffering so today.

With U. S. debt now at around half a trillion dollars, and annual interest payments amounting to more than $30 billion, it is no wonder that runaway inflation is the order of the day, and that the international bank failures I predict could signal the end of an era.

Everyone will simply have to understand that the world's resources are limited, while demands against those resources are infinite. Politi-cians who incite and inflame hopes beyond a reasonable likelihood of satisfying them must be considered irresponsible. All politicians who have acted irresponsibly have contributed to the real roots of inflation. This would tend to include all of Keynes' followers.

To whom does this apply? Listen to those who are willing to "accept" a 5% to 6% inflation rate for a few years, and add up how many years it would take to destroy the U. S. dollar. Every year the destruction of our currency accelerates. Let those whom the shoes fit wear them.

I wish I could be sanguine about the prospect that America will emerge relatively unscathed from the coming monetary storms. There could be a significant decline in America's economic and political free-dom, due to decades of bungling, misconceptions, and downright stu-pidity. After Rome fell, the West went into a decline which lasted for centuries. Perhaps after this decline, all we can do is hope that Keynesian economics will be resoundingly rejected and that people will go back to free enterprise capitalism. This involves much less government, drastically lower taxes, and the return to the economic policies of economists like Adam Smith and Ludwig von Mises.

I was saddened, but unsurprised, when Con Edison spectacularly omitted its dividend for the first time in its history on June 15, 1974. This stock is heavily owned by those greatly dependent on dividend income, so the omission radiates throughout the economy, hitting hardest those who need dividends the most. Widows, pensioners, and small investors seeking above-average incomes are victims of an anti-business, anti-capitalist, anti-ethical mentality which will don the garb of whichever cause is popular at the moment, including ecology. The New York Public Service Commission denied Con Edison the right to build an atomic energy plant in Queens and additional plants along the Hudson River. Instead of allowing business to make the decision, government made it. Partially as a result, New York had a total blackout in 1965, and since then electrical voltage has been cut repeatedly because there is insufficient electricity to handle New York's growth. We are going to have to make more realistic decisions and keep the Washington Economic Establishment's hands off business.

Monetary convulsions will continue until we elect a political leader who is smart, brave, and honest enough to make the American people understand that the business cycle is a necessary evil for eliminating waste and inefficiency in our system. As bad as it is, the alternatives are far worse. I must sternly warn that the concept of continued, limitless, and unending prosperity is a trap and a mirage.

In retrospect, exhausted by the Great Depression and World War II, our legislators made war against the symptoms of that Depression rather than its causes. By enacting the Employment Act of 1946, in a desperate effort to avoid duplicating the trauma of mass unemployment, there was a bland disregard for the havoc this would wreak elsewhere in our society. The Employment Act did not confront the problem of the built-in, long-term inflationary bias that permanent full employment breeds.

I am calling for a repudiation of the way we have been living these last few decades. No one can guarantee permanent prosperity and full employment without creating other monsters like inflation. Politicians who promise these things without pointing up the other side of the coin are charlatans and knaves. If the American people insist on voting for this type of leadership, they have no one to blame for the consequences but themselves. The encouragement of debt must cease, and government red ink will have to be a thing of the past. This is so because governments can no longer raise enough money for their basic needs. They have run out of debtors to gut. By inflation, we have raised our prosperity to an incredible height, which will undoubtedly induce some to demand this forever. It was a temporary heroin-like high from our addiction to inflationary policies, and we must understand it as such. Total and permanent prosperity is unsustainable at the present level of civilization. Instead of demanding full employment, let us try to get as much employment as possible within the framework of stable prices. It does not help the little man if he is given more paper money while prices are rising even faster. Let us turn the clock back to a sound monetary and credit system as it was before the twentieth-century madness replete with cruel wars and misery-breeding ignorance of true history took hold. Regardless of the price that is about to be paid, let us determine not to go through it again in three decades.

With 1984 less than a decade away, where is our national purpose and old-fashioned patriotism? Has inflation already done its dirty work by bringing under fire all of our basic institutions, such as family, religion, the military, police, and so on? The working class has become much more militant than the meekly unemployed of the 1930s. This new generation has never known hard times; it is anti-authoritarian and profoundly iconoclastic. From this pressure cooker, this seething cauldron, could come the type of social upheaval which is written about in history books.

Even a Paranoid Really Has Enemies

The U. S. government makes a big thing out of its anti-trust laws, aimed at preventing companies from getting "too big." However, there must be something wrong with government standards which stress size rather than efficiency. When a company gets so large that it becomes inefficient and almost goes bust (like Penn Central and Lockheed) the government guarantees loans with your tax money instead of letting it go bankrupt, a process which would open up opportunities for the little guy. This is why business gets bigger and bigger in this country. Instead, in 1974, super-efficient AT&T, Xerox, and IBM become targets of anti-trust suits; the red inks versus the black inks. I regret having to continue returning to government as a source of problems but apparently the U. S. federal government is becoming the ultimate guarantor of failing big businesses.

It is said that heroin is dangerous because it makes the user feel like a hero, and there are no limits to a heroin user's imaginary powers. The United States government does not need to take heroin to have an unbelievably egomaniacal concept of its ability to do anything in the world, to save anyone, and most dangerous of all, to become the ultimate guarantor against failure. Unfortunately, the cost of this exercise falls on the American taxpayer, while the politician collects a salary for arranging such dubious transactions.

Now that bankruptcy is increasingly common, our government is taking over many companies. This is hardly surprising in view of the fact that corporations face double-digit wage boosts and borrowing costs, more frequent strikes by workers who are hurt by inflation, plunging stock prices which remove the option of raising capital on the equity markets, wild fluctuations in commodity prices which make planning increasingly difficult—not to mention the instability of international currency rates which undermines long-term contracts and, eventually, leads to declining corporate volume. The more workers are affected by inflation, the less they buy and the more corporate sales decline, the more other workers are laid off. Labor union leaders generally seem to be Keynesians. This is because their superior bargaining power facilitates gains for them at the expense of non-union workers. Something for nothing. Do we really want plumbers to earn more than physicists?

A return to gold would not solve the problems of monopoly union power or excessive governmental interference in the economy. These are separate problems. Everybody should be confronted with the necessity to earn his keep in fair competition with all others.

The United States, having already bailed out Lockheed and Franklin National Bank of New York, even becomes, in some cases, the guarantor of other nations. It has assumed the dubious distinction of guarantor of such precarious states as Italy and Israel. I hasten to add I have nothing against these two countries. Any free individual may help countries if he so wishes. But *The Dines Letter* rails against the use of American bank depositors' money without their knowledge or consent.

On June 26, 1974, the governor of the Italian Central Bank, Guido Carli, stated that Italy looks to the United States for help to pay her oil bills. Carli wants the United States to become "lender of last resort" to nations having difficulties paying oil bills, and suggests that the U. S. Federal Reserve System become a "world central bank." Heaven forbid! The U. S. Federal Reserve has done enough damage to the United States internally, so imagine what it would do to the entire world.

In mid-1974, Israel borrowed $300 million from the United States through an unusual debt instrument guaranteed by the United States government. The City of New York invested $30 million of its employee pension funds in the new Israeli notes which mature in 1994. Israel will use this money for weapons, which is not a very productive investment in a country with a precarious future. I wish all the people well who are involved in this arrangement, and hope that they will prosper. However, I do not think that solutions of this nature are in the best interests of unknowing Americans who deposit money in their banks. Harrison J.

Goldin, Controller of New York City, expects many other states to invest in these new notes, causing American civil servant pensions to become dependent on Israel. If Israel defaults, or has runaway inflation, Americans will share the tax burden. Goldin defended the purchase of these notes as "a legal investment because payment of principal and interest is guaranteed."

Both Italy and Israel are deeply dedicated to socialist dogma, and a lack of concern for debt by inflating it out of existence. What most socialists fail to grasp today is that this is not 1949. The United States can no longer print all the money it wishes to, to get itself out of trouble. The Fed has become increasingly impotent because the Keynesian system has come to its natural end. The City of Rome fell to the barbarians in 476 A.D., and I hope there is no 1500-year long-term cycle ominously pointing to 1976.

The American people should become less like enthusiastic Sancho Panzas chasing a tilting Don. Before a government commits itself to increased spending, however worthy the projects, it must check its resources. Calculations on income should not be based on theoretical textbook definitions which duly impress harmless and cowed students. "Growth rates" based on average rates over a century ignore the stench of the 1930s, which is just a small part of a hundred-year chart. The 1930s were long and hard times for those who lived through each day of those bitter years. Now I view the spectacle of a nation as great as ours getting a margin call. If the United States is to have a recession of the depth and magnitude the market seems to be predicting, the national income will be far below planned and committed expenditures. Such a strain, even for the short-term, could break our once-proud and eagerly sought dollar. It is of little comfort to know that we are watching history being made. Indeed, we might be watching the beginning of some of the greatest ironies of the last few centuries. If President Ford is the victim and future scapegoat, no harm will befall him that will not be inflicted on the rest of us tenfold. We are all in the same American boat. What I have criticized in *The Dines Letter* is germane to the stock market. Political considerations seem to affect the current stock market climate more than economics. What I point out is not exclusively limited to one party. It is a national rot which needs to be expunged. Wild spending engineered by people who never developed a callus or met a payroll is only part of it.

Why is a new political party not formed? A party which would combine charity with fiscal responsibility. Clichés like "right" and "left" should be assigned to the same purgatory to which market terms like "long-pull" are fast descending. I predict there might be a new political party or a drastically changed old party in 1976 or 1980, and it could win perhaps by another man on a wheelchair.

I recommend original thinking, and fewer professors acting like security analysts (sometimes based only on experience with their ten shares of AT&T and four shares of IBM—both bought at the highs). I have an idea which I have never seen in print and which probably will never be heard by America's "brain trust." There is much talk of raising or lowering margin requirements, but I suggest a compromise: specifically, separate margin requirements for buying and holding. This would be superior to present arrangements. For example, I suggest lower holding requirements in order to slow the forced liquidation which harms many innocent people. Buying requirements would be raised to prevent investors from becoming overextended. This is how we can prevent overextension by an overly enthusiastic public and prevent another 1974. In mass margin liquidation, the weakest destroy the weak, and the weak, in turn, make the strong weak. Best of all would be an anti-margin educational campaign, but it would only work in a climate of sound money without encouragement of debt, instead of the sad process we have whereby each problem creates a score of others.

Someone undoubtedly will claim that my pessimism helped bring the entire stock market down. To think for one moment that my subscribers could conceivably affect over thirty million investors in such a way would be akin to declaring that a pebble dropped in the ocean caused a tidal wave, a flattering but impossible idea. No one has that power. Any system brought down by a pebble was not very stable to begin with. On the other hand, if I had not written about the coming stock market crash and gold crisis in recent years, if I squeezed out dishonest smiles as indeed some have, the few whom I saved also would have been sacrificed. Who are our enemies if not ourselves?

I cannot help but wonder what direction the historical shift which lies ahead will take. This country went left after 1932 and it is hard to say whether there will be a turn further to the left, or a move back to the right.

The Sad Crash Ahead

This book bristles with my predictions, many of which will be borne out. Remember, I am not an amateur printing a "scare" book. My "track record" is appended in the "Odyssey" chapter. Many predictions, as you will see, already have come to pass.

No type of government in existence today could conceivably take the steps necessary to forestall the disaster I envision ahead. To obliterate decades of persistent and consistent inflation requires deflation. Not only is it probably too late, but the current political climate makes this option difficult to swallow. During the coming depression (while a generation educated in Keynesian economics is in a state of shock) the public will be much more amenable to the strict measures which will need to be taken. Until then, there is no hope of restoring sound money, secure credit, a gold-backed dollar, much lower taxes, the end of inflation as a national policy, and so on. Today, no politician, even if the thought occurred to him, has the following and support to effect such changes, and therefore the destruction of the once-proud dollar, given these present conditions, is a certainty.

A massive restructuring of wealth is coming. Those who had money before and did not pay attention will be removed from a position where they can affect this country's future course. This is a Darwinian law of the jungle. Some who were once poor will suddenly acquire money. The very act of acquiring it will qualify them to run the next show. They will not be frightened by a climate of volatile interest rates, plunging stock markets, tidal international monetary flows, sporadic but wild currency speculation, and a roller-coaster gold price. The new millionaires will understand that all these phenomena are related and represent the end of an era.

Here are some predictions of what probably lies ahead. Do not expect every one to work out literally, because no man has that prescience. It is the thrust of this book that I believe in, rather than the weight of any particular prediction. Furthermore, after this book is written, foresights will appear to me. Writing a book does not possess the advantage of updating as new input becomes available.

I am positive that in some quarters the Arabs will be considered the "cause" of current economic ills. That will be incorrect. The Arabs, responding to our outrageous inflation, have reacted in self-defense, although they could have been friendlier about it. As I stated earlier, the energy crisis was a catalyst for the inevitable.

At my New York City Gold Seminar in February 1974, one of the more shocking predictions I made was that oil stocks would collapse. It seemed hard to believe then, with the oil crisis going full blast and sky-rocketing oil share earnings rampant. However, oil shares were in Downtrends* when I made that prediction.

Higher oil prices factored into everyone's equation in all sectors of the economy will, undoubtedly, force prices higher. Many individuals will be enveloped and unemployed because manufacturers overloaded

---

* See *How the Average Investor Can Use Technical Analysis for Stock Profits,* James Dines, 1972.

with inventory will begin mass layoffs. Look for skyrocketing unemployment, far higher than the 6% level generally predicted by the government, and probably in the 10% to 20% range by the end of this decade.

The Arab threat to raise oil prices again even from these astronomical levels (because of high U. S. inflation rates) is the beginning of the end. The Arabs no longer trust paper. The jig is up, Keynesians. It is the end of the economic "system" we now have.

What is still unclear is whether the current situation will force a massive shift back to a simpler and less automated society. Will bicycle and buggy whip manufacturers be the stars of the next bull market? The possibility exists, and open minds will be duly rewarded. Remain alert for new business opportunities.

The Coming Infression

The topic of inflation is returned to at this point in an effort to tie things together. Infression, a word invented by *The Dines Letter*, suggests that inflation and depression will occur simultaneously: an inflationary depression. This is a new virus cropping up in the international monetary system. Its feature is a flood of government money which attempts to inflate (because such inflation worked in the past), but the real purchasing power of people—adjusted for inflation—will decline. That is the deflationary aspect of the situation leading to a depression. Some features of this system began to appear in 1974 when people received more paper dollars while the amount of goods they bought with those dollars declined. This will lead to a lower standard of living for many who do not now think that that is possible.

The U.S. economy is out of control. The government has, apparently, long since abandoned worrying about inflation, or the dollar's standing. Having to meet oil payments and trying to obtain prosperity and minimize unemployment could lead to the biggest paper printing binge in history. One can never predict how irrationally people will react, but the risk of runaway inflation is now greater than ever. If President Ford resorts to "reflating" the economy, if and when it comes, it will lead to an economic collapse unequaled in American history.

When Gerald Ford became President, he declared inflation Public Enemy Number One. Indeed, only war is worse than inflation. Can Gerald Ford, who was not even elected by the American people, arrive on the scene after forty years of government attempts to do more than our resources permitted through ruinous inflation, and actually make people understand how dangerous inflation is? Sad to say, some will need empty bellies to clarify their thinking.

People are being conditioned not to trust paper money. The depreciation of our dollar, accelerating annually, is approaching a terrifying velocity. Each new crisis exposes the dangers more clearly to those who would see. It would take a brave man to predict that the trend toward worthlessness, having been carried on for so many years, will by some miracle reverse itself. Perhaps it can, but I do not see how. And it must lead to a devastating deflation somewhere ahead.

Debt can be extended, paid off, or bankrupted out of existence. Since our inflated debt could not possibly be paid off (barring a lengthy depression), a massive debt liquidation must lie in our future. Until it occurs, this crisis can not be considered terminated. This will be the primary cause of most of the distress in the immediate future. During inflationary times, many undercapitalized corporations survive when they should not, owing to the "good times" of inflation. Most of these corporations will eventually disappear. Excessive stock market speculation will have to be liquidated by margin calls. As people withdraw money from banks to pay for the effects of inflation and then deflation, there will be less money available for business in terms of venture capital, construction and so on. Furthermore, low bank liquidity could lead to some spectacular banking failures. The loss of venture capital will hit Wall Street even harder, where there could be some surprising bankruptcies. Banks might have to call loans to meet depositor withdrawals, causing those who can not repay their loans also to be thrown into bankruptcy. The unprecedented high levels of borrowing by individuals will also have to be brought back in line. All this suggests prosperity for bankruptcy lawyers.

Inflation was intended as the euthanasia of the middle class, due to misguided socialist doctrine, favoring the working class. When employees do not take wage cuts, their wages are simply inflated lower. Unfortunately, some unions then gain at the expense of other unions, since unions negotiate differently. All unions gain at the expense of non-union labor. Unorganized labor can not negotiate for higher wages the way powerful unions can. The only way to have sound full employment is to permit wage flexibility. Until we get over that residual hang-up from the 1930s, there is no hope that this solution will be tried. The U. S. government must learn it can not solve anything by printing more money and throwing it at the problem. Nearly all New Deal legislation would have to be dismantled to cure the disease now afflicting the world's currencies, and it is doubtful that this could happen in the present climate. The concept of the minimum wage needs to undergo thorough reconsideration, particularly since it legally forces some to be unemployed. Idealism must be tempered with practicality, and compassion must be tailored to what is best for the long-term because the short-term is always brief.

Each year confidence in the currency diminishes as the flight from paper accelerates. Sometimes the economic climate improves for a while but this is illusory; inflation returns worse than before. It is likely that this low-key panic will increase until it swamps the ability of the monetary authorities to cope with the coming crisis. The public needs education on the merits of a sound currency. Instead, they are buried by a never-ending stream of greenbacks, a wealth created and backed by an order to run the printing presses for another hour. It cannot be that easy to live that well on this earth.

The question remains: Are you ready for massive debt liquidation? If not, perhaps you do not deserve to keep your wealth.

This infression will not end until the government realizes that running printing presses will not work this time, and until it realizes its actions are increasing the rate of inflation faster than before. Just how far along the line governments will awaken to this realization will tell you whether or not inflation will move into the hyperinflation phase, and the subsequent destruction of present paper currencies. Someone will then at last attempt to halt the presses. At that point, the depression will bare Frankenstein's teeth, uncosmeticized by inflation. The process will then shift from infression to deflationary depression.

Few agree with me, but a depression already appears to be starting in the realm of stocks, real estate and advertising. After all, do not expect all areas to topple over all at once. I see a spectacular commodity bust by 1975 or 1976.

A gut cause of the coming downturn will be high interest rates. Lenders demand higher interest rates to compensate them for loss through inflation. These high rates are the seeds of self-destruction spawned by inflation. *The Dines Letter* was one of the first to warn that high interest rates are one of the symptoms of inflation's final phases, and in 1973 I wrote a prediction of prime rates between 10% and 15%. Rates of 12% arrived in 1974. This is not the stuff of which booms are made. Sound economic upturns occur when interest rates are low, and people are cautious and conservative, not when they recklessly pay 20% interest to engage in dubious ventures. As people become more fearful of lending money long-term during inflation, it becomes more difficult for business to expand and, therefore, fewer goods are produced. True, interest rates will have their decline as well, but the broad trend has been up in recent decades.

It is moot to consider at which point inflation becomes runaway inflation or hyperinflation because, historically, every inflation eventually has been followed by a major deflation. Economic conservatism will become fashionable again. Since the government is in charge of the figures, add a few percent to whatever the government declares the inflation rate is (they undoubtedly manipulate them—otherwise the inflation game does not work). However high it is, bondholders look as if they will be the first to be ruined. Those who own bonds, keep money in the bank, or even money in their pockets overnight, are having it outrageously confiscated by self-motivated politicians. Since banks are even more dangerously illiquid than in 1929, a widespread banking collapse could add to America's nightmares. As businesses close down, there will be layoffs and surprising bankruptcies. People will swear off stocks ("never again") and collectively the American people will lose billions, as the bill for a forty-year binge is due and payable all at once. The full impact of the infidelity to gold has yet to be felt. Until it has and true reform takes place, there can be no hope that this crisis will pass soon.

Shoving all this paper money down America's throats like some Strasbourg goose has meant that too many cars are being produced, and too many factories have belched their poisonous pollution into the air and water in order to cater to our collective greed. Suburban sprawl and urban decay might just reverse, and the rest can be left to your imagination. Perhaps, as often happens, much good will come out of this experience.

The "Fabulous '50s" and the "Soaring '60s" Have Led to the "Sickening '70s" but My Path Leads to the "Elevating '80s"

If my predictions come to pass, there will be a decline in the national standard of living in the coming period. Life will become more austere, there will be shortages and rationing, various markets will come crashing down, interest rates will reach Mafia-high levels before engaging in a long downswing (perhaps for the rest of this century). There will be skyrocketing loan delinquencies and bankruptcies, news of entire nations going bankrupt, and an end to the economic control of a generation as blind to gold as other generations were to the earth's revolution around the sun. Soaring commodity prices, currency chaos, and a loss of personal freedoms will lead to a wave of national and international pessimism. Architects and stockbrokers will be driving taxicabs, and plumbers will accept lower wages as unemployment inspires humility. In such a climate of social unrest it is hard to say which personal freedoms will be seized first. Safe-deposit boxes could be sealed and reopened only under government supervision on some pretext, perhaps to catch those who have "secret Swiss bank accounts."

Food riots and a devastating stock market decline could be fueled by the unrest sparked by unemployment. As I have pointed out, the only way to stop unemployment without a police state is by a free market.

We will have to learn to stop destroying the majority of 94% to help the 6% unemployed.

In the years ahead, market rallies will be prematurely halted by "stale overhead longs," as devastated stock owners try to "get out even." Yet, after the establishment of a freely traded gold standard and the restoration of the right to own gold privately, there will be no inflation, interest rates will have declined, and a boom will be ready to start. Hopefully, I will be ready to use the phrase "The Elevating '80s."

The Coming Financial Panic

Adjusting for inflation, a stock market crash has already occurred, as you will see in *The Dines Letters* written in 1974 in the "Odyssey" chapter. Money invested in the stock market in the mid-'60s has already been savaged. Observing this, it is obvious that a full-scale financial panic has already begun, although to the innocent public it is still an "invisible crash." Starting Over-the-Counter, it spread to the ASE, and is moving toward the blue-chips. The recently soaring gold price only adds to the ominous feeling that something horrendous is going to happen. The purpose of this will be the destruction of the individual's capital so that at the next market bottom investors will not have the wherewithal to buy stocks which will be at once-in-a-lifetime low levels. Every instinct you have will caution you against buying stocks at that time. There will be a number of possible reasons for heeding these instincts, such as unreasonable taxation, threatened government take-overs, threats of communism, ruinous wage demands, or merely a poor business outlook. Social unrest, already suggested, could be so threatening at the ultimate bottom that stocks will run out of selling pressures. Major cultural changes at the very heart of our society are to be expected. These changes will be expressed in the long-term political paths we take.

As in any crash, the impact on Wall Street will be profound. The elimination of fixed commission rates by the SEC will lead to numerous mergers and bankruptcies. The lower stock volume from the bear market will cause a brain drain and horrifying attrition from which Wall Street will take a long time to recover. Look for cutbacks in the number of branch offices, and for stockbrokers to retaliate by competing more with banks, real estate and insurance companies. A new, truly international, around-the-clock marketplace could develop. Wall Street brokerage firms should dwindle to a handful of large companies, plus some small specialists.

The accounting profession will likewise undergo tremendous

changes, as will all financial institutions. Accountants have been slow to realize that they made no effort to adjust to inflation accounting. They will be involved in lawsuits from those willing to sue anyone in an attempt to recoup, and some causes of action will be justified.

There could be a new currency in America's future after the dollar as we once knew it is gone and all paper money becomes suspect. As people retire they will discover that their pensions and life insurance are like mirages in the desert, disappearing on the approach. This will add to accountants' nightmares.

The Social Security system could collapse. An August 23, 1974, editorial in *The Wall Street Journal* questioned the soundness of this system by pointing out that as of June 30, 1973, the net unfunded liability of the Social Security system was $2.1 trillion. This means that current working and retired members of the system have been promised $2.1 trillion more (in constant dollars) than will be paid into the system from now on. This suggests radically higher Social Security taxes by the end of the century to pay these obligations. The U. S. government will try to pay for this debt out of general revenue—further embezzlement. Social Security has undermined people's desire to save for the future. Savings are necessary for capital formation and the future growth of our capitalist economy.

My fears and criticisms are not meant in a bad spirit, but in genuine apprehension for the soundness of our economic system. This book is a warning; an effort to help those who can yet be saved, and some hopeful advice to the architects of the next monetary system.

It will be unclear to the public who the scapegoat for the coming crash should be: government, Wall Street, or the Arabs, to name a few. This will be a God-given opportunity to get rid of the Securities & Exchange Commission and return to the rule of *caveat emptor*. The SEC was created by Roosevelt in a misguided effort to force corporations to disseminate sufficient information. Theoretically this would prevent another crash. Prior to the SEC, in the '20s, investors were not privy to sufficient information on corporate earnings, which resulted in the purchase of a great deal of worthless stock. Despite all the information demanded by the SEC, the fact that most of the new issues floated in the 1960s wound up the same way shows that the public was not protected. The few OTC issues which could still find quotes in mid-1974 were down so much from their probable purchase prices that losses to investors must have been horrendous.

The SEC is just one more expense which can be dispensed with, enabling further tax cuts and the return of money to the people. Let investors hire their own advisors privately as was done before the invention of the SEC. Not only have corporations spent staggering sums to satisfy SEC demands for disclosure and information but, by pretending to protect against frauds, the SEC has lulled the public into a false sense of security. Without the SEC, individual advisors would have sprung up to provide warnings to the public. Those who use that service can pay for it, instead of taxing everyone to pay for something politicians think we all should have. I feel that power should be restored to the people by massive tax cuts, which would make them more self-reliant. For example, Alan Abelson of *Barron's* has sent many stocks down the tube by his clear and honest warnings of trouble ahead. Government should not have control over what honest and wise individuals should be selling as a service and paying taxes on their profits to the community.

Our system of taxation will have to be seriously reexamined. Perhaps the solution is to start from scratch, by voiding the Sixteenth Amendment to the Constitution, ending the federal income tax, and preventing politicians from getting more money without a national referendum. Again, this returns power to the people, but this is not meant as a phony revolutionary slogan. Furthermore, any new tax system would have to be so constructed that taxpayers would know precisely, at all times, what they are paying in taxes. It is the only way to prohibit politicians from engaging in irresponsible schemes.

I have long toyed with—let's call it "The Dines Tax Plan." It abolishes all taxation of every U. S. citizen. How does the government get money? Here's the skeleton of my tax plan, whereby a new money could be printed once a year with, for example, 10% more paper money printed. Inflation would be avoided because prices for everybody would go up at the same moment, and it would still be tied to the price of gold, but inflation would be used as a tax collector—with strict limits to avoid Frankenstein. The government takes that extra percentage for its needs, and nobody pays more than a 10% tax. The figure can be raised or lowered only by national referendum, for only one year at a time. There are no "deductions" for anyone, so all special interest lobbyists can leave Washington. Everybody in the country would pay an equal, fair tax, including the Church, farmers, the Mafia, and even oil men. There would be no way to evade this tax. All wages, rents and every price would go up by the same amount on "conversion day." My plan would limit government's financial ability to foul up the economy. The whole expense of the Internal Revenue Service system could be liquidated, and income tax jailbirds could be released to productive work. Best of all, none of us would need to fill out a confusing tax form again. The adjustment period should be spread out to minimize the impact of a large jobless bureaucracy, while some sort of "national charity" for the poor on a purely voluntary basis should take care of the lower end of the spectrum.

Will There Be a Bust in Pension Funds, Banks, or Social Security?

It is difficult to pinpoint where the crash will start. It will probably be on Wall Street, where mutual funds have already been devastated. Although it could come from real estate, which is especially vulnerable to a possible banking collapse; banks are deeply in hock to speculative land interests.

First of all, banks are not used to handling foreign currencies. The new rules involving floating currencies have resulted in a number of banks suffering shocking losses because they are not accustomed to volatility in that area. Banks are encountering currency problems without realizing it, and it would take only one big bank failure to create a domino effect where the weak drag down the strong. It might even come from overseas. A wave of English bank failures would hit U. S. banks very hard. The FRB member bank reserves are more illiquid now than in 1929. Hence, any such increase in deposit withdrawals by panicky savers might lead to the door of disaster.

The crash could come from the bankruptcy of a major pension fund. A man who works all his life for a pension will not be amused to discover it is not waiting for him when he completes all those years of loyal labor. This kind of man could reach for a rifle or vote for a radical political ticket. How about those who, dreaming of an inflation-resistant pension, have put their pension contributions into a fund entirely invested in common stocks? In those days, stocks were thought to be a good hedge against inflation because as inflation continued, the rising dividend payments—which would vary with the income of the pension fund—would theoretically rise to offset the ravages of inflation. But skyrocketing inflation and a stock market slump have been a disastrous nightmare for the variable-annuity dream. Many retirees are watching in anguish as their pension checks slump monthly and they come closer to forced dependence on public assistance.

Approximately three-quarters of all pension funds are privately funded. The remainder are government funded. In 1950, around 83% was invested in conservative bonds, government issues or government paper. Only 17% was invested in common stocks. By 1972, two-thirds of private pension fund money was invested in the stock market. In 1950, 46% of the assets was insured, but by 1972 (with most of the money in the stock market), only 31% was federally insured.

In 1972, these funds paid an enormous $8.3 billion in benefits. With a possible stock market collapse ahead, pension funds might con-

Private Non-Insured Pension Funds

Year	Percent In Common Stock
1959	29.8%
1960	32.4
1961	35.6
1962	37.6
1963	38.9
1964	40.0
1965	42.4
1966	43.9
1967	47.1
1968	50.2
1969	52.8
1970	53.3
1971	59.0
1972	63.5
1973	63.7

Source: SEC—Statistical Series dated 12/12/69 and 4/17/74

ceivably aggravate that decline by being forced to sell stocks to meet the huge benefits to be paid out. A depression, prompting widespread dividend cuts, could leave many pensioners literally penniless in their old age. The victims of this fraud, perpetrated by the U. S. Treasury, could become a destabilizing social factor.

It would not have improved the situation if fund managers had stayed entirely in bonds. Even before distortions, the government tells us that the value of the U. S. dollar has declined by over 70% since 1950, so although pensioners get the same number of dollars they are promised, their purchasing power is drastically reduced. This is why inflation inevitably destroys the society it infests. Frankenstein has never yet failed to destroy his creators.

Again, I accuse Lord Keynes, and the nest of rabid and flint-hearted Keynesian economists who succeeded him. The one thing of which I am confident is that the people responsible for this outrage will never be punished. When are we going to get a sound currency? Sadly, the public expects the Keynesians to be the creators of the next monetary "system," and I am under no illusion that I alone will be able to stop them. But there is at least one angry man left in the world who asks quietly, "What are you doing?"

The Face of the Future: Dollar Dumpty Will Take a Great Fall;
All the King's Horses and All the King's Men . . .

I was among the first of the well-known security analysts, if not the first, to stridently warn of this inflation and gold crisis because of an unsound currency. It has been my singular displeasure to watch Washington's economic actions over the last fourteen years. First, God made morons. That was for practice. Then He made the people who have ruled the international monetary system and took the dollar off the gold standard. These are the inescapable conclusions of all my years of Treasury watching.

One of the most important events of recent decades was the repudiation of the convertibility of the dollar into gold on August 15, 1971. This event was drastically underestimated by people who should have known better. When Nixon announced, after many solemn promises to the contrary, that foreigners could no longer get gold for their dollars, he showed how quickly monetary changes can occur. The lesson that it could happen again was not lost on foreigners.

By now it should be clear that unless the dollar is convertible to gold, there can be no limits on inflation or deflation. Chronic gold crises will recur. This is central to my theories. The dollar must become convertible into gold before lasting solutions to our problems can be found. Until then, the action taken that fateful day in 1971 will continue to be a major destabilizing political force throughout Western civilization, and the problem will not be solved until a blatant economic smash shatters the complacency and drives from office the people who got us there. Meanwhile, as I have shown, inflation will diminish peoples' confidence in paper currencies, in their institutions and, in the end, their politicians. A resurgent nationalism could lead to tariff wars, competitive devaluations and, possibly, violently antisocial acts. Perhaps even urban guerrilla warfare. All paper currency will decline in value in relation to gold. Look for proliferating exchange controls and tariffs (despite the deceptive current trend by President Ford to dismantle them), competitive devaluations, and an all-out smash in the huge Eurodollar market, probably sparked by a major bank failure.

American ownership of gold became legal in 1975. Even now, our government does not understand the implications of this and thinks that they can continue embezzling citizens. Little do they realize that free gold is a silver bullet aimed at the heart of an inflationary Frankenstein. All the ingredients of an economic bust-crash are present, invisible though they might seem now. For this reason, I have advised *Dines Letter* subscribers to invest heavily in precious-metal-related investments.

I predict golds and silvers will be the big movers of the 1970s.

Hopefully, in the coming debate and recriminations, people will come to grips with the true problems facing us, including the query "What is money really worth?" There are other provocatively simple questions like "What is money?" and "Should governments be permitted a monopoly on the production of money?" Perhaps competition should prevail by allowing the people to choose the currency they want. This idea should be explored.

A massive dollar devaluation looms ahead, and all the king's men will not be able to put Dollar Dumpty together again. Instead let us hope there will be a new system somehow linked to gold.

What the Coming Currency Will Be

I predict some country will produce a special gold currency within the next few years. It will be accepted internationally, and its reputation will endure. It will sweep before it all artificial forms of currency, particularly Special Drawing Rights. The world will continue stumbling back toward a gold standard of some type, and in it gold will remain central. This will hasten the bankruptcy of paper monies, so look for frequent monetary convulsions. As confidence in paper is increasingly shaken around the world, "the little man," rather than the so-called "gnomes of Zurich," will break the·dollar and other leading currencies. Only the next currency system will be able to revive rapidly deteriorating international trade.

John Exter, Senior Fellow at the American Institute for Economic Research, aptly points out that currencies are now saying, "I do not owe anybody anything at a fixed price," and therefore calls all paper currencies "I-owe-you-nothings." Since the explosive increase in U. S. paper money is not related to anything tangible, especially not to gold, it has become impossible to have a fixed exchange rate system. Exchange rates are floating. For the first time in centuries, perhaps in history, inflation is world-wide. This is an international attempt by governments to substitute paper for real money, just as John Law did in France nearly three centuries ago. Exter is especially contemptuous of Special Depository Receipts, which are not even IOUs. The SDR has no obligor, and therefore is a "Who-owes-you?" Since nothing is owed, it is a "Who-owes-you nothing?" It has no maturity date, hence "Who-owes-you-nothing-when?" Exter describes the SDR as "the most preposterous credit instrument ever invented . . . if indeed we can call it a credit instrument." SDRs will have no place under a gold standard.

This enormous world-wide flood of IOUs is the main problem

today. Central banks are totally committed to the expansion of this paper money which, in effect, wipes out previous debts. There is no way to pay it off. Central banks are prisoners of inflation, and as they try to restrain it they will plunge the world into a depression.

Fixed exchange rates are unworkable unless paper money is convertible into gold at a sufficiently high price. It would take a radical increase in the price of gold to achieve convertibility, so until then we must become reconciled to floating exchange rates and world-wide inflation. Our collective, accumulated debts, unable to be paid off, will lead to massive deflation, depression, and debt liquidation. Such events rarely occur without blood in the streets—the best time, usually, to buy common stocks before the next boom.

The next monetary system will probably eliminate the International Monetary Fund because there will be no need for it in a world of floating exchange rates. Soon, I predict, some country will break the ice and begin settling its accounts in gold. This will be a revolutionary and electrifying development because it will "unfreeze" all the gold which has been locked up in vaults for many years, and begin to improve the severely deteriorated liquidity of the world's central banks.

These predictions will not occur in a straight-line progression. There will be temporary periods of dollar strength or perhaps even a rallying stock market. Yet unless the basic underlying problems are solved, these remissions should be considered irrelevant. Temporary improvement in the U. S. balance-of-payments should also remain unimpressive until the problems covered in this book are confronted and solved once and for all.

At some point, there will be a national debate between protectionism and the movement toward a free-gold standard. Protectionism will hark back to John Stuart Mill, who basically feared that England would run out of money, and who set up many walls against this possibility.

During this debate, remember that under the gold standard a nation lost gold when it ran a balance-of-payments deficit. This contracted its credit and increased its domestic interest rates. As a result, commodity prices fell and made it a more attractive country from which to buy and a less attractive market in which to sell. The nation with a gold loss then has rising exports and, by necessity, a discouragement of imports. Furthermore, relatively high interest rates not only discourage lending at home, but also attract funds from abroad. This is the magnificent self-adjusting mechanism of a free market, and we do not have to pay politicians to run it—it is free. Fire the politicians running our economy and give the taxes saved back to the people.

Any effort to resurrect Bretton Woods' fixed-parity currencies, which eliminates the free-floating aspects, will inevitably lead to another build-up of suppressed pressures and another explosion near the end of this century. Hopefully, the world will not return to a "gold-exchange standard" but to a gold standard, which will keep economic downturns minimized and break this vicious cycle of booming inflations and catastrophic depressions. Under the gold-exchange standard the United States merely printed more paper dollars to fill the gap when it ran a deficit, instead of paying in gold. In other words, the debtor nation did not lose what the creditor country gained. All nations will have to realize that their means are limited, though their wants are not. The function of gold is to distribute equitably that which is available to satisfy those wants. A corrective mechanism is desperately needed, and it cannot be left to government. As the experience with Keynes has shown, government leaders are not to be trusted with the currency. An impartial and ruthless discipline like gold must be used as our policeman. Remember, a persistent balance-of-payments deficit is the result of a government's own fiscal and monetary policies. Governments will refuse to admit this and blame everyone but themselves, from bankers and businessmen to labor and management. Never forget where the blame truly lies.

In the era I see dimly visible ahead, it seems fair to expect that foreign aid will be much lower. The outlook for the tourism industry is bleak until currency values get back into line. Waves of devaluations overseas will make international traveling too hazardous and difficult, so stay close to home until the dust settles. An international depression will also alleviate the excessive tourist growth and excessively rapid economic development which has endangered the world's ecological balance.

The loss of prestige the dollar has suffered will cause some of our allies to lack trust in the United States for a long time. Foreigners might react to this lack of trust by dumping their American stock holdings, perhaps from a fear of capital restrictions.

It will be realized, eventually, that a higher gold price would help the United States a great deal, barring a temporary insanity which could induce us to dispose of our national gold stockpile. There is still some talk of demonetizing gold. This is the same story I have heard for over a decade. Governments and a few hundred economists can demonetize gold if they wish, but there are still three billion people who trust and respect gold, and our leaders will never alter this loyalty.

I anticipate more dispersion of international power in the post-crash world. The Arab sheiks with their black gold have made headlines as the coming world power. But no one else seems to see that South

Africa, with its colossal pile of yellow gold, could also emerge as a world power. Currently, no one gives South Africa a second thought, but it is a mineralogical freak with a strong military force. South Africa has diamonds, coal, uranium, platinum, copper, a strategic military location, and much more. If the price of gold gets as high as it could, South Africa will indeed become a force with which to reckon. At some point, South Africa might accelerate this process, ending its long policy of cooperation with the West by actually withholding large quantities of gold from the world market, much like the Arab oil cartel. This would send the price of gold skyrocketing. South Africa could defend its position by claiming dwindling gold resources, fashioned after Arab claims of dwindling oil resources. Should this occur, it would bring the whole house of cards down immediately, if it has not already collapsed by then.

Dollar Dumpty's fall will lead to a crisis atmosphere in which it is difficult to predict just yet precisely all that will happen. It would be nice to envision calm discussions of balancing the budget, but this is unrealistic. There could be a wave of nationalizations of industries such as railroads and air transport, banks during a money crisis, and oil companies in a petroleum shortage. The nationalization of U. S. gold and silver mines will be discussed, especially by those who specialize in getting something for nothing from others, and who are better at "redistributing" wealth than *creating* it. There will also be a large wave of interest in silver which, if the price of gold gets high enough, might be described as "the poor man's gold." The ramifications are endless. Each person should consider how this crisis could personally affect him or her. The time to prepare is in advance. If you are not ready now for the crisis that might come tomorrow, what are you waiting for?

It is not difficult to imagine, during a gold hysteria, that some will demand that individuals' gold shares, even gold coins, be subject to government seizure "for the good of the people," and compensation given in fiat currency. This possibility definitely exists. Roosevelt nationalized gold in 1934, so there is ample precedent for such a disgusting solution. Most assuredly, tourism and the outflow of money to overseas will be severely limited by government in a belated effort to temper the outflow of dollars. Right now, it is legal to get paper dollars out of the United States. Perhaps this money should be left legally in gold or silver coins in a Swiss bank, and those who draw no interest, and do not deduct the charges, have no income to report. They should leave their options, such as emigration, open. In the event of another devaluation, they will be able to switch those gold coins back to paper dollars at a much more favorable rate. They will benefit both from the devaluation (which hurts others) and from the increase in the gold price.

How High Will Gold Go?

The long-term history of gold is a fascinating one. The accompanying chart was composed by Wilfred C. Krug, Canada's gold dean for nearly forty years. It was compiled originally in the 1930s and included in his booklet *The Future of Gold*. The chart was updated in the 1960s by *Investment Guideline* of Ontario, Canada. Mr. Krug sought me out many years ago when I was a security analyst. He was among the first to whom I was able to communicate my ideas about gold, and from whom I received sensible answers. In those days, goldbugs were few and far between, and patronizingly relegated to the "kook" category.

From 1250 to 1700, the price of gold rose at the rate of approximately 1.25% per year. In that period, payments were made in cash and credit was extremely rare. From 1700 to 1914, gold prices remained stable. The creation of the gold standard by The Bank of England established a new trust in paper money which in turn permitted the creation of banking and the birth of credit.

In the 1920s, the gold-exchange standard led to an unstable boom and a crash. Roosevelt's decision to raise the official price of gold 69% to $35 was, I believe, an effort to allow all paper money printed to finance World War I to "catch up." The next gold price increase will catch up with all the superfluous paper printed since 1932. The arrows in the chart show how I perceived the future in 1960.

Soaring gold prices instill in many people the feeling that something is very wrong. They are driven to bad-mouth gold and even suggest that it will decline, rather than accepting gold and trying to solve the underlying currency problems. Some say that raising the price of gold would unleash a whole new wave of inflation. I have already discussed the true causes of inflation and how gold would control, not accelerate it.

A longstanding prediction of mine was vindicated in 1973 when currencies were allowed to float freely. Yet ahead is the "official"price of gold floating freely with no monetary restrictions. For the moment the interesting point is that the world could smother the gold price at any time because of this free-float policy. The gold market is a small one, and any determined central bank could, for the short-term, artificially depress the price of gold. Gold is a unique commodity. The fact that it has so many potential sellers, and yet no one sells, is truly amazing. Considerations like this beg the question of what would happen if the Arabs invested a portion of their tremendous oil profits in gold. The results would be even more astronomical if Arab money were invested in the few gold-mining shares available. The possibility of an

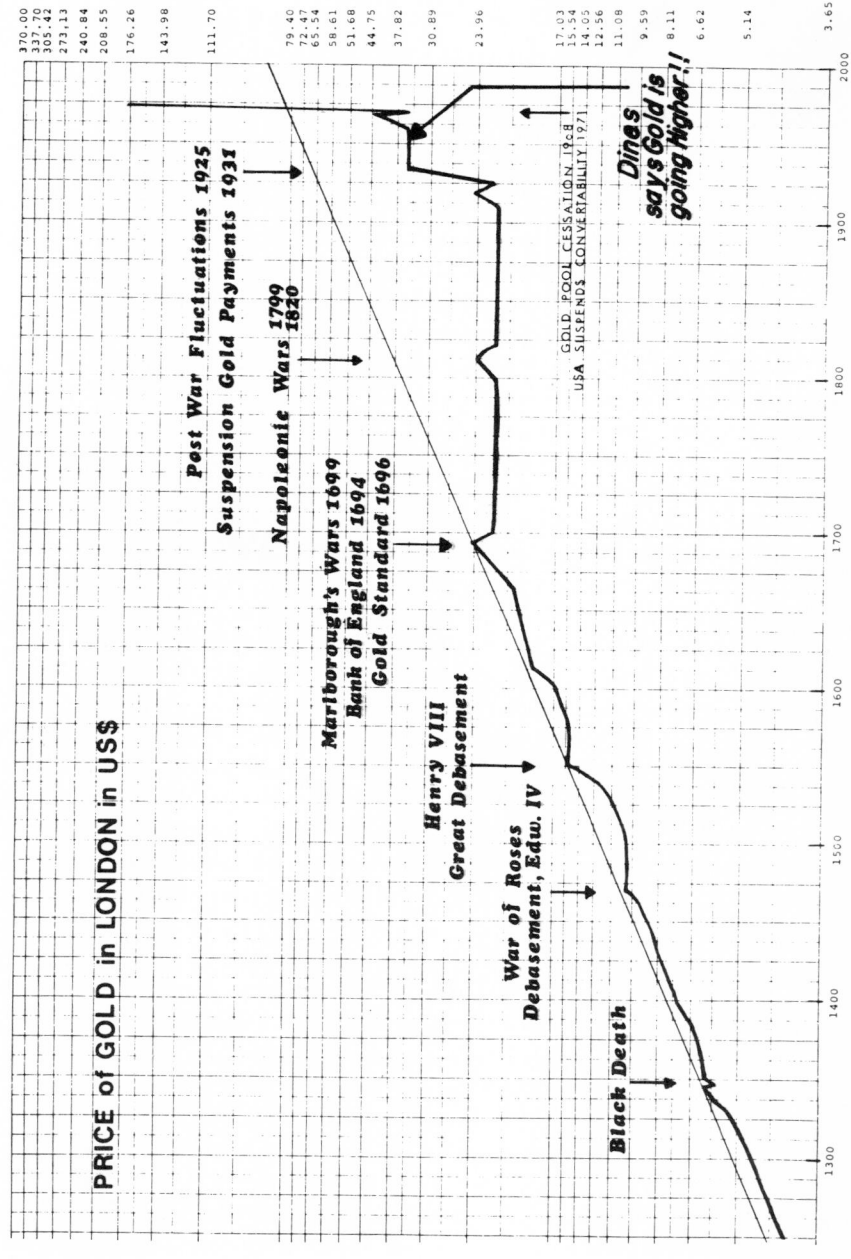

PRICE of GOLD in LONDON in US$

explosive and historic rise in the price of gold and gold-mining shares is therefore high.

Twisted logic might reason that gold rose only because of inflation. According to this line of thinking, during the shift from inflation to deflation which I anticipate, people will sell their gold and move into the security of bonds. This will provide the investor with an income and, at maturity, return his original capital increased in real terms by the amount of deflation over the holding period. In contrast, gold returns a negative income due to the cost of carrying, storing, and insuring. It is said that during a depression industrial demand for gold should contract, since there is then so little industrial consumption of gold.

Actually, costs decline during depressions. Once stability returns, the relatively fixed sale price of gold usually means higher profits for gold-mining companies. For example, gold shares went up, not down, during the first Great Depression. The government could conceivably guarantee dollar convertibility at a much higher level, to avoid another gold decline. In the 1930s, gold remained stable not because of any inherent stability, but because the United States, with its unlimited talent for printing paper money, bought all the gold offered to it at $35 per ounce. Without U. S. intervention, the price of gold might have declined below $35. What if the government did the same thing again at $1,000 per ounce? Either way, this will benefit gold-mining shares, which could be your salvation.

To quiet those who would suggest that raising the price of gold would have a powerful inflationary impact, I maintain that those billions would not be added to the world economic system, but instead used to wipe out the extra paper printed in recent decades.

As for any possible argument that gold prices move with all commodities, as occurred in the early 1970s, I think this was coincidental because the free-floating price of gold is a relatively new phenomenon. People buy gold for reasons very different from the ones which prompt them to buy wheat, for example.

Actually, people might withdraw their bank deposits in fear of a banking collapse and convert their assets to a hard monetary asset, such as gold coins. This would likely provide support for gold during a depression. In the past, gold shares were held for income by widows and orphans, and certainly not for capital gains.

When I first became interested in gold shares, the Swiss and South Africans with whom I came in contact were uniformly pessimistic about them. They will understand someday that gold is for protection of capital. People will take less of a yield when holding gold shares, and much higher prices will be possible for them than for most other stocks. Finally, there has been less monetary freedom since Franklin Delano

Roosevelt, and gold coins are desirable because they are easy to hide. Why hide it? Because inflation has been an informal, annual levy on all wealth, all capital, and all income. Later, in the event of a formal capital levy, gold is portable and leaves no spoor.

How high will the price of gold go, and by when? The United States is still haunted by the sins of the past—enormous deficits and foreign aid which piled up billions of dollars abroad. If the United States stays out of expensive wars abroad, if foreigners begin to move capital into the United States, if we can export enough food to earn foreign exchange and yet not cause a revolution in American grocery stores, if America somehow earns enough money to pay for increased oil prices, and if American inflation can be temporarily slowed down, the burden of our national debt still will not be sufficiently alleviated. The United States owes at least $100 billion overseas. Since the dollar will continue to collapse until it is convertible into gold, our $11 billion stockpile will have to be increased tenfold to begin to cover our liabilities. This suggests a gold price around $300 to $400 an ounce. However, Alan Greenspan figures the amount of money overseas at around $200 billion. Jacques Rueff suggests it is $250 billion. It is difficult to know whose estimate is more accurate, but if they are in the right vicinity, look for a price of gold closer to $1,000 an ounce.

Euromarket volume of around $200 billion, rising by around $10 billion annually, might be as large as Germany's entire Gross National Product by the end of 1974. The growth of Eurocurrency deposits in 1973 alone was $50 billion. At least that much is scheduled for 1974. The total narrowly defined money supply of West Germany is $54 billion. This gives some idea of the dimensions of the problem.

Worse, there are no reserve requirements, which means one dollar of reserves can be multiplied endlessly. If any bank wanted to do so, a simple dollar could be multiplied into millions of dollars of bank deposits because there is *no gold limit.* In other words, banks can use reserves to make loans, which create deposits that can be transferred to other banks, which in turn become reserves for further loans, and so on. A London bank can create dollars when it makes Eurodollar loans.

Those who are alarmed by the rapid creation of money in the American system absolutely ignore the incredible creation of money taking place in Europe. A dollar created in London bids for goods on the world market in the same way as a dollar created in New York, and it has the same inflationary impact.

That Damoclean Sword of paper dollars poised over our tiny gold supply is our real probelm. Those deficits should never have been allowed to happen.

After every crisis was allegedly "solved" by Washington's standards, *The Dines Letter* accurately predicted worse to come. Until the dollar is fully convertible, and gold is traded freely in the world without restrictions of any kind, I see no end to chronic monetary crises. Brand the word "convertibility" into your mind.

There are moments in history when an older order is finished but not yet dead, while the new order is not quite strong enough to be visible. I think the world is at such a watershed right now. A massive restructuring of old alliances is taking place, such as the U. S. shift away from Japan toward China and away from Europe toward Russia, the impact of which will be felt for years to come. The mistakes of the old order must be completely liquidated and that process has just begun. From its ashes will come a tremendous new bull market. Imagine the impact of the end of the era begun in 1932. If the United States no longer subscribes to inflationary policies, then the whole concept of granting wage increases which are immediately nullified by printing more money will bring about a tremendous social upheaval. Tell organized labor that wages will have to level off and watch the fur fly. Inflation is such an integral part of our lives that withdrawal symptoms could be terrible. Hopefully, we will see the repeal of the Employment Act of 1946 with a sharp reduction in taxes and government in general. Look for slashes in the welfare rolls, the end of "new economics" and the return of many American troops from overseas.

Look for increasingly desperate politicians attempting to seize the public's wealth, in order to conceal the results of their financial follies. Historically, governments have rarely hesitated to seize foreign assets during crises. Who knows which ingenious schemes they will think up to take what money you have left? That is why it is likely there will be limits on the amount of money which can be removed from the United States, oppressive red-tape for Americans wishing to travel abroad and, conceivably, confiscation of domestic assets of those leaving for extended overseas trips. Although foreign bank accounts existing at the time of the crisis will be allowed to stand intact, new foreign bank accounts will probably be outlawed. Eventually, even these might be "nationalized." A word to the wise is sufficient.

Despite talk of forcing the dollar higher, world-wide anti-gold propaganda, government actions, bad press, the Wall Street and Washington Economic Establishments and a lower stock market, gold shares made new *all-time* highs in 1974. That should say something loud and clear to sensitive people who are listening. Gold has beaten all its enemies, but they do not know it yet.

The gold mining stock with which *The Dines Letter* is most identified is ASA, first recommended at an adjusted price of $5, and priced at over $100 in 1974, an increase of nearly 2,000%. In 1968, hat in hand,

I visited numerous brokerage houses in an attempt to start a gold fund. They were not interested. Some of those brokerage houses no longer exist.

The world is about to learn the difference between credit and real money. Only precious metals and short sales will survive the coming holocaust unscathed.

Which Is Better, Gold Coins or Bullion?

Smart individuals will have the foresight to invest in some gold and keep it close at hand. Some prefer bullion, others prefer gold coins. For large investors who do not wish to be involved with numismatics, bullion is probably preferable. For the average investor, gold coins are superior because they can be spent immediately without further processing, and when you wish to sell them, an expensive examination is not required. A gold coin is accepted at face value, while bullion must be checked by an expert to attest to its validity, at further expense that could take a chunk of your profits. There is a slight numismatic premium when coins are purchased. Take care to ensure that the premium does not exceed a small percentage. At selling time, the premium should not be too far from that percentage, thus minimizing the possibilities of numismatic loss. Premiums exist due to the scarcity value; gold coins of the old dates are no longer made. Gold coins assure you of access across any border in the world, with or without papers. There has never been anything quite like gold. Gold coins are far easier to store or ship than art, antiques, or land. Someday, numismatic values on gold coins will probably go up a great deal because pre-1934 gold coins will never be made again.

On the theory that it is a stupid mouse that knows only one hole, a prudent course is to spread out geographically, as well as between gold and silver coins, and precious-metal stocks. This is the time to invest in gold mining shares in Canada, South Africa, South America, and so on. So far, few institutions have golds, even though there are several new gold funds on the drawing boards at the time of this writing. As people scramble to board the bandwagon, golds could become the glamour stocks of the 1970s. Since so few security analysts understand gold because of the ignorance surrounding it for so long, there will be much misinformation bandied freely about. Later, the reckless and unwary will be selling gold shares short. Beadlets of perspiration will develop on their foreheads during the ensuing pyrotechnics, as demand for these shares from new funds and latecomers arrives. In this period of run-away inflation, when the values of real estate, art, stocks, diamonds, and bonds are declining, proceed on the principle of maximum caution. Those who understand the problems and take sensible steps will not be engulfed. Be certain your bank account is kept under the federally insured limit, particularly because cash is vulnerable to inflation and the possibility of bank failures exists. *The Dines Letter* has been a long-term pessimist on bonds because of their vulnerability to inflation. Embezzlement of bondholders has been a primary source of wealth of the welfare state. Those who refuse to sell their common stocks should have a properly placed stop-loss under each stock owned. By careful study of my first book, you can learn where to place your stops properly, beneath their uptrendlines. At least if you are wrong, you will be stopped out of your stocks automatically. Of course, the option to simply sell out is always there. With respect to golds, there will come a time to sell them.

International Ramifications

It is difficult to tell precisely what is going to happen during the coming period of upheaval and readjustment, other than to state that there will be unforeseeable ramifications—as there were after 1929. There is the possibility of exchange controls, and the power of gold could smash efforts to build a Common Market. Floating currency rates have already occurred, no doubt a mixed blessing.

As mentioned, *The Dines Letter* has long been a proponent of floating currency rates. These rates have been criticized on the grounds that they make international trade impossible. This is untrue. Fluctuations in currency should be handled the same way Hershey Chocolate handles the fluctuating rate of cocoa, by buying futures on a forward market. Then, while Hershey knows months in advance precisely what it will pay for its cocoa, speculators deal with risks. That is one of the valuable economic functions of speculation.

Currency hedging is time-consuming and entails a slight cost, but it is worthwhile because it maintains the stability of the international monetary system. Currency futures markets lack depth now, but if the world adjusts to floating rates, new institutions will spring up to provide that depth, since international trade, after all, must continue. As I write this, solid forward markets exist only for the Canadian dollar, British sterling, German mark and Swiss franc, where coverage is available for one year or more. Fairly good markets exist for the Japanese yen, Dutch guilder, and Belgian franc.

It cannot be emphasized too strongly that these institutions are brand-new. Eventually, international trade could be financed quite easily without the twisted logic which favors large balance-of-payments deficits by the United States. This is true particularly if the international monetary scene stabilizes to the point where risks are minimized and therefore forward currency premiums could afford to be lowered.

The only benefit derived from the lack of understanding of the gold crisis by the government is that, in breaking down, old institutions (i.e., those set up at Bretton Woods in 1944) have been replaced by free markets and not by new institutions. Therefore, I have high hopes that the next system will be the one presented to the world by the free market. It most likely will be the best one available at the time. One can do a lot worse in this world than institutionalizing what the free market has shown to be the correct way. By this, I mean evolution rather than revolution.

A primary problem will be the danger that a breakdown in international trust in everyone else's paper money will lead to such a proliferation of tariffs and other controls that there could be a return to the seventeenth-century mercantilism so thoroughly discredited by Adam Smith. Perhaps there is a three-hundred-year cycle involved here. Therefore, this would appear to be the time to buy foreign items which you were planning to buy eventually. If the hyperinflation I discussed does occur, U. S. interest rates will go far higher than anyone dreamed possible. International raw material prices could skyrocket, reversing the historic trend of expensive manufactured goods in relation to cheap raw materials. With such high interest rates, it is difficult to see how an international depression could be avoided. The question of whether it will be an international recession or a depression reminds me of a comment made by Eddie Cantor in 1932 when he said, "A 'recession' is when the guy next door loses his job. A 'depression' is when you lose yours."

Politicians who know virtually nothing about gold because of the baleful influence of Lord Keynes, will have to educate themselves quickly. Some politicians still declare that there is "not enough gold to support modern commerce." Apparently, they fail to grasp that all one has to do is raise the price of gold sufficiently and there will be enough gold for all the commerce needed. As a matter of fact, raising the price of gold will probably be the only way to pay for Arab oil. The question in my mind is, Are the Arabs enterprising enough to rush into gold before the industrialized nations slam this door by officially raising the price to $300 or $400 an ounce?

For years, food was underpriced because it was fully competitive. A kernel of wheat does not have its own distinctive fingerprint, and everyone's wheat competed with everyone else's. Farmers were vulnerable to capitalism, but when farmers tried to spend their earnings they had to buck cartels, labor unions, government boondoggling, and welfarism. This will end domestically and internationally. The percentage of income people must pay for food will become more realistic. After all, it is the farmers—the real producers—who should receive these profits, not the government. Farmers should not be penalized for providing us with sustenance.

Likewise, the Arabs were duped into accepting fixed rates for their oil for years, while Washington accelerated its printing press spending spree. The more money they printed, the less real money the Arabs received. In the 1970s, fortified by the laws of supply and demand, Arab oil prices rose from $2 per barrel to well over $15 per barrel, in some extreme cases. As with gold, food and oil are beginning to wipe out some of the incredible price inflation of the last thirty years.

Consider the ramifications. Low fuel prices spawned lavish waste of energy, including increased auto production and excessive driving, second homes, snowmobiles, and all the production that prepared the steel for the manufacture of these engines of choking pollution which fouled our country under the cloak of progress. New York became undesirable and, to some, unfit to live in, not only because of air pollution and the subtle taste of Poughkeepsie sewage in the drinking water, but because of the filth, crushing taxes, and the crime and drug problems (partially related to people with no family structure and too much time on their hands while they collected welfare). While some deserve welfare and relief, it is difficult to believe that over one million people in New York City alone are incapable of working.

Also due to excessively underpriced fuel, airlines held an advantage over buses, ships and trains, so the aviation industry grew unfairly at the expense of other means of transportation. It became convenient and inexpensive to commute from the suburbs, and this led to the dispersal of people from cities and urban decay. Even the automobile was too cheap, so we choked on the excess carbon monoxide and sulphur dioxide.

Here is how gold abuse causes slums, and it is not the standard sociological answer. Landlords are particularly vulnerable to inflation. Indeed, they are probably one of inflation's primary targets, considering inflation is usually instigated by those with a collectivist mentality. Since depreciation allowances are inadequate to maintain or renovate homes when costs escalate rapidly, houses begin to deteriorate because landlords cannot afford their upkeep. In some cases, entire city blocks are abandoned by owners and what was once decent housing becomes a slum with all its attendant evils. Those who inflate in the interests of the little man, or obtain federal or state funds to clear slums are, in effect, aggravating the problem rather than curing it.

Businesses which are not given adequate depreciation are further invisibly taxed on inventory profits due to inflation. These taxes are paid out of capital. The businessman is squeezed and is forced to lay off workers, which adds to unemployment and again only aggravates slum conditions.

Employees who are squeezed are practically forced to join or form a union in the never-ending fight for "more." One can never be satisfied because to stand still during an inflation means to perish. Thus, much of the strength and aggressiveness of the labor unions probably comes from underlying inflationary forces.

In earlier times, monarchs "clipped" gold coins by shaving off the edges and returned the coins to circulation. That is why coins have fluted edges—so a clipped coin is immediately apparent. However, there is no need to clip coins today because most are made of worthless metals. Instead of clipping, our monarchs have taken the whole coin! Even the penny, which contains some copper, underwent severe shortages in 1974. That contributed to the stress on paper, and this printing press mania is extremely similar to that of mass counterfeiting. With a flood of new paper coming in, citizens are robbed of goods although they are generally unaware of it.

This massive embezzlement takes place at the highest levels. Politicians and their bank collaborators claim to deplore inflation. Yet, they do not find a satisfactory solution to repay those who have been robbed of the purchasing power of their savings or of the loss of their life insurance by the deflation of the economy. Their concern is not to make good or to stop the embezzling, but to urge those whom they fleeced to continue their trust in paper money while they continue their inflationary tactics. I predict that the problem of slums will worsen until gold is re-enthroned. It would be refreshing if a top government official spoke out as I have, even if it cost him his career. There was a time when men of towering integrity ran this country.

The next bull market could involve a tremendous resurgence in rails, cruise ships, and the coal industry, at the expense of airlines, auto, snowmobile, and pollution control equipment manufacturers and suburban real estate developers. The wave of the future is almost always clearly visible to anyone who is paying attention.

Because of the interrelation of currencies, and the fact that all countries are so deeply in debt to one another, the failure of any one nation's currency could lead to a chain reaction collapse, unless tight money and competitive deflationary tactics lead to a depression first. The coming turmoil will lead to major shake-ups of international political relationships. I pray I am inaccurate, but even war is a possibility. No one would benefit from a nuclear exchange. However, in perilous times people do not always work in their own best interests.

Amid the waves of futile and competing devaluations, in an effort to improve balance-of-payments at the expense of others, look for protectionist legislation and federal controls, depending on the business you are in.

Look for a national bankruptcy. Whether it is Italy, England, or some other country depends on what happens next. The role of the dollar will decline as the Arabs leave dollars and turn to gold. A national bankruptcy could mean that U. S. troops will have to pull out of Europe unless the European nations agree to foot the bill for their own protection. The 1960s were considered tense. In retrospect, we should have been happy. Perhaps Oscar Levant was right when he said happiness is not a thing you experience, but something you remember.

In the coming upheaval in international political thinking, there could be the elimination of foreign aid, a new export consciousness, less military aid, no more giveaways, a halt to buying "friends," less tourism, subsidies to gold miners and, incredible as it sounds, a movement to permit individuals to mint money, ending government monopoly. Hopefully, there will be no international Federal Reserve System or international bank. Although it would be more efficient and acceptable someday, world development is still too adolescent for this step yet. After the water is painfully squeezed out of the present system, money with intrinsic value will be minted once again, fully backed by and convertible into gold. When this occurs, the United States and the world will launch on one of the most incredibly expansionary booms ever seen, with a tremendous growth of trade between new partners, perhaps China and Russia.

Look for an end to so-called "soft" loans. The Eximbank will go bankrupt and, hopefully, the Federal Reserve System will be realistically altered or rendered impotent. International bonds will be backed by gold. Gold clauses will be included in most major contracts. Attorneys should prepare immediately for the understanding and use of such gold clauses, which were common before FDR's reign. Major creditors, such as the Arabs, will in all likelihood insist on gold clauses in the future.

The IMF is finished. Its function was to fix parities, so it is obsolete. Since no one takes dollars for gold, there is no unit for international settlements, and this will send the IMF into a well-deserved oblivion. It will recede into insignificance also, because it has become an emotional and political organization like the League of Nations and the United Nations. The 1973 Nairobi boondoggle proved that the world is already divided into monetary blocs. By 1976, at least one country will try to settle its deficits in gold at the free market price. It will then be in the interest of those nations to maintain a stable gold price. The first country to use gold in its money could have the currency of the future. Then, the Group of 20 top industrial nations, the IMF, Keynesian economists, the diehards and leeches, and those who wander from meeting to meeting with no real function, will be forced to seek productive jobs.

As inflation and deflation hover on a razor's edge, those who readily understand what is written here will see clearly that what I predict for the United States applies to the world as well. This is a unique nexus in time and space, where the national and international scenes have degenerated simultaneously. And why not? The United States has corrupted the international monetary system to the point where we could all go down together. It is not difficult to see that high U. S. interest rates will have to be maintained in order to attract dollars to prevent them from flooding overseas bankers and weighing the dollar market down. To my knowledge, no one has calculated the impact of the return these dollars will have on the U. S. money supply. This factor might be partially responsible for skyrocketing prices. The United States recklessly pumped dollars overseas for years, and Washington will panic at these now-unwelcome returnees. Furthermore, these dollars will accelerate the inflation rate, as more paper chases the same amount of goods. We really lived high off the hog.

Like most, the solution to this problem is deceptively simple. Nations, like individuals, must earn their keep. It is a pity that blood will probably have to be spilled before this basic truth is realized.

As will be described in the "Odyssey," in July 1974 Italy was allowed to use its gold, its last remaining asset, as collateral for another loan. Italy could not really have sold all that gold on the open market instead of using it as collateral because that would have sharply depressed the price of gold, with such a drop now in Italy's worst interests. This marks a transition because the initial discarding of gold allowed politicians to print more paper money and to manipulate it in various ways. This can no longer be done to gold. Germany lent Italy the needed money, so she cannot allow the price of gold to get very much below $118 to $150 (the collateralized value). As more countries understand that by writing up their national gold reserves, even if they are only tempted to it at this point, the incentive and vested interests surrounding the maintenance of a high gold price will become overwhelming. Even the United States will find it in its best interests to uphold the gold price to prevent more countries from going bankrupt or deflating. As understanding of this develops, investors will cluster to golds as an investment or a speculation. Italy's bail-out underlines the bullish power of gold.

The world needs new financial leadership. It is unfortunate that as foreigners shun the dollar, the United States is losing its financial primacy. All foreign contracts were once written in dollars, and most international businessmen spoke English. We could lose that distinction. Unfortunately, I do not think Washington politicians would understand this even if they did listen, so my confidence in Washington is close to zero.

One would never know it from reading the uninformed press, but the world is actually in the throes of another monetary convulsion. At my annual London gold seminar on November 10, 1973, I told a startled blue-chip audience that the market in London appeared to be on the verge of an immediate collapse. It began the following Monday and, incredibly, still looks as if it could go lower. Class warfare simmers in Lord Keynes' homeland. This is partially due to the monetary strains on their society. In 1973, France surprised the world by permitting the franc to float. This caught most of Europe off guard, and sent the Japanese yen reeling.

World banks are ill-prepared to cope successfully with current international economic strains. So one must be completely ready for the element of surprise at a peculiar time like this. The winds of deflation are in the air, as we scale the crest of inflation. Paper currencies are disintegrating, and those unprepared for the consequences could be financially wiped out.

What Do I Think of the 1974 "Economic Summit"? Will It Spark a New Bull Market?

President Ford's "economic summit," which ended in September 1974, was, in my opinion, an orgy of confusion, and an endless stream of "gimmes." All those genius economists assembled under one roof, with endless hours of hot air crossing their vocal cords, could not even agree on the cause of inflation. The only thing they agreed on was that the problems facing America would last for a long time.

It does not faze anyone that the same people who got us into this economic mess are being trusted to get us out of it. Those great economists in Washington appear frustrated because they do not or will not understand that this is the end of the system they have constructed. No thought is being given to a new system, or a return to the old capitalism.

Our economy is weighted down with government interference and controls like a corpse riddled with maggots. The only thing economists suggest, noting that government interference does not work, is more government, more tampering with the free market, more interference with interest rates, more minimum and maximum prices, and more regulatory commissions.

Our economists do not recognize that the invisible internationally crashing stock markets are dire warnings that government can saddle just so much nonsense on business before business collapses. The government should get off the economy's back, because its prosperity, in turn, will help the workers and the little man. Instead, the demagogues are trying to help the little man directly, at the expense of business. Both business and labor will suffer because of this error. This is a subtle form of the class-warfare ideas prevalent in socialism, which have caused economic failures all over the world. People are reacting to the symptoms of a rotten system—inflation, high interest rates, stubborn unemployment, and the rest. These symptoms stem from four decades of government over-regulation. While it is natural for uninformed people to react to such symptoms, and request government action to lower interest rates, since they have no concept of the intricacies involved, it is absurd that informed economists who should know better also clamor for a messianic government's interventions.

One summit conclusion was that the government should raise taxes on oil to discourage its use and establish controls, such as lowering national speed limits. This, of course, involves hiring more unproductive politicians. There is an easier and cheaper way. As I previously stated, I advocate letting the price of oil rise to any level it wants. If it gets high enough, alternative energy forms become practical. Shale oil producers, now in need of assistance, could then compete on their own. But the government distorts the free market by raising taxes on oil, and thereby gets its hands on more money to spend on any project it thinks worthwhile at the moment. Excluding the government from the oil picture would sort things out quite nicely. Of course, the suspicion arises that Washington holds domestic oil prices down in order to conceal from the public the true inflation rate—at the expense of oil company stockholders.

I got the impression from watching the economic summit that our most important national need—a reexamination of our national attitude toward business—will be ignored. Instead, the message is clear that the government wants to run business. I, therefore, look for an unprecedented increase in government interference. The fact that no one at the summit paid much attention to the crashing stock market—no one even asked pejoratively, "Why are the rats leaving the ship?"—suggests that it does not matter to economists what the market does. I think we will see nationalization of industries, moratoriums on debt and interest, all kinds of "freezes," a revival of the discredited government's "makework" proposals, and profit "guidelines" to circumvent discredited wage-price controls. As anti-profit fanatics come out of the woodwork, there will be a dangerous drift toward dictatorship. The more power drifts toward government from increased intervention, the greater that danger will become. I would not be surprised to see food rationing, as our national food-producing mechanism gets whipped back and forth by the vagaries of first inflation, then deflation, and finally depression. During this time, the flight from cash will have made almost everyone illiquid.

People are afraid to hold cash, since its value declines with inflation. Later, as deflation takes hold, cash will increase in value, and the flight will be from tangibles (art and stocks) to cash. If you have cash at the next bottom with liquid buying power intact, then wise investing at that point could make you rich indeed.

Sad to say, it seems Sewell Avery (ex-president of Montgomery Ward) was right, after all, when he foresaw a depression in the 1950s. But he was decades premature.

Can you imagine an economic summit where gold was not discussed, where the word "capitalism" was not used, where no one seriously considered that the best solution might be to dismantle much of our economic legislation? These issues will be strongly debated before this bear market makes its final bottom. Unfortunately, few in power realize that these concepts are the only "silver bullets" that will work on this, the king of all bears.

Since President Ford took office, the Dow-Jones Industrial Average has virtually collapsed. Almost no official mention is made of Wall Street's invisible crash, but obviously investors see in Ford a continuation of the same old tired policies which have led the economy to this yawning canyon. The pressures which will accumulate around President Ford will undoubtedly cause him to make big changes in his retinue. Conceivably, he could resign during the middle of a full economic collapse if we had one. Some will claim that Nelson Rockefeller engineered the whole thing to attain a Presidency he could never have gotten by popular vote.

Although Ford has declared inflation our "number-one enemy," there are no weapons being used against it except tight money (and even that is kept on a tight leash). The "war on inflation" looks like the way John F. Kennedy invaded Cuba at the Bay of Pigs—inadequate determination and insufficient weaponry.

In 1964, the federal budget was $118.4 billion. By 1971, the $200 billion level was passed. Fiscal 1975 shows Washington's spendthrifts at the $305.4 billion level, and they are having extreme difficulty cutting that sum back to $300 billion.

In the last decade, there has been deficit spending in every year except 1969. The year Lyndon Johnson recorded a $25.2 billion deficit, the news was greeted with yawns. So staggering is this sum that it is more than double the official value of all the gold in Fort Knox—and it was blown in only one year. The public is to blame for this crash because they have not paid attention to whom they have been electing. To give some idea of the enormous sums involved, in the five fiscal years since the small surplus of 1969 was registered, total budget deficits of $66 billion have been recorded, and the federal debt is now at the staggering $500 billion level. There is no way that that can be paid off. Those who buy Treasury bills should keep this in mind. It is not outside the realm of my imagination that the U. S. government could repudiate this debt—for instance, to anyone who has, say, more than $10,000 worth of such debt. This absence of fiscal restraint is being battled with the ineffective weapon of "tight money," wielded by Dr. Burns. The pressures on him after the 1974 economic summit were such that the 12% prime rate began to plunge mysteriously, suggesting government tampering.

The complex question arises as to whether or not there will be easy money in the future. Sooner or later, the economic deterioration I have been predicting will become obvious even to Washington. At that point, the clamor to "reflate" the economy will become intense. You have already seen an example of this at the economic summit, where the pressure was almost universal to force Dr. Burns to ease up on the tight money situation. This manipulation of the interest rates was one of the features of the mid-1920s and, in my opinion, was one of the primary causes of the 1929 Crash. Those who manipulate gold or interest rates are tampering with profound levers deeply involved in the capitalist processes. Generally, such tampering is done in a clumsy and unsophisticated fashion, and places more emphasis on social needs than is economically feasible.

If Ford reflates the economy, there will be an immediate stimulation, as there was during the early stages of inflation after Roosevelt's election. But it will ultimately lead to triple-digit inflation, which would lead to a complete destruction of our currency and the horrendous preconditions which spawned Adolph Hitler.

On the other hand, if we decide to take our medicine immediately—probably politically impossible—and develop a sound currency, permitting interest rates to go where they want, and free the economy from the myopic socialist shackles weighing it down, then a genuine decline in interest rates becomes a possibility. During inflationary times businessmen are not eager to borrow money, since business is bad. The demand for money will drop and if there is a sound currency without inflation, I can easily envision interest rates getting below 3%, which is where they should be if the economy were properly run. Interest rates are like a thermometer, so 1974 prime rate levels of around 12% showed a very sick patient.

Thus, you have two scenarios: the first leading to sky-high interest rates and a complete destruction of our currency, bonds, and all forms of paper debt instruments, the ruination of insurance policies and annuities, catastrophic disruption of an ability to write long-term contracts, and so on. In the second, an immediate depression will be allowed to run its course, which will make workers happy simply to have jobs, and there will ensue a slow and sobering return to sanity.

The choices are suffering an immediate hangover, or getting drunker to avoid an immediate hangover, thereby compounding miseries. If reflation is chosen, it could lift the stock market for as much as a year, into 1976, and give the nation a false sense of prosperity. But stocks will remain surprisingly sluggish, unemployment stubbornly high, and interest rates amazingly resilient to declines. Overall, it will merely be a cosmeticizing of all our national problems. When the problems break loose again, there will be no stopping them. It will be a killer bear like nothing yet seen on Wall Street.

I have discussed Italy's economic plight. England is another candidate for catastrophe. An incorrigible balance-of-payments deficits, ever-accelerating inflation, declining productivity, a conviction on the part of the working class that they are somehow at war with management (echoing the socialist doctrine of class warfare) point to some form of economic and political convulsion for England. Her leading political parties are identical, so the public is given little choice. All leaders are terrified of the unions, and those who pretend to control them act like men holding a tiger on a tissue-paper leash. A possible takeover by the military or paramilitary forces is conceivable for England. Perhaps this points the way for America which, in recent centuries, has followed England's political evolution. These problems pave the way for a "man on a white horse" for England, perhaps even someone like Enoch Powell, former member of Parliament.

The main issue in the next few American elections will be the U. S. economy. Will inflation end? At the economic summit, everyone provided evidence of suffering special hardships from inflation and, therefore, they argued that they needed more government spending and more creation of money—the very "solutions" that got us into this situation. On the contrary, in the first few months of the Ford Administration, there was a new $25 billion education bill, an $11 billion mass transit bill, and an $11 billion housing bill, unbudgeted additional millions in veterans' benefits, and small business loans, subsidized loans for cattlemen and banks, and a refusal to put off federal pay increases. The government can only talk of "holding" federal spending at $300 billion (which is $25 billion higher than last year) while there persists the increasing pressure for spending more for creation of jobs, food stamps and welfare. The interest alone on our national debt already costs us $29 billion a year, and is rising steadily. What can you expect from the group that "held" federal spending from $118 billion to almost triple that sum in a single decade?

The Dines Letter has forecasted a devastating bear market. Unlike the barely visible post-World War II dips, this will be a genuine bear market, a consequence of limitless paper money.

In the coming period you must avoid being brainwashed. The people who have seen this coming: Ira Cobleigh, John Exter, Edson Gould, Percy Greaves, Jr., E. C. Harwood, Tom Holt, C. V. Myers, Richard Russell, Harry Shultz, and many unnamed others, have proved an ability to lead you to high ground. People on both sides of the gold fence have been quoted throughout this book. Remember the names involved. Read *Atlas Shrugged* by Ayn Rand, if you want an idea of the kind of societal breakdowns which could come.

This is the time to disperse your funds in a wide selection of banks in different countries. Stay as flexible as possible to cope with changing conditions. Mobilize your portfolio between bullion and gold and silver stocks. Keep safe deposit boxes in different locations. A new world with a new set of politicians is coming.

In the past, bonds, preferred stocks, cash, blue-chips, and utility shares were the ports of refuge. Not any more. This is the Golden Era. David has picked up his sling and is moving toward Frankenstein.

I reiterate my stand that you should buy precious metal-related assets. There will be scares from time to time, such as U. S. Treasury threats to sell gold to drive the price down. This will be an empty threat; remember, they have scorned the rise of gold all the way up. Their track record does not suggest they are to be taken seriously. Events have moved from arenas in Washington, Paris, and London to the free market. All currencies will be valued in gold, like it or not.

At my July 1973 New York gold seminar I read the table below to the audience. It was a list of all the popular "growth stocks," and showed their percentage increases from January 1962 to date. I threw in a curve ball including ASA, my favorite gold stock. To everyone's surprise, only Xerox out-performed ASA in that eleven-year span. I updated the chart as of September 17, 1974, and ASA had beaten them *all*. I had picked the most outstanding stock of the decade, which should dispel criticism that gold did not keep pace with growth stocks in the 1960s. It kept pace even at the market high in 1973. All it took was patience. Many people still do not know much about the company, which is a good indication that it has a long way to go before it reaches its final Top.

Stock	Price Jan. 1962	Price 7/16/73	% Inc.	Price 9/17/74	% Inc.
Xerox	11	154⅛	1,300%	75¾	589%
ASA	5	51	920%	67¾	1255%
Tampax	17	111¾	557%	27¾	63%
Polaroid	27	137½	409%	18¼	(32%)
Avon	25	119¾	379%	21⅜	(14%)
IBM	123	319	159%	161⅞	32%

Adj. for splits & stk. dividends.

What You Should Do Now to Protect Your Assets

On June 23, 1974, *The Dines Letter* wrote, "Can you hear the faint creaking sound made by a bone under pressure as it is about to break? That is the sickening reaction we get."

The speed with which the market collapsed in 1974 suggests that something horrendous is about to happen. Maybe it is a national bankruptcy or a repudiation of debt of a country like England or Italy, or even a large institution. Crashes like this have always preceded catastrophes. Clearly, dividends are not going to stay at these high levels for long, which is one reason why stocks moved down so urgently. Investors are beginning to understand that this is the longest and most vicious bear market in our history. It is an unraveling of monumental debt, which could lead to a collapse of the house of cards of international promises written by illiquid banks on meaningless paper. Suspicion of financial institutions is at levels not seen since the first Great Depression.

The bear is moving on a personal level to assure that none of us survives. This is an interesting concept. One must ruthlessly question every method of storing assets. Even paper cash is suspect because it might be wiped out by inflation or repudiated by government. Gold bullion and coins appear difficult to fault although they will have wild shakeouts and could bankrupt those who hold them on thin margin. Will Treasury bills be safe? Superficially, they should be. However, what if there is a national repudiation of all debt of more than say, $10,000, to any individual or a wave of bank failures, or a freezing of bank accounts? All these might be in the cards. There is far too much debt in the world to repay, so some repudiation is a possibility. This scene represents the biggest challenge to investors ever, the biggest redistribution of wealth since communism. Communism is a danger in at least one country, perhaps in Italy, Germany, France, or Portugal. While not a probability now, a depression could shift the balance.

I have been predicting exchange controls for years. So if you have not gotten money out of the country yet, do it immediately. And do not make excuses like "I don't know how."

In August 1974 there was still $4.51 billion in margin accounts, most of which will likely be wiped out before this bear market is finished. The main point to remember is that it will be a debt-liquidating bear market. So common stocks, except silvers and golds, look like dubious investments during this invisible crash.

No measure of safety in the next two years is too extreme to be considered. It is difficult to get hurt by being conservative. If I am right, those who have not prepared according to these dictates could be bankrupt. And when IBM is at unbelievably low prices, those who are unprepared will salivate but will not have the cash with which to buy shares. Those who do buy at that time will be rich indeed, because after this depression I see a most incredible bull market.

Almost everyone now admits that the economy has entered a recession. I predict a depression will follow—no hedging, ands, ifs, or buts. There is too much water in this economy, too many underutilized people, too many leeches and loafers, and not enough real producers. Most excesses will have to be wrung out forcibly because no one will volunteer to do it, and politicians are too spineless to demand purification. That is what our collapsing stock market is telling us. It is a message loud and clear to anyone who is sensitive to Wall Street throbs. The parallels to the 1929 Crash are not pleasant and are even terrifying, but you must know them because forewarned is forearmed.

This is a time for maximum caution—caution on a level most investors are not accustomed to. If I am incorrect, the worst that will happen is that you will be stuck with cash. After all, I am not recommending anything crazy. What I am recommending is that this is the time to dump your property and art, anything for which you can still get a decent price. Prices are coming down. Ignore those who have inflation fixed in their heads. Sometime soon I expect a collapse in currently sky-high commodity prices, which could come out of the blue. There will be massive inventory liquidation and a desperate movement from things into cash. If you wait until the mob packs the door, you will never get out. You must move realistically ahead of time and not try to get the top price.

Across the valley and up in the hills, I see paper money tied to gold and, therefore, sound. Inflation will then, at last, be under control. There will be very low interest rates, almost zero unemployment and a great boom.

Will Computer Growth Save Computer Stocks?

Some people have listened to my advice and sold all their stocks, desperately clutching their computer stocks as a last resort. They feel that with high projected computer growth, there can be no way to lose.

Remember, computers are labor-saving devices. Behind development of these devices in recent decades was the ever-rising cost of labor. If I am right in my opinion about an oncoming depression, wages will decline (big unions notwithstanding), so the cost of labor will decline and begin to compete ruthlessly with computers.

There are some uses for which computers are indispensable, such as guiding missiles. Furthermore, computers will show fantastic long-term growth because labor will again someday become more expensive. There will be further economic ups and downs.

However, for the coming period, computer stocks can continue to take the kinds of vicious spills already seen in 1974.

Will Silvers Rise from Here?

Unlike gold, industrial demand for silver mushroomed with the technology boom following World War II. No wonder. The metal is outrageously underpriced for these inflationary times. After all, it is a precious metal and, as such, should have far greater value than other metals because it is a store of value. The need for silver transcends its usefulness as a mere commodity.

Silver rarely occurs alone. Around 78% of mined silver is a by-product of copper, lead, and zinc mining, and is why silver production cannot be expanded or contracted rapidly. Silver production is now lower than it was fifty years ago. If economic conditions deteriorate as I expect, the decline in copper, lead, and zinc production will result in much lower silver production. This will more than offset any prospective disadvantage to silver if a depression should also cause a contraction of its industrial demand.

For more than fifteen years, total demand for silver has been far larger than new mine production (by 234 million ounces in 1973 alone). Since 1958, the U. S. Treasury has supplied the difference—around 2 billion ounces. This source of supply is now ending, since the U. S. government only has around 140 million ounces left.

I am probably the only goldbug to strongly recommend silver-mining shares now. If the price of gold goes as high as I predict, those who feel they have missed out are going to look for a substitute. That substitute will be silver and perhaps, to a small extent, platinum and palladium.

The primary reason silver interests me is that silver gains should be even greater than gold, on a percentage basis. Not only are silver stocks currently at depression levels—but the ratio of the price of gold to silver shows that silver could have a much bigger advance from this $4.00 an ounce level than anyone now believes.

In my opinion, gold will lead the way among the precious metals on the upside, and eventually silver will join the parade. The percentage gains from these levels will be stratospheric. Therefore, I think some 20% or 30% of your investment in precious metals should be in silver, either bullion, coins, or stocks. When gold stocks are up and silver stocks are down, the percentage of silver can be increased. Purchase bullion only on weakness, with not greater than 30% to 50% margin, and preferably pay for it in full with delivery taken. Those who purchase bags of silver coins are probably even better off, since you can always spend the coins without paying to have them assayed, in case of depression or other national emergency.

Constructing a gold portfolio is a very individual thing. Large investors should certainly get professional help in working out details. Smaller portfolios should place at least 70% in gold- and 30% in silver-related investments. Those who are in high tax brackets or do not follow the market closely should, after setting aside cash-equivalents, move into an investment trust. These are excellent long-term holdings. Those who need high income should buy shares of the individual South African gold mines. The bulk of any investment should be placed in South Africa because they have the best gold mines in the world. Those with more money should branch out into some of the Pacific mines and also consider investments in mines which deal in both gold and silver. Those who are undecided should consider owning some gold coins as a starter.

May you prosper and may good times return soon!

7
An Introduction to the Odyssey

You have come to what could be the most important part of the book. What follows probably has never been done in a book before, but its uniqueness has a function. It might be a challenge to read and study it carefully, but I promise that when you finish, you will be a different person. It is akin to the casebook method used in law schools, whereby actual experiences are studied.

The primary function of this book is to help make sure Great Depressions do not happen again. This is written for another generation.

You will see the pathos between the lines as a young security analyst stumbles across a depression which was invisible to just about all the experts. In my initial comments on gold you will see how guarded I was, because I was surrounded by people who hated gold. If you are one of them, remember that behind every hate is a fear. Find it.

I learned to cherish the ignorance about gold, for ignorance is the wetnurse of originality. I was continually amazed at what people believed. As my knowledge of gold and people grew through the 1960s, I learned that if I do not believe in myself, I will believe in something a lot worse. I came to learn that even cynics are sometimes right.

Now that you have read the history of gold, you will more clearly see in the coming pages how many of the old errors are being repeated today. The actual quotations from *The Dines Letter* are reprinted here with no effort to edit. These letters were written under pressing deadlines, and I was quite alone. At first, you will see some fairly immature comments. Compare them with my blistering commentary as you move into the 1970s.

The "Odyssey" chapter was written for the uninitiated; it is a basic explanation of the various crises we have in the world today. Bad gold policies are truly the root of all evil. Remember, it does not matter what is engraved on a gold coin.

As you read through the "Odyssey," watch my predictions for the future, not all of which have come true—yet.

Note how many "experts" of the day were quoted, and how wrong many of them were. Some of them are still in the U. S. government today. I named people as I went along, and I think you will be shocked to see who said what, and when. This book could be the raw material for a scandal far greater than Watergate. True, much was done in the name of a wish to eliminate hardship for people. But try to eliminate depressions incorrectly and we wind up with inflations; then depressions follow anyway. Everybody will be hurt somehow by the coming depression, including you and me.

Before you begin the "Odyssey" chapter, the next brief chapter will show how I figured out the gold crisis. It will enrich your reading of my early gold comments in the "Odyssey" chapter, for it explains why I was so terse at first.

8

How a Young Security Analyst Blew the Whistle on the U. S. Treasury

I was one of the first, if not the first, of the well-known American security analysts to detect the gold crisis. It is really nothing more than the story of the Emperor's Clothing.

I am not telling this story for ego purposes, or I would have done it sooner. Its purpose is to convince readers to become more conservative economically for the near-term, and perhaps to serve as an encouraging example to other security analysts who develop opinions which conflict with the Wall Street Establishment.

Many people who followed the advice of *The Dines Letter* made killings on gold, and others who did not were hurt badly. It should be stressed that I do not envision the end of the world approaching soon, but I consider this a surgery, or catharsis if you will, a prelude to a boom in the 1980s which could see the Dow-Jones Industrial Average soar to the 2,500–5,000 level, or even higher. I will have to wait until we are closer before making my final decisions.

Not long after being hired as senior security analyst with A. M. Kidder & Company in 1960 I read an article by Jacques Rueff in *Fortune*, in which he warned of a monetary crisis. It was the first mention I had ever seen of the possibility and frankly, while it seemed extreme, it made enormous sense to me. The few people to whom I mentioned the possibility sloughed it off as if it were a complete waste of their time. However, a casual glance at gold-mining charts* convinced me that there could be something brewing in the gold shares. Little did I realize that it would cost me my job and lead me on a seemingly everlasting path of

* See *How the Average Investor Can Use Technical Analysis for Stock Profits* by James Dines, 1972, for an explanation of chart reading.

uncomfortable situations. At that time, the price of gold was government-fixed at $35 an ounce, and had been locked there since January 31, 1933. The price of silver was, likewise, locked at 90¢ to 91¢ an ounce, with no apparent hope of an increase in sight.

When I began to perceive the light ever so dimly in 1961, I could not read about gold anywhere. No hint of documentation existed, and gold seemed invisible. I had to analyze the money question carefully, disagreeing with the entire intellectual structure of the country. It was a painful yet exhilarating experience.

In 1960, my reward for accurate predictions was permission to call A. M. Kidder's weekly market letter *The Dines Letter*, the first time they allowed personalization of their market letter in ninety-four years. However, the company was unwilling to allow me to state my views regarding gold and silver. I increasingly understood that precious metals were destined to go very much higher, and that the rise would probably be accompanied by a crash. As a compromise, A. M. Kidder permitted me to print only one sentence regarding my particular views. You will see this sentence in the "Odyssey" chapter, dated March 24, 1961.

I correctly foresaw the crash of 1962, during which gold and silver shares skyrocketed, which firmly locked into my mind that I was on the right path, and provided a "dry run" of what was to come in the 1970s. I repeated the same sentence week after week throughout 1961. Specifically, I recommended ASA at $2½ (adjusted for splits through 1975) and it is most interesting to note that in 1974 this stock went over $50, illustrating a concept which I refer to as the "Dines Theory of Gold Contracyclicality," which posits that when the stock market crashes, golds and silvers skyrocket, and vice versa.

Eventually the censorship at A. M. Kidder affected me to the point where I either had to sell my soul or get out. So, in frustration, in October 1962 I left with *The Dines Letter* as a separate entity because I envisioned a new bull market. The crash I had envisioned for the market had come in 1962, but it did not break the country wide open, so I concluded that the market had one more long lease on life. I started my letter at a fortuitous time and grew along with the economy, never faltering in my repeated insistence that golds and silvers were destined for huge rises.

Sam Coslow, then owner of *Indicator Digest*, helped me a great deal in starting my service. In gratitude, I taught him all I knew about the gold situation, and convinced him in one long afternoon that golds were due for a great rise. To his credit, his became the second advisory service to become as bullish on golds and silvers as *The Dines Letter*.

It was still nearly impossible to find other analysts' or economists' write-ups on the subject of gold because Wall Street interest was zero.

In fact, mentioning gold-mining shares was a good way to get a laugh. So I rolled up my sleeves, and for months dug through musty tomes after office hours to learn as much about gold-mining shares as I could. I concluded that the best gold mines were in South Africa. However, I was concerned about the political situation there, which was then receiving chilling newspaper headlines after the Sharpesville riots. By sheer chance in 1964, I met a South African at a party who told me that the U. S. press was distorting the situation. This was a great shock to me, and after much agonizing, I decided to go to South Africa to see for myself. When in doubt, look. I then wrote some articles on South Africa for *Barron's*,* one of the earliest publications to recognize and appreciate my work on gold. I returned convinced that there would probably be a revolution and mass violence in the United States before it would occur in South Africa, which nearly turned out true in the late 1960s. In those days, some South African gold-mining shares were yielding over 30% per annum because of the Sharpesville riots, and in retrospect, it is probably true that the world's best buying opportunities occur when there is blood in the streets. Remember that if the market crashes the way *The Dines Letter* expects it to. At any rate, I decided that ASA (then called American-South African Investment Trust) was probably the best vehicle for a diversified, supervised portfolio in South African gold-mining shares, and it became my première gold recommendation.

While I was researching these gold stocks, I found some very old charts dating back to the 1920s, and was almost knocked off my chair when I saw that they had skyrocketed between 1929 and 1934. I proceeded to go through other bull and bear markets and noticed the astonishing fact (which I had never seen in print anywhere before) that when the market went down, precious metals went up and vice versa. I was then struck by the logical conclusion that if gold shares were destined to rise greatly, then the market had to crash on a long-term basis. That was one of the many bases of my long-term bearishness on the market, which you will see in *The Dines Letter* quotes in the "Odyssey" chapter.

I was badly shaken in 1962 when President Kennedy announced on Telstar that the dollar would never be devalued. The next day, my new company was struck by a wave of cancellations. One subscriber called me and said, "Mr. Dines, I have just one question for you. Do you mean to tell me that *The Dines Letter* is right and the President of the United States of America is wrong?" It took me a few seconds to catch my breath, but I had added the numbers up many times and kept coming up with the same answer. If that made Kennedy wrong, then he

* June 1965.

would be wrong. At the time, ASA had plunged sharply overnight. Do not bother looking for that drop; it is now only a tiny jiggle on the charts. But at the time, subscribers experienced big losses on ASA, and all I could do was to urge them to hang on. It was good preparation for the tremendous gold plunge after 1968 which shook out many gold investors. All *The Dines Letter* could do was insist that investors stand their ground. I am telling you this story to urge you to be as stubborn as I was. Do not listen to people who tell you to take profits each time golds go up a little because there is really no adequate replacement for them in case you are tricked out. If they wish to have little dips, let them. Nowadays, with more and more advisors picking up on precious metals—and now that I meet so many who not only agree with me, but who have actually gone further—it is easier to get into gold shares because there is more of a mass-herd acceptance. However, so many people are voicing their opinions on gold these days, that it is becoming a bit more dangerous in terms of being tricked out of the group too quickly.

In the early days while educating myself on gold, I had a great deal of difficulty finding anyone with whom I could discuss the topic. Nowadays, most people own gold shares at big profits, but in the mid-60s at a picnic given by the New York Society of Security Analysts, I asked a number of analysts what they thought about gold. They were all uninterested in the subject, save one analyst who told me that he would not buy a gold share for someone even if they demanded it. I said what if they nonetheless insisted, and he replied, "I'd throw him out of my office." It was then that I found the confidence to persist in my belief.

When others tell you to sell prematurely, stand your ground. Someday it will be time to sell gold- and silver-related assets, and that choice will be as difficult as buying and holding them all the way up. Read the "Odyssey" chapter carefully, for with the benefit of hindsight you will see it was the correct path for everyone. Perhaps you will spot it again in the future.

9

An Odyssey Through the Gold Comments in <u>The Dines Letters</u>, 1961-1974*

ERRATUM

On page 147:

"Homemaking Mining" should read

"Homestake Mining"

ırea 18-21.

ʒold specu-

ːh 24, 1961

ining. It is
ɔull market
m 1929 to
ing Mining
ıte in 1960
iding, gold

The gold stocks are obviously a unique group and cannot be measured by conventional yardsticks. Unlike any other companies, they have a guaranteed market to the government at a fixed price of $35 per fine ounce. A disadvantage in a period of general bullish growth, it is a major advantage in times of recession because earnings and dividends are more secure than any other group. Golds should be looked on as

* *Note*: To enhance readability, I have not used ellipses to indicate deleted passages. No changes have been made in any of the quoted material that appears.

** Because it seems to me to be useful in relating my comments to the market conditions at the time, you will find the Dow-Jones Industrial Average ("DJI") and the American-South African ("ASA") market price for the first day of each half interlineated in capital letters. ASA underwent three 2:1 stock splits through early 1975, so the prices have been adjusted accordingly. For instance, the January 1, 1961, price was adjusted by dividing 20 by 8.

income stocks with exceptionally secure dividends irrespective of the economy, plus capital gains possibilities in a bear market. On top of all this is the hotly debated possibility of a rise in the price of gold, which would put gold stocks through the roof if it occurs. The amount of gold holdings behind a nation's currency is what fixes the value of that currency in relation to the others. When a nation's gold holdings decline excessively because of chronic deficits in international trade, the amount of gold behind each dollar shrinks to the point where an increased number of dollars is needed to represent a fixed amount of gold and it amounts to an increase in the price of gold. This is called devaluation of a currency. If this country continues to lose gold in the next few years, devaluation will become a strong possibility.

Gold ore around a mine is normally uniform in gold content, which accounts for uniform earnings over the years. South African mines, especially in the Orange Free State area, are the only exceptions in the Free World, and are unquestionably the most fabulous in the world because the richness of ore increases with depth. Were it not for the political situation in South Africa, we would recommend no other gold stocks, but prudence dictates that no more than 30%-50% of the entire investment in gold be placed in that area. My favorite is American-South African Investment listed on the NYSE, which is sort of a gold mutual fund.

March 30, 1961

*　　*　　*

In preparing for the bear market expected by 1962, continue quiet accumulation of gold stocks: *Homestake Mining* (HM - 43⅛) and *Teck-Hughes* ($1.70 bid OTC). For high value and high income in South Africa, balanced by political risk, *Free State Geduld* (13 bid OTC) and *Western Holdings* (16¾ bid OTC). For those in high tax brackets, *American-South African Investment* (ASA - 20), a gold trust.

April 14, 1961

JULY 1, 1961: DJI 689.81; ASA - 2¼

*　　*　　*

Market commentators barely mentioned silvers this week and, according to the Theory of Contrary Opinion, this is good news indeed. Also, low priced Canadian mining stocks continued higher this week. Last July, we strongly recommended to our readers an article in *Fortune Magazine* by Mr. Rueff, a leading French economist. This article discussed in depth some inherent dangers in the present international financial gold structure.

Gold stocks were firm or slightly higher. There was some relief over the lack of a large gold drain last week. This relief over a lack of loss is a dangerously defensive attitude, and we should be disappointed when our country does *not gain* some gold in a week.

December 8, 1961

JANUARY 1, 1962: DJI 724.71; ASA - 2¹¹⁄₁₆

*   *   *

Watching the golds scoring new highs these days makes us think back to the spring of 1961, when A. M. Kidder & Co. was the only leading brokerage house to recommend a shift out of "growth stocks" into golds. Nobody agreed with us then; that is how we knew we were right. While a few are jumping on the bandwagon now, there has not been enough of a stampede for us to sell them our Amer-So African bought around 19, Homestake bought in the low 40s, or the Dome bought in the low 20s. Buying of these three is still recommended, anytime, to average down on weakness. *Amer-South African* (ASA - 30) is by far the world's greatest gold bargain. This closed-end mutual fund not only still sells at a discount from assets per share (last reported around $35.00) but the stocks it holds are all deeply depressed. Furthermore, these Kaffirs have been picking up recently, and it would not surprise us to see the net asset per share reported over $40. ASA is especially attractive to those in high income brackets because the dividend is kept deliberately low. Stocks in ASA's portfolio typically yield 10-15%, and nearly all of the high income is immediately reinvested in golds. The American investor thus pays no taxes on the high income, and the reinvestment makes for a built-in growth factor. Recent strength in Kaffirs could be indicative of increasing political stability in South Africa. In fact, we would not be surprised if there were an announcement, in the near future, that currency restrictions were being removed. We do not intend to argue as to whether the U. S. will devalue or not, but we bought golds to protect ourselves from a government that has such little respect for a paper dollar that it feels 3% inflation a year is good. Nor will we be dissuaded by the fact that men in high places say that devaluation is impossible. Put yourselves in their places and you can see they never will be able to say anything else. The proof is in the pudding. ASA made a new high at 30 yesterday, and Homestake over 53 the day before, when pitifully few new highs were being registered. We might be in for fantastic fiscal deficits, despite all the talk about "the end of inflation," and record prosperity. While the modern world places great strain on America's assets, we must do first things first, and then only what we can afford. Even we must learn to live within our means. I specifically think of the $5 billion spent annually to help rich farmers glut our storage bins, a sum that would be better spent among underdeveloped areas in our own country. Or some of the waste and inefficiency in our foreign aid program. You and I can do nothing to stop the situation, other than protect ourselves and pray that someone in power begins to understand. People say gold is useless. Not true. It is demonstrating its function right now. Gold is the ballast for the printing press used in making paper money, and gold relentlessly punishes offenders.

June 8, 1962

JULY 1, 1962: DJI 573.75; ASA - 3⅝

* * *

In the week ended Wednesday the Treasury's gold stock fell $50 million after staying unchanged for three weeks. This made the year's decline $792 million, leaving Uncle Sam with $16.098 billion (on which we owe around $24 billion). Investors are urged to view purchase of gold stocks not as "dirty speculation," but as self-defense of one's hard earned savings from a government which seems unwilling or unable to stave off trouble by changing its policies.

August 31, 1962

* * *

Get ready for a barrage of propaganda during next week's annual meeting of the International Monetary Fund and the World Bank, replete with additional denials that the price of gold will be raised. Too much protesting, seems to us. Also prepare for implications that gold stock buyers are somehow "Unamerican." Ridiculous. Most people invest in gold stocks in self-defense against government policies that could undermine paper dollar life savings, just as the dollar has lost half its purchasing power since World War II (which was at least partly due to Roosevelt's 1934 devaluation). But this is a surprisingly emotional topic, and facts are rare. One fact is that in the week ended Wednesday, the Treasury's gold holdings fell another thirty million to $16.068 billion, making this year's decline $822 million so far. Another fact is that gold securities (Homestake, Dome) are undisputedly in long Major Uptrends. In other words, there are bigger buyers than sellers, and big money usually does not get big by being wrong. As for non-government expert opinion, Per Jacobsson (chief of I.M.F.) says it is time to worry less about devaluation and more about deflation, and, balance in the United States international payments is "in sight." Leslie Gould of the *Journal American* acridly replied "What Per Jacobsson is expressing is a hope rather than a fact. President Kennedy, in pledging there would be no devaluation, expressed the same hope, using the same figures." I agreed when Mr. Gould added that foreign countries who owe their post-war prosperity to us should now share more of the foreign aid burden. Why doesn't our government do something about this, instead of worrying about furtive "speculators and raiders" like a shadowy paranoid? Speculators could not raid our currency if it were truly sound.

September 14, 1962

* * *

Anti-gold speculation propaganda this week was buttressed by Treasury Secretary Dillon's pledge that the dollar will not be devalued. However, history shows that official government denials have occurred before devaluation (Roosevelt denied it a few years before he devalued).

September 21, 1962

* * *

Studied unemotionally, remember that Homestake is the last important gold mine to survive in this country, despite federal policies that have been proved to be less than wise. However, it is not impossible that somebody in Washington will awaken and try to save our gold mining industry, if only by means of subsidy. Furthermore, should Homestake stop producing gold it would only aggravate the world shortage, further endangering our dollar internationally, and aiding other gold stocks.

[A]s a group there is no better hedge against deflation and/or devaluation, or even as a simple protection of bond positions. Wise portfolio planners, more of the investor type, should plan to retain all gold selections until around 1965, through thick and through thin.

Do not wait for gold stocks to make headlines again before buying. Patiently retain *American South African* (ASA - 30⅜), *Dome* (DM - 25⅛), *Homestake* (HM- 48).

November 9, 1962

JANUARY 1, 1963: DJI 646.79; ASA - 3⅜

* * *

The *long-term* position taken in the Spring of 1961, and repeatedly reaffirmed since, is unchanged. We believe this whole post-1949 bull market will end, as it already has for many individual stocks, sometime within a few years. We believe the 1962 crash was the handwriting on the wall of a grossly overpriced situation, and not just an accident.

March 1, 1963

*   *   *

While the American public as a whole is still fairly ignorant of the gold situation, professional accumulation continues and the golds were noticeably firm this week. Our favorite, *American-South African* (ASA - 33⅛) surged into new 1963 high ground and is now challenging its 1959 all-time high of 34½. Combining the features of a mutual fund with a fat 30% discount from its underlying assets, with an income growth rarely found among golds, and providing one of the few hedges against devaluation available to Americans, American-South African is an outstanding candidate for every investment portfolio despite the political uncertainties probably exaggerated by our press. America's gold holdings are shrinking, and gold cannot be dispensed with as the common denominator of all currencies. The penalty of profligacy must come just as it did with silver, which we predicted early in 1962 based on precisely the same type of reasoning.

March 22, 1963

*   *   *

We are not going to say de Gaulle must show gratitude, for true giving should never expect anything in return.

We should remember that France could ask for the gold we owe her, and thus conceivably start an international financial crisis single-handedly. We, on the brink of a gold crisis, should expect neither help, gratitude, nor mercy. Nor should we be angry with de Gaulle, because a France that would ask for her gold would surely think herself doing the right thing. (De Gaulle is probably a sincere man, agree with him or not.) A far better vent for anger, than at people who owe us nothing, would be the civil servants of the United States who got us into this rotten position of vulnerability.

This nation must begin to live within its means, for "too late" is far too late to worry. Not much that ordinary people can do, other than to place 10-20% of all portfolios in gold and silver stocks. Our outstanding favorite is *American-South African* (ASA - 33½) which brushed against its all-time high of 34½ this week.

March 29, 1963

* * *

JULY 1, 1963: DJI 701.35; ASA - 4³⁄₁₆

An article in the *Wall Street Journal* this Wednesday was entitled "Gold Gloom—South African Mining Boom Near Its End." The article implied that South Africa's gold production is petering out. It seemed to ignore the boom in the Orange Free State area, and concentrated on the bad and nearly depleted mines. We flatly assert that South Africa still has the richest gold mines in the world, and is second to none. *American-South African* (ASA) is still the best gold investment; it provides a balanced investment medium for anybody seeking gold representation. Gold is a rare commodity in that it is not abundant. But that is precisely why it is the common denominator of all currencies. When all are evaluated by the amount of gold backing them, international trade becomes simple (the other solution would be a unified world currency). Any rare and indestructible commodity such as platinum, silver, or diamonds could replace gold. There is no point in changing to these because the problem of scarcity would still exist. We counsel you to ignore dire warnings of gloom, and maintain investment positions in gold stocks to protect against the inevitable devaluation of the dollar which most authorities vow will never occur. All of you should own ASA (30) and *Hecla* (HL - 23¼) by now.

October 25, 1963

* * *

A hearing by the House Senate Committee turned to no outside monetary expert who would urge repeal of the 25% gold-backing rule to back U. S. currency. Senators Douglas and Javits favor the repeal and both disagree vehemently, for it would remove the last vestige of meaning to the paper we call money. Granted, in the history of nations money has never lasted long without gold to back it up. Nobody has ever been able to resist the temptation of going to the printing press to solve fiscal problems when there is no restraint such as is provided by the limited amount of gold in the world.

More rational is a proposal to form a new unit of international currency. While it would not solve our own particular gold problem, the western world would not depend so much on the state of our own gold reserve. If there could be some form of absolute control over the amount of paper printed, the suggestion would indeed be helpful and we would favor it. But as we said above, never in the history of the world has any paper currency lasted for very long without the backing of gold. The U. S. balance-of-payments deficit showed a spectacular decline during the third quarter this year, and undoubtedly reflects the Administration's policy of increasing short-term interest rates while at the same time discouraging long-term investments to foreigners. We still have grave doubts about this nation's capacity to live within its means and we therefore feel that no citizen should entirely trust his government's paper promise.

November 15, 1963

JANUARY 1, 1964: DJI 766.08; ASA - $3^{11}\!/_{16}$

* * *

CONGRESS'S JOINT ECONOMIC COMMITTEE WANTS REPEAL OF THE PRESENT LAW REQUIRING 25% GOLD BACKING FOR MOST U.S. CURRENCY

Translated into simple English, the attrition of U. S. gold holdings from a peak $24.8 billion in 1949 to its present level of $15.5 billion will be permitted to continue all the way down to zero. As things stand now, the present rule would force us to devalue by halting attrition at the 25% figure. What terrifies us, frankly, is the prospect of a currency not tied to gold at all. Historically, such moves have always preceded the phenomenon whereby paper money becomes worth exactly the paper it was printed on (see Brazil). America's complete disregard and ignorance of the gold problem since 1949 has led us into a serious situation. We do not yet see any real awareness by either party of the catastrophic implications negligence could bring, and we do not believe we are exaggerating. Needless to say, we are vehemently against the release of gold-backing from the U. S. dollar for historical reasons, and we strongly recommend you protect yourself by including in your portfolio some precious metal investments. Our favorite in the golds remains American-South African (ASA - 33½). We know there are not too many advisors who would agree with us on ASA, but in this situation we suggest you just trust us.

April 17, 1964

* * *

Needless to say, we are vehemently against the release of gold-backing from the U. S. dollar for historical reasons, and we strongly recommend you protect yourself by including in your portfolio some precious metal investments. Our favorite in the golds remains American-South African (ASA - 33½); the net assets per share of this holding company on February 29 were $46.73, up from $44.16 on November 30, 1963. Continuing its internal growth, irrespective of the price of gold, net income was up over 30% from the previous period.

June 17, 1964

JULY 1, 1964: DJI 836.06; ASA - 3¹⁵⁄₁₆

*   *   *

The United States reported a loss of gold in June, placing the nation's gold holdings at $15.623 billion and indicating the long attrition begun in 1949 is nowhere in sight. We've been howling about the situation since 1961, and unless Federal authorities make the public aware of what is really happening, the dollar could well be broken.

July 31, 1964

*   *   *

A distinctly ominous note was sounded by a report that the balance-of-payments deficit in the April-June quarter widened to an annual rate of $2.968 billion after seasonal adjustments. This compares with the loss of $856 million in the first quarter, and the total 1963 deficit of $3.286 billion.

[P]ay attention to our flat assertion that *you must protect yourself from disastrous federal policies by buying one or two selected gold mining issues for storage in your vault as permanent protection.*

Incidentally, Britain's trade gap continued to widen in July with a substantial excess of imports over exports. Such news items indicate to us that interest rates on the Continent will continue to rise, and sooner or later money that is now invested in U. S. bonds will be shifted toward the higher rates available in Europe. *This will aggravate the gold outflow.*

Our gold problems, which we estimate only one out of several hundred thousand Americans (at best) are even aware of, derive from the idiotic notion that a planned federal deficit on a permanent basis is a good thing.

The gold cover, or the requirement to maintain an absolute limit of gold behind our currency, was deliberately placed there as a stop-gap measure against profligacy and spendthrift folly. To remove it, is to stick a penny in the fuse box. It removes the last vestige of protection American citizens have for their currency. Not only have Keynesian economists failed to understand the problem in recent decades, but they still don't. Simply stated, nobody can live beyond his means forever— not even governments, not even governments as big, powerful, and profitable as the United States of America. There is a day of reckoning and there are just so many times it can be put off. To invade the last hoard of gold we have left is not to run this country into the ground, but it is to complete the job and run it underground. When, oh when, is the American public going to wake up and see what is being done to them?

August 14, 1964

* * *

One subscriber points up a possible repudiation of the gold standard rather than Federal acceptance of devaluation, which would make gold stocks almost worthless. This is another typical effort (and one that has occurred to many rulers throughout the world's history) to get out of a gold squeeze; it also occurs when a little boy loses almost all his marbles and threatens to stop playing altogether. Such sulking will not last long, and a currency not backed by gold will become worthless so fast that a new currency will later be developed based on gold.

You have nowhere else to go but to precious metals, for everything else will drop during a period of deflation and devaluation (including commodities, real estate and other general mining stocks—which, on the contrary, fare well during deflation).

Frankly, the aggressiveness and ferocity of buying that recently poured into the gold and silver stocks scared the daylights out of us. (American-South African made an all-time high, and Dome almost did.)

Maybe France has just demanded gold for all the dollars she holds —but we do not know, and theorizing without fact can be dangerous.

One of the world's leading financial publications had this to say about strength in golds: "Part of the strength in gold stocks, some brokers said, was the result of buying by Europeans who were acquiring the issues as a hedge against inflation." We are saddened by the ignorance of some who claim to be financial writers. Even a minimal intelligence would realize that inflation (involving higher labor and other costs) when matched by a fixed selling price (as exists for gold) will result in a classic profit squeeze. *If inflation were the only factor affect-*

ing the golds, they would be collapsing. Golds do well when costs drop, such as during deflations. (It is incredible what one finds in print.)

October 29, 1964

*   *   *

Here is your chance to buy American-South African on weakness, and in our opinion few of us will ever live to see ASA's present capitalization in the 30's again, barring revolution.

November 13, 1964

*   *   *

[I]t is dawning on many people that to defend the dollar, *U. S. interest rates will have to go up*; else, money will be transferred from the U. S. to England to take advantage of higher interest rates, and a dollar crisis would ensue. However, *if interest rates go up, this might choke off the boom* in our economy. What a dilemma!

December 4, 1964

JANUARY 1, 1965: DJI 869.78; ASA - 5$^{11}\!/_{16}$

*   *   *

We were dumbfounded by a brief note in the May 17 *Wall Street Journal* which reported that President Johnson has given the "go ahead" for the government to start turning out $45 million worth of silver dollars. With a fantastic short squeeze on silver, turning out more silver dollars is simply not defensible. (In future years, people will look back on this action and compare it with the folly of minting Kennedy half dollars during the orgy of grief that followed his assassination.) President Johnson said, "They will be distributed in the areas of the country where the silver dollar has traditionally been used as a medium of exchange." Well, don't expect to see any in circulation because the hoarders will be jumping on this bargain of a lifetime.

May 21, 1965

JULY 1, 1965: DJI 871.59; ASA - 6¾

*   *   *

International monetary troubles lurking in the background might come out into the open at next month's Washington meetings of the members of the World Bank and the International Monetary Fund.

Much of what is really happening is at least partially obscured by considerable talk these days of inflationary pressures. Basic items such as containers, aluminum ingots, and probably steel are slated for price increases. Furthermore, the House Labor Committee approved a rise in the minimum wage to a $1.75 an hour. Investors seeking inflationary hedges have been seeking haven in the stock market, especially in electronics, airlines and defense-oriented issues. We predict that those people who are hedging against inflation will have prepared for the wrong enemy. Just as England is raising interest rates in a deliberate attempt to cause a recession and deflation, so the United States will be forced to raise interest rates to defend the dollar. By making it harder to sell at home, the theory goes, businessmen will be forced to sell more overseas and thus correct international monetary imbalances. In other words, the gold stocks are not a protection against inflation (which is their worst enemy) but against deflation. Let us make this crystal clear.

Because of this expected deflation, we are again alone in remaining bearish on bonds, which will have to drop so that their yields are competitive with higher interest rates elsewhere. As if to bear out our frequent predictions along these lines, bonds slumped on Wednesday, and yields increased to a 5-year high. You can now get around 4½% on highest quality U. S. government bonds, and there are plenty of corporate issues crowding the 5% level.

August 20, 1965

* * *

We predict deflation, rather than inflation. If you have been watching what England is doing, you will notice that in an effort to defend her currency, she has been forced to raise her bank rates. They are now 6%, and there are rumors of a possible jump to 9%. *Can you imagine what would happen to the United States' economy with a 9% bank rate?*

FOWLER TO TOUR EUROPE

American-South African once again blasted into new all-time high ground at 78¼, and a few days later it was announced that Secretary of the Treasury, Henry Fowler, would leave on August 28 for a 2-week tour of Europe.

Ostensibly this trip will be a prelude to this fall's conference on the international monetary system.

The situation is coming to a boil; September and October look critical. Ignore public statements, for it is discussion behind closed doors that will count. You should be able to deduce what is being said, however, by simply watching the gold shares. As of this moment, our observation of the ticker tape shows that golds are under very aggressive accumulation. If this continues, or accelerates, you should be able to guess what is happening and what November headlines will be. Don't wait for the official announcements, because responsible officials are severely circumscribed in what they can responsibly say in public in a situation like this. Also pay attention to an *absence* of news on the monetary situation. For example, President Johnson called the price of gold "immutable" early this year, but has been resoundingly silent on the subject ever since. As a possible prelude to a shift of President Johnson's position, he said on August 25th "that the United States was not wedded to any particular procedure, nor to any rigid timetable in its search for improvements in the international monetary situation," as reported in *The New York Times*, August 26th.

August 27, 1965

* * *

In 1516 A.D., a Bohemian valley named after St. Joseph (Joachimsthal, or Joachim's Valley) discovered silver. They minted coins the size of current U. S. silver dollars, which were called Joachimsthalers. These pieces became popular and were eventually called "thalers." As the coins spread across the Continent, their name was changed to "taler," to "daler" and then to "dollar."

Residents of that Bohemian town could hardly have foreseen the destiny of the word they created, in the same way that American monetary officials as little as a year ago had not the vaguest notion that the dollar was being challenged to its very foundation. Experts were equally misguided this week, in our opinion, when a wave of optimism hit world monetary centers after announcement that the United States and nine other nations had come to England's rescue once again. These nations, with the conspicuous exception of France (which has obviously decided to use a form of "stoploss" in a refusal to throw good money after bad) have accumulated money, the amount of which they refused to disclose. This is a stand-by fund for Britain's use so that she may buy offerings of sterling when selling threatens to break the pound.

Perhaps to forestall the possibility that the above news could spark a renewed wave of speculation against the pound, the Labor government announced a national economic plan to put England back on its feet financially by 1970, and a determined bid to break out of the vicious circle of on-and-off controls which have plagued Britain for twenty years. Conservative Party opposition leader Edward Heath promptly described the project as "the biggest publicity gimmick which the government has so far produced." Secretary of State George Brown wrote, "Britain has one over-riding economic necessity. We must pay our way in the world and to do this we must increase our production. We must sell a greater amount overseas and hold steady and even reduce the prices at which we can profitably do this." This is bad news for runaway labor, and it remains to be seen whether England can respond to this note of sanity the gold crisis is forcing on her socialists. It is even debatable whether the Labor government will survive, since they came to office partially on a promise to end the "stop-go" policy of monetary restraint. Trouble is, instead of all "go," it now looks as if Labor will be all "stop."

Secretary of the Treasury Henry H. Fowler reported to President Johnson, after his European trip, that the ten leading financial nations have agreed the first stage of negotiations for improving the world monetary system should be started this month. Apparently realizing that the pound could not come back to health before 1970, but could get much sicker quickly, Mr. Fowler said, "ways will have to be developed to expand international liquidity after the balance-of-payments deficits of the United States no longer exist . . . such a time is rapidly approaching because the United States' deficit is already under control." We respectfully submit (1), our deficit is *not* under control, and (2), that U. S. deficits are not the sound way to finance the world's international trade.

So far, everyone agrees there is at least one problem, but at the IMF conference the differences between possible solutions should break out into the open. The best that can be hoped for, is an agreement to assign a group to study the problem; thereby keeping the sparks out of public view.

There is a spectrum, a rubric of indicators, which reveals the quiet but well-informed final assessment of people in the know. The action of the gold shares, for example, illustrates how far ahead a crisis lies. Golds are slightly below their highs, but without determined selling. There is much hullabaloo about a rally for the English pound, but the prices are not even above the May highs. The international oils rallied, and they should be declining if a crisis is imminent.

Should these events become urgent trends, we would consider putting off the probability of a monetary crisis within the next few months, and begin to think that instead of devaluation there would be a last ditch effort to halt the monetary deterioration by raising interest rates substantially. As if to underscore this point, U. S. government bonds fell again and now provide yields of 4.29%. *Should sharply higher interest rates be in the cards, you can expect a zipping economic recession and bear markets throughout the world. Thus, you have our conclusion, that either devaluation or recession are imminently in the cards, and possibly both.*

September 17, 1965

* * *

MONETARY WOES—AND THEN THERE WERE 8

The big ten free world countries are now concluding their International Monetary Fund annual meeting in Washington. The members are the U. S., Great Britain, France, West Germany, Italy, Belgium, The Netherlands, Sweden, Canada, and Japan.

The members agreed to report in the spring on the "scope of agreement that they have found" among themselves in informal discussions to take place between now and then. The first meeting, in late October or early November, will probably be in Paris. The dominant mood favored getting the issues into committee as quickly as possible so that uncertainties would not disturb the present system. This would then be followed by a "second phase" review by the Executive Directors of the IMF, so that 12 months hence it can be discussed before the IMF's 102 member countries.

France predictably lay down the law by insisting that any new world payment system must provide unanimous approval of credits to hard-pressed countries, a direct link between a new international reserve unit and gold, control by a small group of leading nations, and credit operations outside of the IMF. The U. S. is in direct conflict, asking for majority voting, no tight gold ties, and operations within the framework of the IMF. The U. S. wants surplus countries, including France and West Germany, to supply more money to underdeveloped countries, non-tariff barriers cut, elimination of existing curbs on international movement of capital, and reduction of interest rates so that funds are not bid away from the U. S.

The big news of the conference was the movement of West Germany over to France's side, thus sharpening the split we envisioned months ago. We still predict that other nations will probably join France and West Germany against England and the U. S. Carl Blessing, president of West Germany's Deutsche Bundesbank, said, "Let me say quite frankly that in my opinion a sustained improvement of the balance-of-payments situation of the U. S. and of the U. K. is of greater importance than all the efforts to reform the international payments system." Mr. Giscard d'Estaing endorsed the German official's statement and called it "brilliant." And so, there are now eight little indians—as the old song goes.

We were especially watchful of the gold shares, for sudden strength in this group would imply that something not yet made public had been decided at the IMF conference behind closed doors: perhaps devaluation. However, golds are drifting quietly a few percent under their all-time highs in an apparent wait-and-see attitude. Meanwhile, the pound showed considerable strength on heavy buying from Paris (not the U. S.) and in fact rose to a two-year high at $2.803. This was partially due to short covering and England's Chancellor of the Exchequer's expectation that second quarter results would be good. But Britain will soon announce her September gold and foreign exchange reserves, and after the 24 million pound decline in August, there had better be an increase. England might squeak through again, as it has several times in recent years, but the country remains in deep trouble. Even the most optimistic English Socialist sees no relief before 1970 while, on the other hand, things could get much worse.

Meanwhile, the U. S. announced its gold stock dropped by $54 million during August, bringing the decline this year to $1.556 billion.

The high risk of a gold crisis remains real. If Germany begins to aggressively cash in dollars for gold, or if any one of a number of factors occur, the whole monetary structure could cave in. Retain your positions in gold stocks, and do not use stops—since gold shares have given false Downside Breakouts throughout their long advances since 1960. To protect your portfolio, we continue to strongly urge retention of selected gold shares. As soon as they halt their quiescence, and begin to spurt upwards again, you may safely deduce that something is going on behind closed doors; meanwhile, just stand pat and be patient.

American-South African announced the highest net asset per share in its corporate life: $55.39 as of August 31, 1965.

October 1, 1965

*   *   *

Lenin was certainly right. There is no subtler, no surer means of overturning the existing basis of society than to debauch the currency. The process engages all the hidden forces of economic law on the side of destruction.

Lord Keynes

[G]old shares moved lower, but not as much as perhaps could have been expected considering the massive propaganda campaign mounted against any possibility of monetary crisis. You have had these various propaganda bursts, and corresponding weakness in gold shares, periodically throughout the long upward sweep in the golds. We urge you to resist being lulled, just as we urged the same so many times at lower levels. Historically, gold shares movements are slippery and hard to catch—thus we warn against trying to trade this group. Few analysts really understand golds, because they defy capricious prediction. Traders risk missing the real and inevitable move, and are urged to stand pat while averaging down on weakness.

Conclusion: With the pound and dollar floundering beneath a facade of bravado, you should exploit any weakness in the gold shares to take additional positions. As we have pointed out on innumerable occasions, American-South African is a fine investment whether or not the price of gold goes up—giving you two-way protection. Individual gold shares, on price concessions, provide yields approaching 10%, which doubles your money every ten years.

October 8, 1965

*   *   *

Gold shares continue to drift lower, and are now behaving in more contracyclical fashion with the market than a few months back. The decline was led by the South African shares, which previously had advanced the most. If normal patterns are followed, the moment the market turns down, the golds will turn upwards again.

We hope we will not be called prophets of doom for maintaining that we still envision a monetary crisis ahead, and we can only submit in evidence that when we first began to discuss golds in early 1961, we were called far worse.

The big news of the week was the announcement of Britain's trade figures with the rest of the world. There are two sets of figures: one is seasonally adjusted, the other not adjusted. The big headlines were given to the adjusted figures since these were more favorable to Britain. We hope you are learning to read the monetary fine print.

October 15, 1965

*   *   *

NYSE volume on December 1 hit 10,140,000 shares with little noticeable price change—such high volume has been reached only 9 times in history and *never before, during a bull market!* It occurred twice during the 1929 crash and twice during the 1962 crash. If this isn't the century's first "buying climax" we can't imagine what one would be like.

December 3, 1965

JANUARY 1, 1966: DJI 968.54; ASA - 7¹⁄₁₆

* * *

8 REASONS WHY GOLDS WILL RISE AGAIN

Many buyers of gold shares expected the pound to be devalued by November. While we thought the likelihood was indeed high, it mattered little to our *long-term* bullishness on these shares. The short-term traders are now discouraged, and their selling has driven the gold shares down again to attractive levels. Do not wait until the golds are popular before taking your position—if you have no position in them yet. There has been no change whatever in the underlying justification for purchasing golds; to the contrary, our reasons have become increasingly confirmed: 1) The United States balance-of-payments continues to deteriorate, with exports level and imports rising. Further, England's balance-of-payments is improving far less quickly than projected, and their wage spiral cannot fail to keep rising in the face of full employment; the credit squeeze is not working. 2) As the United States prints money, more short-term claims are accumulating abroad. 3) The rise in our discount rate was too little too late, for the Europeans are simply raising their rates to neutralize the boost. 4) The U. S. gold reserve ratio continues to decline. 5) As foreign nations continue to accumulate gold, they have more and more of a vested interest in linking currency to gold. 6) deGaulle's re-election implies that French resistance to American monetary policies will continue. 7) The Vietnam war continues to attract away from this country dollars which will find their way into European banks. 8) Private hoarding is sufficient to absorb all new gold production.

January 14, 1966

* * *

The U. S. balance of trade for 1965 was announced this week, and the picture is just as bad as we had warned. Imports leaped to a new high, above $2 billion in December, while exports dipped. The foreign trade surplus for the full year narrowed to $5.2 billion, down a walloping $1.5 billion from the previous year. There is no change in our rather daring prediction that, in due course, U. S. imports will actually exceed exports—plumping an intolerable strain on the dollar. However, this is not the type of specific surprise event which would now be triggering precious metal strength.

Perhaps the reason lies in the current Paris meetings, where officials are trying to negotiate a new international monetary system. France represents the "hard money" group, insisting on a link between paper and an immutable commodity like gold, whereas the United States represents the "soft money" countries, wishing to set its own inflationary prosperity as an example for the whole world.

Obviously, something is brewing behind closed doors, and whichever—it is favorable to the precious metals.

Again, we urge you to take a position in the golds.

January 28, 1966

* * *

U. S. Treasury reported that the nation's gold stock dipped by $100 million to $13.634 billion—the lowest level since September 1938. The dollar hemorrhage still shows no sign of ending, and we will not burden you with how rotten a job we think Republican and Democratic Administrations have done with U. S. currency since 1949. We would be hard put to describe which of the two parties is worse since they are both infected with a distorted Keynesian approach that is sure to debase our currency and lead to inflation. The new silverless quarters are simply a manifestation of a much larger rot. The only way to prove our politicians are wrong would be to go through an economic depression—a bitter way to be right.

March 25, 1966

* * *

American-South African leaped 9 points in 5 consecutive trading sessions, clearly challenging its all-time high at 78½.

Attention is being refocused on the gold shares for two primary reasons. First of all, the English election has unsettled the pound, which is fragile enough.

The second reason for gold strength concerns meetings for "reform" of the world's monetary system to take place in Paris and Washington during the first two weeks in April. It is an open secret that the group's attempt to negotiate an agreement had a violent falling-out, behind closed doors—as we had expected. You will remember that we also predicted the "sound money men" would ultimately prevail over the "funny money boys" for the simple reason that Europe has suffered all the inflation and monetary disasters it can stand. We also expect that Europeans resent our running the printing presses to turn out dollar bills that are promptly used to buy foreign industry. (Why can't Ghana run the printing press and buy out General Motors?) We are not saying that France has no political motives for opposing us—that is not the function of this Letter. What we do say is that European countries have sound *economic* reasons for demanding that we balance the dollar inflow and outflow—with a dollar completely backed by gold.

England is now undergoing a self-enforced recession in an attempt to balance the pound's inflow and outflow. We expect President Johnson will have to do the same thing, and it is precisely such a recession that prompts our long-term pessimism for the stock market.

The American public has been led to believe that France constitutes the sole problem in international monetary negotiations. We believe otherwise, and the first time we have seen these sentiments in the press was a small passage in *The New York Times* on March 28, 1966, where the last (but not least) sentence was also important as the first "public balloon" suggesting that the plan for a new reserve unit is doomed to failure. Here is the quote:

> The fact that France has virtually walked out on the negotiations is only part of the trouble. The other Continental powers are also unhappy with the United States position, nervous about the much-discussed idea of creating a new monetary reserve unit, and opposed to letting the whole affair pass into the hands of the 103-nation International Monetary Fund. The nervousness about a new reserve unit leads many participants to feel that this idea is dead, at least for now. And the distaste for the IMF indicates that this may not be the route for reform either and, thus, that there may be little or none.

The very next day, in the same publication, the other shoe was dropped: suggesting that after a full year of propaganda to the effect that the U. S. could balance its international budget "anytime we wanted to," the balancing act was doomed to failure. This fulfills another of our predictions in that the United States has gotten so fat and lazy that it is incapable of balancing its dollar flow without much more drastic measures—which we believe will come. Here are the first two paragraphs of an extremely important admission from Washington:

> The Government is pondering with deep concern a new analysis that indicates the nation's balance-of-international-payments, far from improving this year as expected, will actually show a larger deficit than last year. The Cabinet-level committee on the balance of payments was given a report last week estimating a deficit in international payments for the year of $1.8 billion, it has been reli-

ably learned. Last year's deficit was $1.3 billion, and the target for this year was, in effect, to eliminate the deficit.

The New York Times

We still envision a monetary crisis. It is a terrible thing to learn to mistrust your own government's comments on international monetary finance, but at this point, we can only say that we have learned to watch numbers and not words. Until we see improvement in the numbers, we refuse to change our opinion.

April 1, 1966

* * *

It is useful to study England's actions, because their history has typically foreshadowed the history of the rest of Western civilization by as much as a century. (Heaven help us.)

This week the British Treasury barred individual ownership of more than four post-1837 minted gold coins. There are exceptions for genuine coin collectors, but from now on they will have to go to the Bank of England and prove their case. It is apparently intolerable to the British government that Britons, in their disgust over the rape of their paper currency, are shifting their savings into gold medallions and coins. Coin dealers have been churning out commemorative gold coins for nearly everything, including St. George's slaying of the dragon and the closing of the Ecumenical Council.

Further, England joined the United States in banning importing of such gold commemoratives. Thus, citizens alarmed over inflation, continue to lose outlets of self-protection. Pressure will build up when they insist on trading increasing supplies of paper money, which are depreciating in value at a frightening rate, for something that will survive the politicians. At some point, it will dawn on them that the ideal switch is into gold and silver stocks. Perhaps that process has already begun.

April 29, 1966

* * *

If the U. S. cannot balance its payments at will, as it long has pretended it could, then the Europeans will refuse to sever the link with the discipline of gold. And, if the United States' dollar is tied to real money, namely gold, the constantly depreciating dollar does not stand a chance. *For, when you tie real money such as gold to an ever-increasing supply of paper money, then the price of gold must rise (devaluation of the dollar).*

We consistently export more than we import, but we blow the

benefits on foreign aid, overseas military expenditures, and the like. Trouble is, we spend even more than that beneficial surplus created by the excess of exports over imports; hence, the persistent *net* deficit. To make matters worse, our exporting ability—the primary item now keeping this country's currency even remotely afloat—is being seriously undermined by the inflationary domestic trends. If prices continue to rise here (and even President Johnson is openly alarmed at this prospect), it will be increasingly difficult to export competitively. Conceivably, our exports could shrink closer towards balance with imports, driving us even deeper into the red, internationally. The strains are building up in the dollar, and far-sighted investors will protect themselves from something they cannot affect or stop by buying selected gold and silver shares.

May 6, 1966

* * *

It is no secret that we expect, within 5 years, a monetary crisis that could drive the Dow-Jones down to around 400 (we hope we are not told later that we hedged that statement). It is difficult to time the exact break because it will be the effect of a political decision, subject to manipulation by politicians. We will say that a prime signal will be a tremendous upsurge in gold and silver shares, whose earnings are quite immune to a business downturn. Eventually, we hope to switch you out of golds and silvers with tremendous profits, into industrials which will be at the buying point of a lifetime. So much for our long-term plan.

May 13, 1966

* * *

The probable reason for recent strength in gold stocks rocked the international financial community this week: the United States announced that *exports plunged 10% while imports rose 3% in April.*

The figures show that the favorable United States trade balance is now at a seasonally adjusted annual rate of $4.1 billion, down sharply from last December's expectations of a $6 billion surplus. Secretary of the Treasury, Henry H. Fowler, cited the Vietnam War and rising imports as factors in the Administration's inability to make progress towards reducing the payments deficit (the first quarter deficit was $582 million, seasonally adjusted). In the first four months this year, exports were up about 7% from last year, but imports jumped by 14%. Thus, the four-months' $4.1 billion surplus is down from last year's $4.8 billion and 1964's $6 billion surplus.

Here is one of our long-standing predictions coming home to roost. As inflation increasingly prices our goods out of world markets, it invites lower-priced goods to come in from overseas; thus, our exports drop and imports rise. We have already pointed out that our deficit derives from the fact that we not only spend our trade surplus, but *more* on foreign aid and what have you. A narrowing of our trade surplus will aggravate the net deficit even further, and the makings of an ear-splitting international monetary crisis remain present.

May 27, 1966

* * *

A devaluation of the pound is now closer than at any time since last summer's crisis. Britain will not muddle through one of these crises and the pound will be devalued. With the dollar so inextricably entwined with the pound, we would expect a dollar devaluation to follow promptly.

We have long predicted increasingly "tight money" domestically, which could not be effectively countered by our government because of a looming monetary crisis. This week Belgium and West Germany raised their interest rates, which we expect will, in turn, put upward pressure on both English and American interest rates. High interest rates are traditional boom-killers, which we predict will cause a very serious economic downturn in this country. We remain bearish on long-term bonds, only reinforced by this week's new lows for the Dow-Jones bond averages; obviously, money is continuing to tighten its stranglehold around our economy's neck.

And, to top it all off, the United States lost $70 million of gold in April, following a $73 million loss in March; so far this year the U. S. gold loss is $138 million.

We predict the downside gold loss will accelerate to the point where even the Europeans get it through their skulls that the U. S. is broke— and has been broke since 1959, when dollar liabilities exceeded the amount of gold this country owned. The day they awaken will come to be known as the day the run on the bank started.

In sum, we believe golds & silvers have embarked on Major Uptrends, implying Major Downtrends for the rest of the stock market. Within that framework, the enormous precious metal advances deserve a temporary Correction, suggesting a short-term rally for industrials. We have been looking for an extension of the bear market rally begun on May 17. Yet, in the teeth of the largest short-interest in the recorded history of the stock market, an aggressive short-covering rally cannot even get started!

June 3, 1966

JULY 1, 1966: DJI 877.06; ASA - 10¼₆

* * *

[A]s things stand now, English reserves are $3,276,000,000. Of this, $2,517,290,000 is Britain's published foreign debt. The difference between the debt (the bulk of which is owed to the IMF, and scheduled to be repaid soon) is $758,710,000. Thus, the loss of $442.4 million in the last five months is serious indeed. Do you wonder why the price of gold has been strong in London, and the price of the pound in a downtrend?

In our last Letter we pointed out that France and West Germany were not satisfied with Wilson's measures to bring on a recession in England—the prime reason being his omission of a restriction on wage increases. We do not have any facts, but we think it a fair guess that instructions were given to Wilson behind closed doors, for, during the week, he came out with a strict limit on wages, prices, and dividends.

The above developments induce us to believe that a devaluation has been temporarily postponed. We feel this way because, having been forced to take the "ultimate step" against labor, England will be given a chance to see whether these final measures will work. Also clear to us: this is the last crisis England will be permitted. England played her ace of trump by clamping down on labor, which should explain to you why Wilson avoided adding that to his list of deflationary measures last week (the Europeans understood what he was doing, and refused to give him another chance, they forced him to install a wage freeze). The chips are on the table, betting is closed, and many around the world are waiting breathlessly to see how this final gambit will work.

August 5, 1966

* * *

The international monetary picture continues to deteriorate, and only this week France announced a $146 million increase in reserves in July for the biggest monthly inflow in several years. France now has almost all of its reserves in gold. Their total reserves at start of August was $5.967 billion, as against around $13 billion for the United States (in addition, the United States has nearly $30 billion in short-term foreign claims against it). Would you believe that the franc is now stronger than the dollar? In fact, in the ensuing chaos, the French franc might survive as the strongest currency in the world.

Here is a perfectly stunning paragraph from the August 5 *Wall Street Journal* that might well be a "straw in the wind." The second sentence is the first public admission we have seen in the American press that there are secret contingency plans, obviously not excluding devaluation of the dollar. Watch the press carefully for amplification of this theme.

> Although the U. S. pledges to sell gold at the $35 price, it is under no treaty obligations to do so. A U. S. official, who admits that "secret" contingency plans exist in case there is a devaluation of the pound, declines to "rule out" the possibility of a dollar devaluation at some point.

Furthermore, here are the first two paragraphs from the lead editorial of *The New York Times* of August 9. Study the second sentence of the second paragraph; it is an eye-opener.

WILSON'S CRISIS OF CONFIDENCE

> The failure of sterling to recover strongly after Britain's drastic new deflationary measures indicates that traders and speculators are still betting against Prime Minister Wilson.
>
> Immediate devaluation of the pound is no longer expected. But longer-term prospects remain uncertain. Major trade unions have reacted unfavorably to the wage freeze in the face of majority approval by the central leadership. Confidence is lacking in the Labor Government's ability to go beyond crisis measures to the profound structural reforms the economy needs.

The rest of the editorial is not worth showing since it wanders off with the "solution" as being England's entry into the Common Market. Confusion reigneth at the *Times*. But that second sentence is really remarkable. Notice the word "remain" with which the *Times* weasels out of its prior confidence that the long-term prospects of the pound were sound. The *Times* makes it sound as if everybody was saying the pound was weak all along. As for the last sentence, England doesn't need any more "structural reforms"—just some good old-fashioned capitalism. When *The New York Times* discovers that, we'll do handstands.

August 12, 1966

* * *

Our opinion is still flatly bearish for the primary market trend, and flatly bullish for the golds & silvers primary trends. All we are talking about here is a slight contra-cyclical motion by both groups, in the form of a bear market rally and a Technical Correction in the precious metals.

Escapees from this crash, such as we, are in a very tiny minority, and many analysts continue to spend their time guessing at Bottoms—repeatedly wrong. The same thing occurred all the way down in the 1962 crash. This is an old-fashioned bear market, the likes of which this generation has not seen. A lesson for the investment public lies ahead. Sit back and relax—we see no reason yet to change any of our major policies.

September 6, 1966

* * *

The United States gold stock fell $116 million in July, the biggest monthly outflow since May 1965, and more than double June's $53 million. The decline stems mainly from another purchase of gold by France, bringing U. S. gold holdings to $13.413 billion as against $13.806 billion at the end of 1965. This despite purchase of $200 million of gold from Canada so far this year. U. S. holdings of foreign currencies rose sharply in July, presumably reflecting U. S. acquisition of British pounds in price support operations. (Oh well, blow another few hundred million to save an unsavable currency; it won't really matter in bankruptcy.)

Even the International Monetary Fund joined the call for a slowdown in the U. S.'s economy, in an unusually critical report. Since our drawings from the IMF in foreign currencies so far have totaled about $1.6 billion, our near-automatic right to draw more is down to around $300 million; beyond that, further credit would depend on an IMF finding to the effect that the U. S. is taking proper steps to end its payments' deficit. Perhaps Johnson will find other assets to squander, but he must be rather busy these days, having just pushed through a mighty increase in the minimum wage law which does little more than feed the inflationary fires.

September 9, 1966

* * *

New support arrangements with the British pound were announced this week, raising the total U. S. line of credit available to the British to $1,350 million. English markets responded by rallying exuberantly, especially in view of the fact that Britain's August trade deficit (the excess of imports over exports) narrowed to $193.2 million in August from $294 million in July.

If a near-bankrupt gambler at the race track lost $10 in the last race instead of the usual $20, would this be progress?

Myopic statesmen continue to concentrate on rebuffing "speculators" instead of attacking the root of the problem: *balancing imports and exports.* Nothing short of his goal will alter our conviction that the price of gold is headed higher, despite all protestations to the contrary. Further, there would have to be a substantial surplus over a period of time to make up for past deficits. Since the latter condition is most unlikely, we are more bullish than ever on gold shares. New support arrangements of ever-increasing amounts are strictly a postponing process intended to buy time. But, if nothing is done in this period of newly bought time, then the collapse will be all the worse when it finally comes. To us, devaluation still appears inevitable.

As if to underscore the suspicion with which the dollar is regarded internationally, it was announced that French gold purchases from us in the second quarter were $221 million, up from $103 million in the first quarter. Adding insult to injury, France took physical possession of the gold rather than leaving it earmarked in the vaults of the Federal Reserve Bank of New York, as do most other nations. France obviously fears vindictiveness of American officials when the crisis breaks.

September 16, 1966

*   *   *

The fundamentals have not changed. In the October 17 issue of *U. S. News and World Report*, page 60, an exclusive interview with Jacques Rueff lays down the law. We regard Rueff as the brains behind de Gaulle, and we strongly recommend you read this article; it should answer your remaining questions. Rueff calls for a doubling of the price of gold.

Quite the most interesting news item of the week was reported by Reuters from Canberra, to the effect that the Australian Federal Treasurer, Mr. William McMahon, advocated an international subsidy to increase gold production, and said any new international currency should be based on gold. Replying to questions in the House of Representatives, he said that in the interests of world liquidity and sanity he unhesitatingly supported the French rather than the British view of this issue; Mr. McMahon said that at the recent meetings of world and Commonwealth finance ministers in Washington and Montreal he presented these views. He said he argued that there was a strong case for the belief that insufficient international money existed to support the desired increase in world trade. He had strongly supported the view that any currency created should be tied to gold, because the real danger in international financing was that unless currency was tied to a commodity in short supply too much money could be created.

September 28, 1966

*　　*　　*

The price of gold on the London Bullion Exchange rose to $35.1925 an ounce, barely under its highs reached on August 25 and again on September 1. Buying is probably attributable to the belief emerging from this week's IMF meeting in Washington that gold will not be removed from international monetary finance. Actually, an impasse has developed and there is no resolution clearly visible to monetary leaders yet. There is, of course, the Washington group which claims that gold should be discarded in favor of the printing press—predicting dire things "when" the U. S. brings its payments deficit into balance. Another group, notably the French, who are also probably tweaking the lion's tail in revenge for Suez, challenges Washington to balance its payments and suggests that this nation is incapable of doing so. France seemingly would like to move back to a complete gold standard; this would also include a substantial upward revaluation of the price of gold. As for the majority of delegates at the IMF convention, which made headlines this week, they simply appear confused. However, they have come a long way from just a year ago when the prevailing opinion was that "gold is a barbaric relic of the past."

One of the more important speeches at the IMF World Bank meeting this week was that of Treasury Secretary Fowler. He definitely hinted at stronger steps to control the U. S. dollar outflow—over and above the present Interest Equalization Tax and the so-called "voluntary" restraints on direct capital investment abroad. (We have been warning you about this for several years, and here is the handwriting on the wall; if you are going to get money out of the country, do it now.) With that brand of arrogance peculiar to American officials in trouble, Fowler advised nations with a balance of payments surplus (presumably France) that it makes no sense to use capital markets "as instruments for the accumulation of gold and other reserves beyond their needs." Fowler advised such surplus countries to liberalize capital outflows and foreign aid. He then went on to defend American profligacy by pointing out that our deficit was cut in half last year to around $1.3 billion. (He did not mention that this was a one-shot influx due to American repatriation of capital from overseas, and not likely to recur.) He pointed to the Vietnam war as a cause halting further movement toward balance. He also indicated that the war is currently causing dollar outflows to the Far East at an annual rate of about $1.5 billion.

The second part of Fowler's argument can be speedily brushed aside with the admonition that the United States had no right in the world to run the deficits it has during the last two decades. Many have gone on record opposing that dollar hemorrhage, as for foreign aid to our enemies, predicting that it would return to haunt us. The responsibility is squarely on each administration since 1949. The fool's paradise is over.

Telling nations like France to stop being efficient represents the socialist mentality. Washington refuses to play by the rules of the game as France piles up its winnings. Having been burnt in 1933 by holding English pounds before that country devalued, France is no longer interested in trusting foreign economists. Washington wants France to live the way Washington lives, and like it or not, France just won't.

In conclusion, the way it stands now, monetary leaders see the possibility of an international monetary collapse, and know things cannot go on as they are. They reject equally the solution of severing money from gold and also raising the price of gold, thereby enshrining it—but as yet they have no positive alternative. We flatly predict, as we have for five years now, that they will ultimately come to see gold not as a metal but as a fuse in a fusebox. It is the only discipline that can be trusted to guard against abuses of the money presses. We predict that the solution to which these gentlemen will come will be a higher price for gold. We can only hope they go even further, and develop a "free gold" system whereby the value of every nation's currency would be allowed to fluctuate freely against a fixed unit of gold. Thus, nations living beyond their means would promptly enjoy a currency slump which would tend to correct the problem, avoiding the postponement, and thereby enlargement, of conditions such as exist in the United States today. This would not make international trade more difficult, since international dealers can simply buy futures contracts to hedge against currency fluctuations. This solution would surely be anathema to the "new economists" because it is terrifyingly well-disciplined and unmanipulatable. This plan will surely separate out those who prefer a permissive mother to a strict father.

September 30, 1966

*   *   *

The United States' payments gap deepened in the third quarter with our trade surplus of $2.9 billion annually down sharply from the total of nearly $4.8 billion last year.

[The] Commerce Department said the reason for the slump was that the British sold some of its U. S. dollars to protect the pound, and high interest rates by U. S. banks overseas encouraged foreigners to keep their dollars in foreign branches of U. S. banks. As Treasury Secretary Fowler said, "We must expect some increase in the deficit in the immediate period ahead." Thus, the logic for holding gold shares is utterly unchanged, for there is not yet even the announcement of an intention to balance dollar inflow and outflow.

October 18, 1966

*   *   *

Speculative pressures on the pound are continuing heavy despite the announcement that Britain's gold reserves increased slightly in November for the third straight month.

It was also announced that France's gold and convertible currency reserves declined in November for the fourth consecutive month. Nonetheless, they have practically come out and made it official French policy that the price of gold be raised. They are insisting that raising of the price of gold be included in possible solutions to the international monetary crisis now going on, a subject heretofore excluded from discussion. They are either preparing for a major showdown on a price increase or only trying to prolong the monetary discussions, which they consider premature anyhow. The monetary situation appears to be heating up again, accounting for recent strength in selective gold shares.

Here is a fascinating chart. The trends speak for themselves.

December 9, 1966

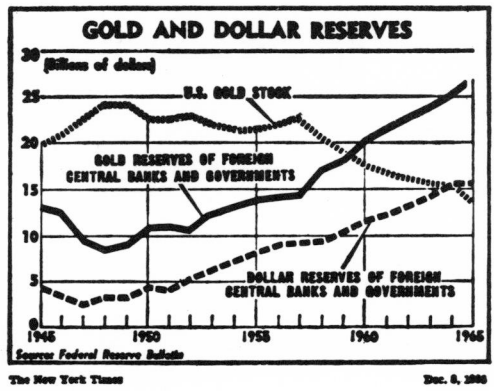

JANUARY 1, 1967: DJI 786.41; ASA - 8½

*   *   *

The gold situation is like a time bomb, or rather, like an underground stream which leaps to the surface briefly before disappearing. In an effort to remove this time bomb, the Treasury has tried to hold international monetary discussions *excluding* the possibility of a rise in the price of gold. However, Michel Debré, the French Finance Minister, in an interview in the Paris newspaper, *Le Monde*, urged a study of the gold price as part of the current monetary negotiations. As the feud erupted in public, the U. S. Treasury summarily rejected the proposal as "completely unacceptable." But the days when U. S. can dictate international monetary policy are coming to an end, and gold prices in Europe soared to all-time high ground. On Jan. 12, Maurice Schumann, chairman of the French National Assembly's Foreign Relations Committee, said Britain could not become a member of the European Common Market unless there is a reform of the international monetary system—including a higher price for gold. Schumann's views frequently reflect official French government policy. And so the struggle goes on. France understands that gold is power, and has increased her gold holdings to their 1938 levels—now ⅛ of the gold in the world.

January 13, 1967

*   *   *

The coming gold crisis began to quicken its pace when Congress asked this week to double the 15% Interest Equalization Tax on foreign securities. The request would not have been made if the 15% tax were sufficient to avoid a gold crisis. Before there was an IE Tax, we predicted there would be one. When it came we predicted it would be doubled. We still predict it will be trebled and quadrupled—until you are blocked from moving dollars out of this country. It will come; we are relatively confident of that.

On the import-export front, more deterioration in December as exports fell 2.7% and imports fell only .1% from November levels. Our foreign trade surplus, traditionally one of the strongest features in the overall U. S. balance of payments with other countries, shrank to $3.845 billion, as against $5.334 billion in 1965. This does not count foreign aid, tourist and investment expenses, which converts this plus figure into a minus figure and causes gold to flow from this country to make up the net deficit. Last year exports rose 10%, but imports leaped 20%, and that should tell you something about the U. S. economy. Keep it in mind.

January 27, 1967

*   *   *

There must have been a worsening in the United States' international balance-of-payments deficit in the fourth quarter of 1966, because the full year's total is now estimated at around $1.4 to $1.5 billion, somewhat higher than the $1.3 billion of 1965. For the first three quarters of 1966, the deficit had been slightly below that of 1965. Imports in 1966 rose to 3.5% of the Gross National Product, the highest in 20 years. This undoubtedly accounts for Fowler's call last week to increase the Interest Equalization Tax.

President Johnson's Economic Report to Congress also disclosed that the United States' gold loss last year was $571 million, indicating that the December loss of gold was $27 million. The Report stressed that deficits in international payments are "clearly not indefinitely sustainable if confidence in the safety and stability of the dollar is to be maintained." Meanwhile, back at the ranch, Secretary of Defense McNamara offered Congressional testimony to the effect that the Vietnam war will add more than $1 billion to the United States' balance-of-payments problem this fiscal year—bringing the foreign exchange costs for defense to about $2.4 billion in fiscal 1967, the highest since $2.8 billion in 1961. The December drain is comparable to a gold loss of $49 million in November and $45 million in October, despite the fact that France did not convert any dollars into gold for the third straight month. The losses brought Uncle Sam's total gold holdings to $13.2 billion on December 31, lowest since 1937 from which point the gold supply was on its way to 1949's peak of $24.6 billion. France took a total of $601 million of gold during 1966, a drain partially offset by purchases from Canada and the United Kingdom.

The gold situation might now rest on the back burner while the market moves up, but anyone who thinks the problem has been solved permanently is enjoying delusions.

February 3, 1967

*   *   *

The low-level gold crisis now gripping the international scene continues. Britain announced that imports continued to exceed exports, but the January trade gap was slightly lower than last year's monthly average. We remain pessimistic on England's ability to balance her accounts.

Meanwhile, back at the ranch, the United States announced that its deficit in international payments expanded to $551 million in the fourth quarter, for a total 1966 deficit of $1.424 billion; this compares with the 1965 deficit of $1.337 billion. Secretary of the Treasury Fowler said this was a "respectable performance," to which we take calm exception. Fowler went on to say that the expanded fourth quarter deficit does not "indicate any worsening in the over-all balance-of-payments position." Why then, Mr. Fowler, are you asking for the Interest Equalization Tax to be doubled?

February 17, 1967

*   *   *

January saw a gold loss of $50 million, up from the December drain of $27 million. For the fourth consecutive month, France failed to buy any U. S. gold, puffing the myth that all our problems are due to France. At the end of January, the total U. S. gold supply was down to $13.202 billion, the lowest level since August 24, 1938. Look, ma, no hands.

March 3, 1967

*   *   *

The gold situation hit the front page of the *Wall Street Journal* on March 20, the lead article in fact. It was entitled "Some Nations Secretly Promise U. S. They Will Avoid Buying Our Gold." As the *Journal* pointed out, this effort probably represents "the greatest change in U. S. gold policy in decades." In other words, since we could not possibly pay off in gold on all the dollars held by foreigners, we have to go begging, hat in hand, that they will not request gold for these dollars. As usual, this panacea does not go to the root of the problem. As the *Journal* pointed out, the real problem is "the persistent U.S. balance-of-payments deficit . . . that keeps increasing potential claims on U. S. gold. This deficit, which rose slightly last year to $1.424 billion from the previous year's $1.337 billion, occurs when foreigners acquire more dollars through American trade, aid, investment, travel and the like than they return to the U. S. in all transactions." Since more dollars keep flowing out than we earn overseas, more and more dollars will keep piling up overseas. Thus, the expedient of asking foreigners to hold dollars can be seen as another way of postponing the gold crisis, but making it much worse when it finally comes.

Still, many confuse the balance-of-payments problem with the size of our Gross National Product. There is no relationship. What if the U. S. ran its printing presses day and night and flooded the world with paper money; there would be no effect on our GNP, but the dollar would obviously be worth less. It is only the number of pieces of paper in relationship to gold at $35 per ounce that fixes the value of a dollar. With more than double the amount of dollars overseas than we have in Fort Knox to redeem them, we cannot understand how foreigners can be dumb enough to refrain from cashing in their chips. How would you feel if your local banker asked you not to withdraw any of your savings because there was nowhere near enough money to pay off all depositors and you might put him out of business? You would run, not walk to the nearest withdrawal window. Yet, this latest ploy will result in yet another postponement of the gold crisis, and continuing loss of gold from Fort Knox. The solution we believe will eventually be arrived at will involve an absolute termination of balance-of-payments deficits combined with a substantial increase in the price of gold to wipe out past profligacies.

March 23, 1967

* * *

The gold situation is boiling up again, this time because of statements by two of the leading bankers in the world. Last week, Chase Manhattan Bank's "Business In Brief" newsletter suggested the possible need for a change in U. S. gold policy. Shortly afterwards, in a virtually identical statement, Rudolph A. Peterson, President of Bank of America, said the U. S. should quit selling gold if the drain becomes oppressive. He said "There is no overwhelming reason why we should sustain the dollar value of gold." Then, Chase Manhattan Bank, compounding the error, "clarified" the remarks by saying that the United States should make it "unmistakably clear in a crisis" that it would stop selling gold. Then, with the kind of double-talk prevalent among some Wall Street observers, they added "no implication should be read . . . that a change in our gold policy is in any sense advocated." As an aftermath, the buzzing among European leading bankers was probably loud enough to be heard on a calm night across the Atlantic.

This stratagem was launched last month when Secretary of the Treasury, Henry H. Fowler, hinted the United States might have to take steps to protect its own interests if it did not get more help from Europe in protecting the dollar. Conceivably, the Chase and Bank of America proposals were trial balloons inspired by the Treasury or President Johnson.

[T]he Treasury Department quickly disavowed any connection with the statements by the two bankers.

What will surely alarm the Europeans is this change in American gold policy. Heretofore, Washington stood ready to exchange gold for dollars anytime at $35 per ounce, so Europeans did not mind holding surplus dollars, even knowing that the U. S. could not possibly redeem all of them. But lately, Washington has been exerting pressure on its allies to avoid converting dollars into gold (see our recent comments on

Canada and West Germany). Europeans who trusted the dollar and American presidential reassurances are suddenly taking a new look at DeGaulle's mistrust of American intentions. The big danger all these disavowals are trying to stave off is that alarmed Europeans will promptly try to cash in some of their surplus dollars as they hedge against the day when the U. S. refuses to permit gold out of the country. Gold prices soared in London yesterday.

The surprise and consternation of Europeans will once again focus their problem on the shortage of gold at its current price, and the ignorance of even top bankers when they confuse international monetary deficits with the heights of the United States' Gross National Product.

None of this should surprise you. In a nation like ours, with only around 14 billion dollars in gold, as against over 30 billion dollars held outside the United States as a potential claim on it, sooner or later somebody will blow the whistle. If the U. S. tripled the price of gold that 13 billion would suddenly be transformed into 39 billion, and, eventually somebody in the United States government is going to discover that. It might be too late, however, to stave off an international monetary catastrophe, and perhaps this is what has unnerved the stock market in general in recent weeks. *There is most assuredly no change in our long-standing admonition to maintain 40% of all portfolios in precious metals.*

Some bankers should study history, and learn that gold embargos are standard in gold crises. When other nations begin pulling gold out of a country in the form of a money hemorrhage, every government has tried the embargo. They then learned to their despair that foreign nations would no longer accept their currency unless it were backed by gold, and the price of gold was therefore raised (the paper money devalued). There is nothing new under the sun.

April 14, 1967

*   *   *

England's high interest rates are attracting an excessive inflow of short-term funds, otherwise known as "hot money," because it tends to flow out of a country when other countries pay higher interest rates. In an effort to discourage this inflow, since it is so unstable, England cut its interest rate from 6 to 5½%. This was no surprise. The element of weakness involved is that the next time money begins to flow out of England, a great deal will leave, and the situation remains unstable.

Meanwhile, Prime Minister Wilson announced to a cheering House of Commons that Great Britain will formally apply for membership in the six-nation Common Market. A similar application was vetoed four years ago by deGaulle, and he will probably be the main stumbling block again. His price? Devaluation of the pound. The final stages of the international monetary drama are fascinating, no? At any rate, action of gold stocks indicates that devaluation is not immediately in sight, although a number of the more important gold shares are within easy striking distance of their all-time highs (many of them are capable of moving several points a day). When new highs begin cropping up in this group will be time enough to expect another of the periodic gold crisis of recent years.

May 5, 1967

*   *   *

While the Silver crisis, which will soon be over, was only a national problem, the coming Gold crisis is far more serious. Last week, we pointed out that Golds and Silvers were starting to move again. Though there is no real connection, other than the general debasement of currency in the United States by inflation, the reasons for the Gold move have been thoroughly outlined by us in recent years. We expect it will explode with precisely the same abruptness as the Silver crisis broke (after the close one day) with absolutely no warning. But the gold situation has a short-term aspect. Golds are contra-cyclical, and they generally rise, or begin to rise, shortly *before* the rest of the market goes into a tailspin. Conversely, when Golds began to weaken in mid-1966, it was an early indication that the market would Bottom out shortly. Sure enough, Bottom was reached in October. Thus, if gold strength accelerates here, it will become highly probable that a *severe* market decline will occur soon. Thus, a new element has been injected into the picture. We must be especially alert for signs of a renewed bear market. We have persistently maintained, all the way up, that you should cash in on the current Uptrend, but that you would have to get out whenever the gold crisis broke, whenever that was to be. As is our custom in bull markets, we steadfastly remain bullish while the Uptrend is intact *and* as long as the majority of Technical Indicators point upward. This is the situation now.

If our Dollar does not fall by itself, it could be tragically undermined by a collapse of the British pound, since Britain's trade deficit widened sharply in April, eliminating any near-term substantial easing of their austerity program. The surplus of imports over exports was $285.6 million in April, from a mere $176.4 million in March. British government bonds softened noticeably. Perhaps President De Gaulle's rejection of Britain's latest Common Market application was partially responsible for the weakness. De Gaulle has been known to secretly favor devaluing the pound before England should be admitted. The Bank of England was forced to support Sterling and their Government securities. As a consequence—elsewhere in England—gold shares began to move up.

Conclusion: Hang on to your Golds!

May 19, 1967

*   *   *

Whenever a leading service comes out against a rise in the price of gold, a number of our subscribers wave such a publication in our faces and say "How come?" Anticipating such a movement, we would like to comment on Standard & Poor's May 29 comments on gold. The *New York Times* editorial page does not seem to be able to get beyond a call for the lifting of the "gold cover" which requires that we hold a 25% gold reserve on Federal Reserve notes. This would increase the supply of "free" gold to meet foreign demands by some $10 billion. However, Standard & Poor's correctly points out that potential foreign claims in excess of $30 billion makes such removal of the gold cover a temporary amelioration. Standard & Poor's then goes on to point out that "stopping the outflow of dollars would correct the situation." And, again correctly, an increase in the price of gold "would only buy time. We would still be losing gold until the balance-of-payments situation were corrected." All of this is correct and we are in complete agreement. Except, that after devaluation we would be losing much less gold. If we tripled the price of gold,* our 13 billion dollars in reserves would suddenly be transformed to $39 billion, more than enough to meet all foreign claims against us. Even total withdrawal would leave us with almost as much gold as we now have.

Standard & Poor's concluding paragraph is shown here and we are in fundamental disagreement with it.

A PANDORA'S BOX

The evils accompanying a rise in the price of gold would be great. The inflationary impact throughout the world might be severe, as it would remove the last restraint on monetary discipline. Confidence in the dollar could be destroyed—there would be no guarantee that further rises in the price of gold would not follow. This could well have an adverse impact on world trade. And, of course, such action would penalize all who had faith in our guarantee to maintain the price of gold, while rewarding hoarders and speculators.

---

* Tripling the gold price of $35 at that time suggested that a gold price of $100 would have been sufficient. Because my warnings were not heeded, the gold price will now need to go much higher than $100.

Thus, raising the price of gold would not solve our balance of payments problems. While international liquidity would be improved by such action, there are alternative methods to achieve the same objective without such unpredictable and potentially dangerous consequences. Our conclusion—the possibility of a rise in the price of gold is at yet remote.

The inflationary impact of devaluing gold hasn't nearly the inflationary impact of the inflationary policies of every inflationary government in the world for the last few decades. And as for removing the last restraint on monetary discipline, we can only ask quietly "what discipline?" As for confidence in the dollar being destroyed, it is rapidly eroding anyway and further increases in the price of gold would follow only if the first devaluation is insufficient. As for an adverse impact on world trade, we expect it will grind nearly to a halt unless there is devaluation. As for penalizing those who have been foolish enough to believe the U. S. government's guarantee, we have always repudiated those who offered such guarantees—anyway, rewarding speculators is called capitalism and we are 100% in favor of it.

We'll see.

June 9, 1967

*　　*　　*

What is really going on is a transparent desire by America to transmit its domestic system of inflationary deficits to finance social welfare to international monetary finance—strongly combatted by France which wants sound money. Thus, France insists that the U. S. stop running deficits, of which the United States is apparently incapable.

June 23, 1967

JULY 1, 1967: DJI 859.69; ASA - 10³⁄₁₆

*　　*　　*

In a major economic White Paper, the American Bankers Association warned that unless the U. S. eliminated its deficit in balance of international payments, we might well be faced with a devaluation of the dollar. To avoid this "unacceptable" possibility, the ABA called on the Johnson Administration to make ending the payments deficit "an objective of U.S. economic policy which claims the highest order of national priority," and said that the international financial system "would be seriously threatened—indeed probably completely disrupted

—by breaking the link between the dollar and gold." We have often said that this possibility could lead to an international depression.

And President Johnson signed the law giving him flexible power to increase the interest equalization tax by up to 50% during the next two years. In other words, Americans who buy stocks from foreigners will have to pay a tax of 22.5%, up from 15% previously. Kennedy initiated this tax and the minor impingement on your freedom has now been expanded. We have predicted this would happen, and we go on to foresee punitive taxes in excess of 100% as the government makes it increasingly difficult for you to get your funds out of this country. We envision upheavals ahead which few now would believe.

Oh yes, England. The British Treasury reported July reserves decreased only $42 million, and it was greeted with hoots of laughter on the London Stock Exchange. The figures were "doctored" to make them look better by including dollars borrowed on a short-term basis from other central banks. We have no sympathy for them if only because they are getting what they voted for, and the inherent immorality of attempting to live at the next fellow's expense.

August 4, 1967

*   *   *

On August 17, a *New York Times* headline read "Payment Deficit Narrows for U. S." while the *WSJ*'s headline was "Payments Deficit Widened From '66 In The 2nd Quarter."

Actually, both were right, with the *Times* indicating that the second quarter deficit was slightly lower than in the first quarter, while the *Journal* said that the deficit of $513 million was considerably wider than the $122 million of the second quarter last year. Either way you look at it, the deficit in the first half is running at an annual rate of $2.1 billion, higher than the total loss of $1.4 billion for all of 1966.

There were indications of heavy purchases by foreign government accounts of U. S. securities maturing in less than a year. Dollars put into such accounts cause an improvement because not only do they not form part of the deficit, but are actually counted as part of our assets! However, they are as real a threat to the U. S. gold stock as the longer term obligations, because they can be converted to gold at any time.

On a more positive note, our trade surplus moved up to an annual rate of $4.5 billion, versus a 1966 surplus of $3.7 billion. Seems American capitalism alone is supporting the whole rotten structure. The continuing series of new highs by gold shares is a damning indictment of the statistical window dressings used along with every other trick in the book to cover the failure to eliminate the deficit in U. S. balance of payments. One day, without warning, as with silver, probably late on a Friday, we expect the dollar to be devalued.

August 18, 1967

*   *   *

In England we have come to rely upon a comfortable time-lag of fifty years or a century intervening between the perception that something ought to be done and a serious attempt to do it.

H. G. Wells,
The Work, Wealth and Happiness of Mankind

Gold stocks exploded on August 24 after *The Wall Street Journal* published an article entitled "Pressure Grows for Devaluation of Pound As Britain Fears Jobless Rise This Winter." The article points out that unemployment will rise about 50%, to 3.1% of the labor force. Britain must pay $347 million in foreign borrowings by the end of this year and another $1.4 billion by 1970, and "some sources say Britain might better devalue its currency."

In recent years Britain has been spending more than it earned internationally, straining the value of its pound sterling, which was shored up by massive loans from the IMF and credits from major central banks. Internally, Britain has been forced to freeze wages, restrict credit, enact price controls and generally deflate. The article pointed out that businessmen and left-wingers alike favor devaluation more than they did a year ago, if the choice is really between recession and devaluation. And, in the last week, the National Institute of Economic and Social Research, an independent panel of top economists, predicted Britain's 1967 BOP surplus would be $70-$140 million, whereas earlier this year it had confidently predicted a surplus of $470 million.

August 25, 1967

*   *   *

On Saturday, August 26, a plan for monetary "reform" had been agreed upon by ten nations after six years of negotiations, and will probably be approved by the International Meeting in Rio de Janeiro on September 25.

While we are still studying this complex accord, these "Special Drawing Rights," or SDR's, will supplement dollars, pounds and gold in settlement of International earnings and balances. This surprising coup over the gullible Europeans, which left U. S. officials jubilant, will essentially permit the U. S. to continue its international deficits indefinitely. James Callaghan, British Chancellor of the Exchequer, said the new plan would make it easier for countries that spend more abroad than they earn, to put their books into balance without going through severe deflation, but he cautioned this is no license for any country to run a continuing deficit. (How's that for doublethink?) Treasury Secretary Fowler called it "one of the great days in the history of international financial cooperation." President Johnson hailed the pact because it "advances the welfare of all Americans." He also said it "marks the greatest forward step in world financial cooperation in the 20 years since the creation of the I.M.F. itself." Richard N. Cooper, Yale Economics Professor, said "gold is on the way out." In agreement, the *New York Times* intoned "probably after the year 2000, economic historians may see the London accord from a different perspective. They may point to that agreement as the beginning of an irreversible process that culminated eventually in a single world currency that would end the reign of metallic gold over international exchange, and make it merely another commodity useful in dentistry, industry and the decorative arts."

But quiet voices of reason began to express doubts. While it was generally conceded that this agreement would check speculative attacks on the dollar and pound, the plan will not go into effect at least until early 1969 and the relatively small amounts of money involved makes the plan's importance "greatly exaggerated." If $1 billion of SDR's is created annually, the United States' share would be 24.59% or $245.9 million.

Clearly, a blow has been struck for gradual international inflation such as exists in the U. S. It seems the height of folly to own bonds (unless they pay at least 10%) or cash, or life insurance, or annuities. The answers seem to be land, common stocks, fine art and numismatics, considering who is running the world. We persist in asking these questions: Why can't the U. S. run its dollar printing presses for an extra few days, and buy France with it? In other words, what is to be the relationship between paper money and reality? What will force the U. S. to balance its outflow and inflow of dollars? At the moment, these questions are being ignored, but will have to be faced some day. It will be a long time before it is realized how bad was the creation of SDR.

We still feel that gold will assert itself as the true yardstick of paper currencies, but that day has been postponed. France's apparent capitulation to the inflationists is a stunning surprise, but facts are facts.

August 30, 1967

*   *   *

SDR's might well be the opening wedge in a runaway world inflation, releasing any restraint on the manufacture of paper money, with a net result that will make the Mississippi Bubble seem small by comparison. However, such degeneration takes years, and the SDR's could well postpone such a crisis for some time. On the other hand, the size of the cushion provided by SDR's might not be enough to cover the deficits of the key nations; and a devaluation might come before the SDR's have a chance to gain their pernicious foothold. It is a race against time.

As we went to press, it was announced that Secretary of the Treasury Fowler was making noises about removing the 25% gold cover for the U. S. dollar. It is this gold cover which limits the amount of paper in relation to gold, and is some restraint on inflation. To remove it means that *all* of our gold would be available to foreign creditors if they sought to cash in the $30-odd billion U. S. debt they hold. The fact that we only have around $13 billion of gold in the kitty makes this something of a puny gesture. If they call our bluff the U. S. will be cleaned out quickly. Fowler seems to be throwing down the gauntlet and declaring that the U. S. plans on having a "fiat currency," like the ruble or the Ghanian pound. The bluff might work, considering the extraordinary inability of the Europeans to grasp what has been done to them. While giving them gold, we have run the printing presses and made major investments throughout Europe with the resultant buying power. In many cases, we have prevailed on them not to take out their gold, but rather to take more paper in the form of short-term Treasury Bills! If they ever wise up and clean out Fort Knox, the result would be too horrible to contemplate. The subsequent refusal of foreigners to accept dollars would cause fantastic economic dislocations not seen in three decades.

September 15, 1967

*   *   *

England's gold and foreign currency reserves showed a $28 million loss in October, but a Swiss bank loan of $103.6 million enabled them to report a "gain" of $75.6 million. Of course, the so-called gain made headlines. Dock strikes and Suez closing continue to hurt, despite the touted increase which reversed five successive monthly declines in Britain's reserves. Devaluation rumors raced across the Continent, or at least an increase in London's bank rate to match competition from other high-interest centers. Meanwhile, back at the once-smug American ranch, a $100 million gold loss brought U. S. gold supply below $13 billion for the first time in nearly 30 years. U. S. now has $12.909 billion, and the patrimony is going fast as foreigners continue to clean us out at 1934 prices. You would think Americans would have learned something about price controls by now.

November 10, 1967

*   *   *

The temporary selloff in the gold group—due to the SDR fiasco—has now been digested. Though alarming to holders of gold shares when first announced, the SDR proposal no longer seems the answer after all.

On November 14 a British trade gap of nearly $300 million was announced, the worst ever reported by the Board of Trade. The Bank of England again had to support the pound from slipping below its official floor. The Board of Trade said it was "impossible to draw any useful conclusions on the true positions and trends" because the dock strikes in Liverpool and London distorted October's trade patterns.

Then, in a dramatic move, the Bank for International Settlements announced a loan to Britain to help repay some $350 million due the IMF next month. This is from 1964's pound supporting operations.

Meanwhile, the Bank of England raised its discount rate to 6½% from 6% (the second such move in three weeks) to help the badly floundering pound. We anticipate additional increases as bait for people to hold on to the increasingly worthless English paper money. The earlier reductions this year from a high of 7% are thus seen to have been "psychological" and therefore premature—they might have actually done more harm than good. There have been rumors of the pound's devaluation since the Middle East war in June, but the discount rate boost rallied the pound. There have also been rumors that the West German currency might be revalued *upward*, which caused heavy demand for marks (often against sales of sterling). There has been strong demand for good currencies since Julius Caesar.

November 17, 1967

* * *

We first predicted increases in the prices of gold and silver in 1961, which we understand is the first of the widely-read market letters to so haruspicate. In fact, we have become rather identified with precious metals. It was therefore of some grim satisfaction to us this week when England devalued by 14.3% on November 18. This is the second serious crack in the international monetary structure, the first having been the disappearance of silver from U. S. coinage and a rising silver price. During this melodramatic week, with a violently fluctuating bond and stock market, we thought back to Lord Keynes' description of gold as "that barbarous relic." So much for Keynes.

Reduced prices for British goods would indicate some American export market loss, reducing our surplus of exports over imports and enlarging our b-o-p deficit. British goods will be coming into the United

States cheaper. A Rolls Royce now will cost 14.3% less than last week. The purpose behind devaluing the pound was to lower imports and raise exports so that the English will earn as much as they buy plus give away overseas. There will also be increased tourism in the countries which have devalued, since dollars are now worth more in those places.

The big question is whether England will devalue again. With the other major financial powers refusing to devalue, and offering a $3 billion standby line of credit, accompanied by additional deflationary measures in Britain, then this small devaluation may have gotten the inflationists off cheap for a while. But basic criticism of the present system is that it is too rigid. When an important nation is in "fundamental disequilibrium," there is no way for it to devalue without upsetting the whole rigid system. If all currencies were allowed to float in the free market (it's called Capitalism) there would not be any strain in the first place.

A lot of big money has been shaken and will henceforth move into gold. This is not over yet, and there are those who will sacrifice potential interest of 8% to sleep well at night owning gold.

Is England Bleeding to Death?

Prime Minister Wilson repeatedly suggested that speculators caused Britain's problems, despite having said he would not fix blame. (Please *do*.) He conspicuously failed to indicate that his own policies might have been at fault. He said, "Even our increasing exports could not earn enough to meet the successive waves of speculation against sterling." Unfortunately, Britain's export-import balance turned unfavorable *before* the closure of the Suez Canal and the dock strike Wilson cited as crippling factors. Instead, Mr. Wilson should have pointed to the dockers, who refused even *to vote* on a proposal to end their wildcat strike. Frank Cousins, when asked whether his union would still persist in its wage demands, said "of course." Is this the same England that defiantly told Adolf Hitler to go to Purgatory? In more than a figurative gestalt, England is slowly bleeding to death.

Other important effects will be increasing pressure on and suspicion of the U. S. dollar itself, which we have long predicted will be broken ruthlessly in the marketplace sooner or later. The devaluation will hurt the U. S. b-o-p, but it remains to be seen by how much. The devaluation, despite solemn assurances to the contrary, makes gold stocks much more attractive long-pull. More people have been awakened to devaluation in the last week than in the previous 100 years in this country. We see a new interest in gold, and we doubt that any new gold production this year will reach official monetary reserves. Why should it at 1934 prices? Look for a boom in the gold numismatic market, since even leading New York coin dealers have no concept of what is coming. Many U. S. double eagles, with around $34 in gold at present prices can be obtained for around $50 each. And it is just about the only legal way for the average American to own gold. Most will be forced into gold shares.

Do High Interest Rates Mean a Depression?

The new bank rate of 8% is Britain's highest in 53 years and the second highest this century. There was 10% paid in 1914, during World War I.

Remember the rejoicing earlier this year when it was reduced from 7%? High interest rates are the immediate result, but do not let anybody bulldoze you into thinking that they will remain high. Any time there has been an *adequate devaluation of a currency, interest rates have collapsed.* Witness the U. S. after the 1934 devaluation. Those who are really for the "little guy" should be working for a sound currency; only the wealthy can hedge themselves against the premeditated inflationary rape of those on fixed incomes.

November 24, 1967

*   *   *

Devaluation is dominating editorial comment throughout the world. *Life* magazine summed it up by pointing out that if one of the two reserve currencies was so feeble, what guarantee is there that the other, more crucial dollar might not also collapse? *Life* says, "The balance of payments problem simply must have more influence on White House fiscal decisions." We disagree, as we have for years, and still feel that international monetary finance is *the primary* (and we mean second to none) area of danger to our freedom in this decade. Anyway, *Life* goes on with the usual platitudes, about the U. S.'s economy being superior to Britain's, and that Britain's retreat will shift even more of the burden onto us. But, nowhere do they grasp that this whole problem has nothing to do with a tax surcharge or spending cuts. We still, as we have for years, get the impression that people in the highest places in this country do not really grasp the importance and meaning of international monetary finance. It is a horrifying conclusion we reached long ago, and are beyond being depressed by it. Particularly when the *New York Times*, in approving the new Chancellor of the Exchequer Roy Jenkins, says, "He is a radical in the best and European sense of the word." What more can we say? *Life* magazine declares, "But even a big run on U. S. gold would not automatically result in devaluation. There are so many dollar-holding countries in the world with so much to lose that they would rush to support the U. S. with loans." The immorality of blackmail aside, a big run on U. S. gold would indeed clean us out quickly.

December 1, 1967

* * *

ENGLAND STILL ON THE ROPES

Sterling plummeted on December 4, and the wild fluctuations that Monday were described by some traders as "chaotic." This came after the British reported that November 30 gold and convertible currency reserves rose last month by $127.2 million, including the Government's portfolio of dollar securities; without this addition, reserves would have shown a *loss* of around $362 million. *This shows that the money which took flight from England has not yet returned.*

AMERICA'S INTELLIGENTSIA SPEAKS

1. George W. Ball, a former Under Secretary of State, characterized General de Gaulle's recommendation to return to the gold standard as that of a "medieval alchemist." Mr. Ball (presently chairman of Lehman Bros. International Ltd.) followed the Party Line about the United States not wanting to dominate others or forcing Europeans to finance our investments through our balance-of-payments deficit. We respectfully disagree with Mr. Ball and would like to point out that comments such as his did nothing to prevent the devaluation of the pound.

2. A *New York Times* editorial on December 4 called for a debate on whether or not the dollar should be tied to gold at all. We agree— so long as there is somebody who knows gold taking the "pro" side and will not simply harp on how nasty is de Gaulle. This he might be, but it's irrelevant. The *Times* again—"The strength of the dollar is not really in the hoard of gold in Fort Knox, as often popularly assumed. The basic strength of the dollar derives from the vast resources of this country and from the talents, industry and production capacity of the American people." No—all that counts is the *net* balance-of-payments. What the *Times* does not grasp (nor most other commentators) is that when we earn a foreign surplus through the talents, industry and productive capacity of the American people, and then spend *double that amount* overseas, the dollar will *weaken*. Get rid of foreign aid and stop running a foreign deficit, we say, and the dollar will indeed reflect the true strength of the American people. The fact is that you must subtract, from the American people, unwise bureaucrats and silly newspapers.

3. And Sylvia Porter announced in the December 7 *New York Post* that things don't look well. She said, "As of today, the potential claims on our $12 billion-plus of gold amount to $28 billion" (Ed: Wrong, Sylvia, it's $30.8 billion, and probably more since England's crisis).

As for what to do, she starts with slashes in Federal Government spending and increases in Federal income taxes. This, of course, would help preserve our balance of trade surplus but, so far, this is the *strongest* element in the picture. She asks for high and even higher interest rates, which of course will tend to keep money here since people will take the risk of dollar devaluation if the interest rates are high enough.

Again, she fails to meet the problem squarely. She then recommends stiffer restrictions on U. S. investments abroad (despite the already stiff restrictions), short-sightedly forgetting that these investments eventually contribute to the inflow, when money is earned. If General Motors builds a plant in Europe earnings will someday flow back to the United States, to add to our surplus.

Sylvia then wants direct incentives to exports, showing the Welfarist-Subsidyism mentality at work again. And restraint of tourism, again the direct control over individual freedom—a typical feature on the trip to Fascism. Finally, she calls for sharp reductions in spending for U. S. soldiers, reminiscent of Harry Bridges' call in 1945 to "bring the boys home."

Nowhere does she mention that the first item to be hit by the axe should be foreign aid. How can she, since she is in favor of foreign aid? And, when will it be understood that it was foreign aid which got us into this mess in the first place? We await the day when even one of the champions of foreign aid will say "we were wrong." We decline to hold our breath until then.

RUEFF, THE BRAINS BEHIND DE GAULLE, SPEAKS

Our first comments on the gold situation, back in 1961, were predicated on substantially what Jacques Rueff said in a *Fortune Magazine* article that year. Even then, we singled him out as the brains behind de Gaulle in monetary affairs. So, we listen closely to his remarks, rather than de Gaulle's, when he spells out his views on gold. We paid particular attention to his interview, as reported in *U. S. News & World Report* on December 11. It shows his position is still unchanged from that "expressed since 1961 about the dangers of the present gold-exchange standard."

He feels that the United States' $13 billion dollars in gold (overdrawn 145% by debts outstanding) must lead to a fiscal collapse which will be triggered by accident—a war in the Middle East, a bad harvest or a recalcitrant trade union. He flatly ruled out devaluation of the dollar, because "American goods are perfectly competitive in the world." Of course, correct. In fact, he termed such a devaluation a "catastrophe" and "entirely useless." Instead he wants an increase in the price of gold in all countries with convertible currency, with no changes in parity. Again, correct, because our problem is what we give away rather than in the amount we are not earning overseas. Our problem is fundamentally different from England's. When asked about a possible financial war developing, he said, "May I protest this term 'war.' It's childish . . . it would be stupid to think we want devaluations of the dollar and pound."

Finally, he indicated he would like to see the price of gold doubled, to allow us to repay "the dangerous part of the dollar balances" overseas. We disagree with Rueff on this point. We think the price of gold needs tripling, which will magically enlarge our $13 billion into $39 billion, leaving us some margin of safety in case foreigners ask for their entire $30.8 billion. Special help will have to be given to England because their $2 billion in gold, even doubled, would be no help against foreign claims of around $12 billion. He suggests a special 20-year loan to them. A transfusion would be better.

Gold shares would be spectacular beneficiaries of a price increase, because costs would not increase at all while the entire price increase would flow into profits. Thus, retention of gold shares is a shrewd move against the kind of surprise that happened when England devalued.

December 8, 1967

*   *   *

Mr. Fabra, financial editor of *Le Monde*, predicted on Monday that the U. S. would ask other members of the gold pool for part of their gold reserves in exchange for certified receipts. There was no Washington confirmation or denial. Fabra also wrote the U. S. would ask for restrictions on potential gold buyers. Meanwhile, former French Prime Minister Pierre Mendès-France said on December 8 that though doing its best to sink the pound, France may well suffer from its own action. He said devaluation would mean French exports now face stiffer resistance abroad, and in any case, "the British will not forgive us for a long time . . . they remember that in the past the franc was much more often in difficulties than the pound and they never refused us help in such a case."

The world was startled by the U. S. gold loss last week of $475 million, leaving us with $12.43 billion; this was the largest single weekly drop in the history of U. S. gold reports, and was to replenish the London gold pool from the November run on the pound. The decline, plus a $325 million increase in Federal Reserve notes outstanding in the last week, pushed down the proportion of gold backing to outstanding notes to a new low of 28.5% from the previous week's 29.9%. Law requires the ratio to be at least 25%, although the FRB has the authority to suspend it under certain conditions. We still grimly believe that the United States is broke, and will need to rent an empty Fort Knox unless the gold price rises soon. Only this week social security benefits were voted to be increased 13%. Keep buying those votes boys, just keep it up.

And George S. Moore, Chairman of giant First National City Bank, doubts the continued link between the dollar and gold, but believes the world still needs "the discipline of gold." The man wants to repeal the 25% gold backing, which we still feel is silly and solves nothing. In a blunt talk, he said that the United States is "over-extended" financially, and that the "dollar is at bay." While he was "firmly" against a rise in the price of gold, he double-talked by saying it was unrealistic to say the price would never rise. He stated there were three reasons why we were in trouble, saying that the Federal budget is "literally out of control," the balance of payments deficit has reached the critical stage, and there has been an excessively easy credit policy by the FRB. He wants higher taxes, lower Government spending and blah, blah, blah.

Then, on Tuesday, the news broke that the United States was indeed trying to set up rules for world gold markets. This is the third major secret in recent weeks revealed by *Le Monde*, suggesting they have extraordinary news sources. Apparently the U. S. wants other nations to forbid their citizens to buy gold, on the grounds that if Americans can be raped financially, why not everybody else? The plan apparently seeks to tie up at least half of South Africa's production, whereas last year all their output drifted into private hands. Apparently, the South Africans are not bright enough to understand that they can refuse to sell gold, and force the price to rise above $100 anytime they wish. Some nations like to sell their patrimony cheap.

Le Monde reported resistance (among European central banks) and outright rejection of the alleged suggestion that they lend the United States part of their gold; also, decisions taken in Basel would emerge only gradually. Meanwhile, the uncertainty caused increased gold buying in Paris, with another new low for the English pound.

Thursday, in the fast-breaking situation, Britain announced a fantastic trade deficit while gold stocks soared to new highs. Gold buying reached near-panic proportions in Paris when rumors indicated the dollar might be devalued any day, or that many would be banned from owning gold. Politicos don't understand that controls breed more controls and blocking sales elsewhere would only boom the Paris gold market or a black market.

Those alerted to the appalling ignorance of gold in our national leadership are now wondering what to do. All government moves represent a scrambling about for a solution, which does not exist given Keynesian premises. We still predict the dollar will be broken, which, considering Washington's bungling in recent years, will probably lead to an international depression. Your solutions revolve around precious metals, whether they be coins (numismatics), commodity futures (platinum, silver), or gold and silver stocks.

Conceivably, the world could go to fiat currency, including paper SDR's, but fiat currencies have never worked and never will. In the interim, gold mining companies can go broke, so you are faced with a severe dilemma. Those who own no golds should remedy the situation instantaneously.

December 15, 1967

JANUARY 1, 1968: DJI 906.84; ASA-13¾

* * *

Cynically, it had already been secretly decided on December 25 to make a stunning announcement because of the gold crisis. Johnson unveiled, on January 1, in the crisis atmosphere of surprise news conferences, a series of measures which contracts bank lending to foreigners, forces repatriation of capital held overseas, provides more financial aid for exporters, and, most fantastic of all, indicates that tourist travel abroad would be restricted.

Backed by the power of criminal prosecution and fines up to $10,000, there will be a new "Office of Foreign Direct Investments." A new set of bureaucrats has been born, and it is significant that it has police functions.

The drastic crackdown on corporate investment overseas, coming less than two months after the "voluntary" program was announced by the Commerce Department, indicates that Johnson was clearly panicked by latest figures. If the U. S. foreign deficit in 1967 reached $4 billion, that would mean foreigners accumulated more dollars than in any other year since World War II. Johnson's actions indicate the figures were indeed bad. To confirm this, the Treasury on January 3 announced a $450 million gold loss accruing from the devaluation of the British pound.

Since $10.6 billion of gold is required as cover for outstanding currency, the latest total of $11.98 billion indicates that the legal gold cover will shortly be removed. U. S. citizens should scream bloody murder to stop the irresponsible squandering of all our gold. The cover was put there precisely as a stop loss on irresponsibility, and these people must not be permitted to stick a penny in the fuse box.

SIMPLY STATED, WHAT IS THE GOLD PROBLEM?

Oversimplifying, every nation's paper currency must be translatable into a common denominator so that international trade can continue. Gold, mink coats, diamonds, slaves, it doesn't matter what, but you must have something. Americans are naive enough, and trust their politicians enough, and believe the controllers of the printing presses when they say they will not abuse their privilege. When tied to gold, paper becomes quite trustworthy since gold cannot yet be created alchemically.

If you know that Upper Volta has a unit of currency which can definitely be exchanged for gold anytime, then this currency will be good anywhere in the world. In 1949, the United States had so much gold behind its dollar ($24.6 billion worth) that the dollar actually became as "good as gold." Touchingly, many began to assume this would be a permanent state of affairs. When Washington poured billions of dollars overseas, and ran the printing presses for worthy causes (Oh, dear, those poor people in Togo just *must* have a new steel mill and dam) the U. S. began to lose gold. From the 1949 pinnacle our gold holdings plunged to $11.98 billion. This is bad enough, but what few Americans realize is that there are additional claims against this hoard *of at least $32 billion*. Actually, the United States became bankrupt in 1961, and to our knowledge we were the only ones in this country screaming bloody murder about it. We received for our pains amused tolerance.

The situation is so bad it is unbelievable. If the price of gold were tripled our $11.98 billion also would be tripled to $35.94, which barely covers liabilities around $32 billion. The longer we wait, and the more gold we lose, the more we will have to devalue the dollar. As it stands now, if the price of gold were *quadrupled* we would at last have enough gold to pay off the paper poured overseas by well-meaning innocents.

GOLD HOARDING ACCELERATED IN 1967

The First National City Bank's annual gold review guesses that $2.5 billion in gold was hoarded overseas in 1967. Around $1 billion was taken from Central Reserve Banks, while the remaining $1.5 billion from newly mined gold. The review said "For the first time in modern history the international monetary system is left to work with declining official reserves." And why not? At 1934 prices gold remains one of the world's biggest bargains. We still feel the only way to handle this is with a rather old device called a free market.

OUR SOLUTION FOR THE GOLD CRISIS

We're in bitter disagreement with what has been done, is being done, and what we understand will be done. No nation this rich should put itself in a position where a De Gaulle can knock it off balance, or where Americans cannot spend anywhere in the world the fruits of their fantastically productive labor. We cannot agree with additional controls, and the check on corporate investment abroad strangles the flow of future earnings which has already been spent by irresponsible vote-seekers. We would remove all controls, and issue orders to capitalists consisting of just one world: "go." We would terminate foreign aid immediately, returning charity to the voluntary arena where it belongs. We would make earning a balance-of-payment surplus the highest national priority. And, most important, we would remove all controls on gold, and let the price seek its own level. "Free Gold" should be the rallying cry of all economically responsible citizens.

DEPRESSION AND END OF AN ERA?

As we have been saying all decade, and now there are increasing numbers who agree with us, monetary questions are the most important issues of this decade. If they continue to be bungled an international depression is probable. Given Keynesian premises, it should not surprise you that we cannot name a single man in government today who we feel understands gold anywhere near the level of Jacques Rueff of France. And the Republican (including Conservative) Party has been completely incompetent as an effective opposition on this point. The few who do understand gold have, for some puzzling reason, not made themselves heard. The press is hopeless. We are amused to note that the most strident voices calling for basic reform are those of the individual market letters. (Is that our fault?) As noted before, market letters have no impact on government. Nonetheless, the market understands what might be ahead, and if the United States curbs overseas spending and tourism or demands repatriation of profits, there could be a collapse of international trade. It is not hard to imagine what that will mean for American firms who depend on trade. An era that began in 1933 might now be in the process of an end. If so, a Major reversal of stock prices is a distinct possibility. This would include a DJI reading below 400. Clearly, L.B.J. is now in the process of building an American equivalent of the Berlin Wall; we have been calling it the Dollar Curtain. (How about the "Gold Curtain.") When tourism is banned for Americans and so-called constitutional questions are brushed aside in favor of public policy, which is ostensibly the custom, many Americans will finally understand that foreign aid has helped create a colossal disaster. *Sauve qui peut.*

WILL ENGLAND DEVALUE AGAIN?

The reason the market is so trendless is its puzzlement over the second most important feature of 1967: namely, the devaluation of the English pound despite the new Special Drawing Rights. Impact is still rippling outwards and has already forced the United States into a number of startling new laws involving exchange controls, clampdown on tourism, and various schemes to stimulate exports. These poor old Keynesians have not stopped to look at England, which tried all these various palliatives, but wound up devaluing the pound anyway. It is a major question as to what a dollar devaluation would mean to our elections—and to the world. It is serious enough for the market to dread another devaluation of the pound, an event widely expected in more sophisticated international financial circles. It would be an achievement indeed if devaluation of the dollar or pound were staved off until 1969.

1968 OUTLOOK FOR GOLD

In last year's Forecast Issue, before France became unpopular, we wrote "France understands that gold is power" and "the days when the U. S. can dictate international monetary policy are coming to an end." The prediction came true with a vengeance when the pound was devalued 14.3% to $2.40 on November 18. We have already chronicled that drama, and would now like to predict that 1968 will see another major devaluation, probably in the United States or England again.

January 5, 1968

*   *   *

We were hardly surprised when L.B.J., in his 1968 State of the Union Message on Wednesday, asked for removal of the $10 billion "Gold cover" required by law as backing for the U. S. Currency. If the Gold reserve law were repealed, all of the $12 billion in gold now held would be available to be squandered in a quixotic defense of an indefensible $35 gold price. Cooler heads prevailed overseas, where there was a stampede out of dollars into gold. If nobody in the U. S. understands gold at least they have read history books and know what is coming. Walter B. Wriston, president of the First National City Bank, asserted that L.B.J.'s plan to bring the B-O-P into line was really a foreign exchange "control" which, "in peacetime never have operated efficiently."

Poor Senator Vance Hartke, the Indiana Democrat, made himself immortal by offering this solution: reduce the buying price of gold to $33.50 an ounce and this would end much of the world gold speculation.

Since his definition of the problem differs from ours, there is not much we can say about his solution except that if few are willing to sell U. S. gold at $35, who on earth is going to sell it to us for $33.50? If his logic is correct, why not reduce the price of gold to $1?

At any rate, we mean no disrespect to Hartke, but a man of his position really ought to know what he is talking about. After all, he does have the alternative of silence.

Let those who underestimate the power of gold consider England's announcement this week that it will withdraw from East of Suez by the end of 1971, cancel its order for 50 American F-111's, charge for medical prescriptions which violates one of the Labour Party's most deeply held articles of faith, introduce the principle of giving welfare to those most in need, instead of to all, and "postpone" until 1973 a plan to make all British children stay in school for an extra year. Pretty soon there'll be no more "brain drain"—less school is one solution.

This comes on top of the November 18 devaluation of the pound, an 8% bank rate, a freeze on additional bank loans, restricted installment credit, corporate tax increases and Government spending cuts of £200 million.

We must conclude that gold is powerful enough to make what 50 years ago was the most powerful nation on earth decide to try to give as much as it takes. By that we mean it will attempt to earn abroad as much as it purchases abroad. Gold is economic sanity.

A year or two ago, we described America's efforts to help the pound as akin to two actors on a stage with only one costume between them, with each appearing alternately to give the impression that both were fully clothed. With England knocked out of the picture, and plainly exposed as a bankrupt, it will become increasingly obvious that now there is not enough for even one. England's cancellation of the F-111 will cut into American exports, lessening even further our ability to bail out England. What a chain reaction!

January 19, 1968

*   *   *

The coup d'etat came yesterday when our gold loss was put at a record $900 million. This was the largest gold loss for any month on record, somewhat higher than the $74 million November loss. (This is the Great Society?) Our gold stock is half as large as the record $24.432 of 1949 and is at the lowest level since the $10.575 billion of 1936. With only $11.982 billion, 10.7 billion of which is required as currency backing, no wonder that LBJ is eager to remove the "gold cover!" To make matters worse, buried in fine print, foreign holders of U. S. dollars —all of which are potential claims on our $11.982 billion—have soared to $32.433 billion. Never has an obvious bankrupt been so completely obviously bankrupt. It is difficult to believe that the dollar will be able to stave off devaluation for another year.

Our Swiss correspondent informs us there was extremely heavy gold share buying in Europe this week, particularly from America. There is also substantial Swiss demand for double eagles ($20) which have risen 13% from $46 a year ago to $52.

February 2, 1968

*　　*　　*

U. S. gold skidded by $100 million last week, lowest since April 1937. At $11.884 billion the proportion of gold backing outstanding Federal Reserve notes plunged to a low of 27.5% (the minimum set by law is 25%).

Such losses have come with such boring regularity that they no longer attract much attention, but somebody in the U. S. Government must have been stung by something behind the scenes because there was a blistering reaction from William McChesney Martin, Jr. In a blend of William Jennings Bryan and *la plume de U Thant*, Martin rejected the possibility of a gold price increase as "highly disruptive and highly inequitable." (TR.: I might be thrown out of office.) Barely pausing for breath, Mr. Martin said, "I am firmly of the belief that a high gold price is neither necessary nor desirable and would "break faith" with nations around the world that have held dollars on the basis of the American pledge not to increase the price of gold.

February 16, 1968

*　　*　　*

Malcolm S. Forbes, Editor-in-Chief of *Forbes Magazine*, lashed out at gold in a lead article in his March 1 issue. Let us anticipate the flood of mail from our readers which usually follows such statements. He said, "The day this country is out of the stuff, that day gold becomes what it's worth as a metal and no longer will have much significance as a monetary measurement. It isn't the gold we have that makes this nation rich. It's what we make, our knowhow, our productivity."

We're not knocking *Forbes*, for it is an excellent magazine that is required reading in our Research Department. We respectfully submit, however, that Mr. Forbes does not understand gold, in common with most Americans. Unfortunately, this type of editorial simply reinforces the naïveté which weakens us in our dealings with people who really do understand gold—like de Gaulle.

Mr. Forbes' basic argument revolves around the idea that ". . . the worth of the dollar relates to our productivity, the application of brawn to brain." To destroy his argument quietly but effectively one need only ask what would happen if the United States ran its printing presses *full time* for a year, filled up several ocean liners with $100 dollar bills, and shipped them to de Gaulle offering to buy France. All of it lock, stock and barrel. It can readily be seen that Forbes failed to grasp that the dollar is not related to our productivity unless the number of dollars is fixed in relation to every other country's productivity—and *gold is the common denominator.*

With the Day of Judgment grinding inexorably closer for a world that has lived beyond its means for too long, Jake Javits decided to speak. The newspapers promptly labeled it the "Javits Plan." Jake suggested that the U. S. suspend gold payments to conserve our remaining $12 billion gold stock, combined with a direct support of dollar in the international money markets. Poor Jake also proposed to free the market price of gold, without realizing that this would be tantamount to a complete devaluation, even though he stressed that his plan would continue an "official price" of gold at $35 an ounce—which of course would not last long.

Now, follow these figures closely: At the end of January, the United States had total gold holdings of $12.003 billion. Total foreign claims on it as of the end of November were $33.812 billion. These stunning figures starkly highlight that the United States is not only bankrupt, but *colossally* so. Of the claims against us, only foreign central banks are eligible to buy gold from our Treasury, and they held $15.958 billion of the $33.812. Foreign banks alone could clean us out, not counting all the rest of the claims by any foreign citizen going to his bank and asking for real money—or gold. The smaller figure is usually used when discussing liability against our gold holdings, but don't believe such fiscal trickery. Javits argued that the United States should see to it that its gold stock remains the largest of any nation so that it will "still have some cards to play." You're too late Jake. Our cards are gone. Where were you when we needed you?

Watching the explosive advances by gold stocks this week (Dome Mines alone was up 11 points) we expect the general insolvency of the U. S. dollar.

March 1, 1968

* * *

The N. Y. Times on March 5 editorialized "it has long been obvious that the present monetary system must be replaced by something stronger and more flexible." Naturally they exclude going back to the gold standard, which is incredibly strong and infinitely more flexible than bureaucratic decisions. But such is the state of the world that people start by ignoring the correct answer and then scramble around trying to patchwork something workable together. *The Times* writes "the need for maintaining the present machinery—despite all its flaws—is more critical than ever," to which we ask "why?"

"The biggest responsibility falls upon the United States" which is of course quite correct, except that for example *The Times* wants a repeal of the gold cover requirement. This is supposed to build up foreign confidence when they realize they can clean us out completely instead of just down to a certain legal minimum. (You know, *Times,* the minimum *was* put there for a reason.) But anyway, they also suggest that "Congress must take swift action on new payments *to reduce the deficit.*" (Underlining ours.) Nowhere does anybody talk about running at a surplus, and the Keynesians cannot get it through their thick skulls that the days of reducing deficits are gone, and that massive surpluses to wipe out past extravagances are mandatory. *The Times* did say something we thoroughly agreed with in demanding more men for our Treasury who are "experienced and sophisticated in the workings of international finance." But this type of person is squelched beginning with Samuelson's first course in economics in American colleges. Seems all the good ones are working for de Gaulle. When it comes to economics, the drain is out of the Anglo-Saxon countries. Maybe we can reverse the brain drain from France.

March 8, 1968

* * *

A dramatic move is coming. One can feel it in the air. The gold hemorrhage being what it is, LBJ will have to do something extraordinary, since his reassurances are presently being ignored completely by gold buyers. Even atavistic cursing of the ubiquitous 'gold speculators' does not seem to help. The current collapse in the bond market carries a strong odor—is it one of sharply higher interest rates (which is another of our old predictions about what would happen when the gold crisis broke) ? The Federal Reserve's discount rate might be hiked to as high as 6%, highest since 1929, with the amount of increase (1½%) the largest since 1931. This would tend to set all interest rates in motion upward, and begin to attract dollars back into this country.

The *Times* says that "the fumbling of the authorities is clearly contributing to the demand for gold. It could well bring on a flight from all paper currencies, which would spell the end of the present system before something better is put in its place." This represents an important change in their thinking, since until now they have implied that the present system is excellent and ought to be supported "since it has served the post-war world so well."

The *Times* proclaims that "there is still time to prevent collapse, provided time is used wisely." Then they at long last stumble on the fact that there is a huge dollar outflow by the Government, but they blame it all on the war in Vietnam. When, oh when, *Times,* will you begin to attack the foreign aid outflow? When will America stop aiding Yugoslavia, or any other Iron Curtain country which then supports the North Vietnamese? If foreign aid is charity, then let charity begin at home.

On March 12 the Treasury released $450 million of gold for sale to foreign buyers by transfer to the Exchange Stabilization Fund. This drops the Treasury's gold stock to just over $11.68 billion, following the February 6 transfer of $100 million and several transfers exceeding $400 million last December. This gold is available to the gold pool to which America provides 59% of the gold sold on the London market. This used to be an eight-nation pool, but when France dropped out the U. S. dimwittedly accepted France's 9%.

The flight from paper money continues unabated and day after day the world is witnessing massive gold losses to those who are gobbling up the metal. Nevertheless an agreement was reached in Basel, Switzerland, last weekend by the seven members of the London gold pool to continue accepting losses from their monetary stocks in order to hold the fixed price. As the *N. Y. Times* put it on March 11, "the meeting represented an effort by the United States to gain time to put its balance-of-payments into order. . . ." Preposterous. We have yet to hear any important Government official say anything except "reduce the deficit." Nothing will stanch this gold hemorrhage short of a direct determination to solve the problem. In that same issue Edwin L. Dale, Jr. said "The steel-nerved decision today of the central banks . . . may have marked the beginning of the end of gold as a monetary metal." We think Mr. Dale's jubilation is premature, and seriously doubt that the gold pool countries will allow all of their $25 billion of gold to be dissipated for nothing. As if to confirm our judgment gold demand in London the next day was again heavy. Where we come from, the judgment of the market place still counts for something.

In the past we have commented that Paul Fabra of the French newspaper *Le Monde* has been painfully right about secret machinations behind the scene. He charged that the U. S. and Britain had arranged a "swop" of gold in which the same bullion was being counted in the holdings of both nations. This is a charge of fraud which places everything in a new light. We didn't for a moment imagine that things could be any worse than they are now, but we are distressed to learn that even our pessimistic appraisal might be too optimistic. *Le Monde* has been right too often for comfort. Perhaps we wouldn't be so annoyed if we did not firmly believe that those responsible for this chicanery will never be held accountable.

On March 13 a *N. Y. Times* editorial struck a really serious tone by declaring "the drama of the gold rush has entered a new and perhaps decisive stage. It is quite clear that the existing international payment system, based on a fixed relationship between gold and the dollar, is eroding. . . . There is little. to be gained by further reassurances from defenders of this system. The blunt truth is that they are no longer believed." But of course the *Times,* as usual, wanders off down the daisy path by rejecting a rise in the price of gold on the grounds that it "will not solve any of the afflictions that now plague the system." They end with "it is time to end the suspense so that a new system can be built on the framework, rather than the ashes, of the old." To which we query, with the way girls' skirts have been rising inch by inch in recent years, why don't they just wear bathing suits and be done with it?

WILL $35 GOLD LEAD TO A DEPRESSION?

Federal Reserve Board Vice-Chairman James Robertson told the Weekly Bond Buyer on March 7th that the Fed will maintain gold at $35 "irrespective of the price we may have to pay." The *N. Y. Post* on March 15 said: "The 'price,' outlined in that interview, would seem to include 1) greater unemployment, 2) a reversal of national economic growth trends, 3) idle industrial plants, 4) downturn in the housing industry and 5) cancellation of many social and welfare programs."

March 15, 1968

*   *   *

After we went to press last week the gold crisis we have long predicted would come, started. As the flight from paper in recent days proceeded first from a walk to the bank to buy gold, to a trot, a canter, and finally a gallop, the Bank of England closed down on March 15 under the strain. Everybody stopped selling gold, even NYC coin dealers. Secret sessions were suddenly summoned in Washington as the full extent of the panic finally began to penetrate the concrete between our politicians' eardrums. Economic illiterates who were weaned on Samuelson and trained to worship Keynes have led the world to the brink of a major disaster. The full effects are at present incalculable, although it is clear that a major turning point in the history of the world began last weekend. The man in the street does not know it yet, although he is uneasy, as would be the inhabitants of a valley where the distant rumble of thunder precedes an earthquake. Those whose financial plans were built on sand will be destroyed, while others, who have prepared, at least will have a fighting chance. You will have to keep your wits about you, for it is certain there will be an erratic and wild period just ahead. Any temporary decline in gold shares should be ignored as ill-informed and irresponsible selling by the same panicky people who have disagreed with our position on gold all along. Dramatic announcements from Washington sparked a wild market rally and a sharp drop in golds. Remember what we told you when President Kennedy declared over Telstar in 1962 that the dollar would not be devalued, and gold stocks plummeted the next day—we told you to step in and buy the bargains of a lifetime, that you would never live to see American-South African sell under 30 again (that was before it split 2-1). Hang on to your golds.

We watched with rapt fascination last Friday when, despite the fact that gold on the Paris Bourse soared to an all-time high at $44.36, gold shares actually declined! This was due to profit-takers who expected a dramatic announcement over the weekend to suddenly cure the problem, as has happened in every minor gold crisis this decade. Now is the time to take major additional positions in golds.

The *New York Times* on March 16 editorialized that "The members of the international gold pool who meet today in Washington are confronted by one unassailable fact: the payments network has broken down. Their task no longer is to avert collapse but to rebuild the structure." We agree, if understood by 'rebuild' they mean radical new solutions (much resembling 'reactionary' old ones) which will avoid recurrence. Unfortunately, the *Times* continued to wander down the daisy path of increased taxes and so on. We did agree with their wish to avoid new tariffs, but the way to handle this crisis is to let the price of gold and all currencies fluctuate freely—which will put an end to 'planned inflation' and other nonsense. Let all currencies be expressed in terms of gold. All contracts and travellers can then simply adjust to the fact that there will be slight fluctuations in their currencies, which is true today anyhow. Then, when the politicos turn to the printing press to get themselves elected, the trouble will show up immediately. We'd like to see the *Times* print *that*.

'The barbarous relic' is amazingly strong for some sort of vestigial appendix. Gold managed to get the U. S. President to promise spending cuts, gold summoned ministers across the world wringing damp palms, it closed banks, and it might just get the United States out of Vietnam. Newspapers described the scene with words like "pandemonium, panic, crisis, fantastic and turmoil."

Edwin L. Dale, Jr. in the March 16 *Times* wrote, "Some Americans abroad learned what an international financial crisis could mean today when they were unable to cash their travelers checks into dollars. But there is no reason to expect any major impact at home, even if the world monetary system collapses." Wrong again, Mr. Dale, and most surprising in view of the fact you even note in your article that "the last time there was a financial panic in the U. S." was in the "early 1930's." Barely pausing over the similarity Dale smugly declared that, "the American economy, the most self-sufficient in the non-Communist world, is probably the most immune to international monetary turmoil." He feels that since Americans cannot own gold legally there can be no scramble here to convert dollars into gold—"a big factor in past panics." This statement is so historically wrong there is no point commenting on it. Later in his article, Dale suggests that the great trading nations (Britain, Continental Europe, Canada and Japan) "could go through some kind of depression." How on earth Dale expects us to go through this unscathed is beyond our grasp. And, he grandiloquently concludes "it is doubtful that dollars will become literally worthless in foreign countries. But the value of the dollar might be less than now, or might fluctuate." (Stocks could either rise or fall, unless they don't.)

In the same *Times* issue a column headlined "Peking Gloating Over Gold Crisis" disclosed that China has been steadily converting its reserves into gold since 1965, something we commented on at that time. Where are those who said that raising the price of gold was bad because it would help Russia? Note, these are the same types who want to 'build bridges to Russia' (good Communists) while China (bad Communists) should be punished.

We did learn something about gold from the *Times:* while it is illegal for Americans to own gold, this does not include unrefined gold. This loophole is to protect miners who stay at their sites accumulating raw gold for a year or more. Apparently gold does not lose much weight in refining, so many traders have been active at "undisclosed mine sites" paying $40 an oz. for the raw material. They have also been buying gold dust from certain Canadian banks—an interesting loophole for those who find it impossible to buy gold coins. Just as we once told you to accumulate silver certificates (which now sell at a 50% premium) and the old coinage which had real silver in it, and just as we told you to buy some gold coins, we now tell you to grab some gold shares before they are gone also.

Over the weekend, news of international reaction to the crisis began to pour in. Gold dealers in Milan stopped selling gold because of an "almost complete lack of supply." Foreign exchange operations were halted around the world, from Finland to Malaysia. Stock exchanges closed around the world, even in South Africa, while hotels, restaurants and stores were refusing to accept U. S. dollars and British pounds. Jacques Rueff, the man we singled out in 1961 as the prime mover in the coming world crisis, was quoted as asking for a "simultaneous increase in the price of gold in all countries with a convertible currency." And so it goes.

As noted, the man in the street still has little concept of what is happening, primarily because he is lulled by the media. The *Times* wrote on March 17 "There is no reason whatsoever for a financial panic in 1968. The turmoil in the gold markets abroad need not have any effect on what a housewife can buy with a dollar bill in the supermarket." Don't be suckered into believing this, because our $800 billion GNP is bloated. We have run the printing presses and given ourselves ever higher wage increases, all of which is increasingly out of touch with reality. No wonder the American economy is sick. It is grossly bloated. So much for the $800 billion.

On March 17 Edwin L. Dale, Jr., again missed the point. He quoted a Zurich banker as saying "Don't they realize, these people who are buying gold, that they are destroying the whole monetary system of the world?" (Perhaps they are just protecting themselves from the rage-of-the-moment mentalities who have been running the printing presses for the last few decades.) Dale says "If the U. S. payments deficit continues, the system will collapse. (Ed: Dale is still in the conditional tense!) What that will mean to world trade or even world prosperity, no man knows for sure. (Ed: He suggested on March 16 that there's no serious problem.) What is known is that the system—with assured rates of exchange among currencies, based on the U. S. gold commitment—has brought with it a prosperity unknown before." (Ed: Has Dale never heard of the Mississippi bubble, or the boom of the 1920's?)

Eileen Shanahan wrote in the March 17 *Times* that "Some of the blunders of the Thirties look simply incredible in the light of present day economic knowledge." Wait until they describe this one in books! Another *Times* article correctly said "It was . . . not a speculators' conspiracy or any occult financial operation—that brought on the gold rush." Edward Du Cann, former chairman of the British Conservative Party and ordinarily friendly to the U. S., spoke last week of "the ineptitude of the Americans . . . the most astonishing folly and lack of foresight."

Lord Keynes once declared that men do not have to be concerned with the long-range consequences of their policies and actions—"In the long run we are all dead." In the short run, however, the price we are paying for adherence to Keynesian economics appears to be economic crisis and the threat of a world-wide depression.

Industrials are now indissolubly linked with gold. When the "two-tier" gold system was announced over the weekend (and we think it will be a "many-tears" system) industrials rallied sharply. But the Monday rally, accompanied by declines in gold shares, progressively receded as the week wore on. Volume contracted as investors decided to let the "gold dust" settle. Thus ended another brief chapter in the sordid drama now unfolding. The "Javits plan" outlined in our March 1 Letter was indeed adopted—Javits was a "stalking horse" as we intimated at that time.

March 22, 1968

*   *   *

Conceivably, the two-tier system could work for a few months, but the basic problem is still there and the gold crisis will recur. The U. S. is hemorrhaging dollars overseas, and the question is: who's going to finance this?

Mad scramble on Tuesday when Government-backed bonds sold at a 6.45% rate, highest since the Civil War. The triggering of public interest is highly reminiscent of the 'magic fives' a few years back, which halted the then decline in interest rates. That this is 1½% more than the 5% indicates how much more people want these days as a hedge against inflation. Underlined by the February increase in consumer prices of 3/10 of 1%—and sharply higher wholesale prices indicate no future let-up. The Labor Department's consumer price index is up to 119, which means that last month it cost $11.90 to pay for what $10 would have bought you between 1957–9. The value of that dollar 10 years ago was 84¢ in February. The index has been rising for the last five months at an annual rate of 4%, and if the trend continues this year, it will be the steepest climb in more than a decade. Next it will be 10%, then

20% and then a runaway inflation and a ruinous collapse. The chart of the consumer price index clearly shows the trends involved.

Charts like these were the ones which led us, long ago, to understand the coming gold & silver crises. Now we point to the galloping inflation that might lie ahead, *unless* there is a monetary collapse. The only thing we can say with certainty at this point is that there is a radical change coming within the next few years. Either inflation will accelerate, or there will be a major deflation. The day of reckoning (and readjustment) cannot be postponed forever.

We said last week it is difficult to know what bargain was made behind closed doors two weekends ago, or the full price. The fact that a bill to eliminate the gold backing for the dollar was signed the following Monday, March 18, having only been cleared by the Senate on March 14, suggests this propitiation was in the bargain. There were also rumors that massive shipments of gold went from Fort Knox to Paris to depress the free price of gold and thus add an element of risk to gold speculation. Some believe that since gold speculation was riskless at $35, *that* was why there was a gold crisis. There was also a small item in the March 21 *New York Times* hinting at 'reports' that the central bankers at the conference had agreed not to draw gold from the U. S. during the period of stress created by the Vietnam war. This rumor was vehemently denied by 'one source' on the grounds that "it would mean the dollar was no longer convertible, the opposite of what the communique said." The communique to which the source referred stated that the U. S. government "will continue to buy and sell gold at the existing price of $35 an ounce in transactions with monetary authorities." It is difficult to say precisely what is happening. However, in Paris on March 21, the chairman of the First National City Bank of New York, George S. Moore, addressed the American Club and said that "the two-price gold market cannot last, and that a rise in the official gold price may be inevitable." On March 22 in Geneva, Edwin J. Stopper, president of the Swiss National Bank, told the Bank's annual meeting in Berne that without a stricter monetary policy it would become increasingly difficult to maintain currency convertibility and fixed exchange rates. He blamed the problems of the international monetary system on "the excessive expansion of the supply of money . . . [which] even exceeded the exceptionally fast pace of economic growth over the last ten years." Mr. Stopper is an astute man who understands the economic imperialism of the printing press.

In the March 24 *Times* a 'gentleman's agreement' was disclosed almost inadvertently. The *Times* pointed out that since 1950 (except for 1957) more dollars flowed out of this country than were taken in. *The cumulative deficit over the 17-year period was almost $40 billion!*

Blandly ignoring the fact that no major politician has called for ending the payments deficit, the *Times* declared "Elimination of the payments deficit has long been a goal of the United States Government. . . . Last year, it soared close to $4 billion which was almost a record." Then, unconsciously slipping back into its normal stance, the *Times* declared that "Policymakers in Washington are now pondering whether, even with the stringent controls on international transactions that President Johnson imposed on New Year's Day, it will be possible to bring the payments deficit *down to an acceptable size* [underlining ours] while the Vietnam war is continuing." And here is the coup de grâce: they talk about the U. S. inability to continue the war without full wartime controls and "for this reason, the agreement—or perhaps more properly the 'gentleman's understanding'—reached by the central bankers at Washington last weekend not to draw further from United States gold stocks at least for the time being, is of the utmost importance."

Paris newspapers on March 11, apparently relying on government sources, estimated that U. S. balance-of-payments deficit in the first quarter would reach a record $2 billion. This was based primarily on the $1 billion outlay to meet the recent gold-buying rush, and also on the sharp decline of the U. S. trade surplus in January and February. Promptly denied by U. S. Government economists who declared that based on preliminary estimates the deficit will be around $1.2 billion, compared with $1.8 billion for the last quarter of 1967, and $533 million for the first quarter of 1967. If Paris was right, the U. S. would end the year with an $8 billion deficit as against last year's total of $3.6. Forgive us for becoming slightly cynical, but we find it difficult to trust a Government which withheld from us, for several months, the news that France had dropped out of the gold buying pool.

Over this weekend one of the most important monetary conferences since 1944 will occur. They will attempt to reach final political agreement on creating artificial reserve assets to be used along with gold, dollars and pounds. There is little doubt that behind closed doors other important issues, such as the strength of the dollar and the pound, U. S. and British economic policies, and the recent gold rush will be thoroughly discussed. France's Finance Minister, Michel Debré, opposes the SDR plan and favors a return to the gold standard. He is pressing for a major conference, like the one at Bretton Woods in 1944.

March 29, 1968

*　　*　　*

The new two-tier system appears to be working, and the American strategy of forcing all newly-mined gold onto the free market, in the hope that such action will depress the price, also seems to be working *so far*. Last week's meeting in Sweden, which approved the new 'Special Drawing Rights,' has concluded another phase of the gold crisis. Each time this decade that the international monetary system has been on the verge of collapse, there has been a desperate last-minute 'solution' which has temporarily allayed the crisis. This latest crisis has left the world with a free gold price (in fact quoted by Mr. Charles Englehard), so the initiative has passed from U. S. Treasury Secretary Fowler to a knowledgeable individual. The next time there is a scramble to buy gold, without official suppression of the price, it should soar. In relief over this additional postponement of the gold crisis, gold stocks declined in classic contracyclical fashion with the rising market.

Since the earliest date for SDR activation is probably Spring 1969, and since there will only be $2 billion a year created, the U. S. share of $492 million is a mere raindrop poured into a bottomless barrel, namely the staggering U. S. deficits of over $3 billion per annum.

It is hard to know precisely what happened at the secret conferences in Sweden last weekend. It is known however that the U. S. yielded greater voting influence in the International Monetary Fund to the six members of the European Economic Community if they vote as a bloc. But the U. S. and England retain the right to veto any change in the official gold price. The U. S. won a key point by limiting activation of SDR's to a simple statement of a 'better' balance of payments equilibrium rather than to any quantitative improvement in the U. S. B.O.P. The efficacy of the SDR's was undermined by France's refusal to approve this new 'paper gold,' but experts say the new system could work without France, although they doubt it could work without the participation of a second major country.

A *New York Times* article on April 2 by H. Erich Heinemann declared "Paper gold, even if it succeeds beyond the hopes of its stanchest [sic] supporters, will not solve the basic problems created by the persistent balance-of-payments deficits of the two principal reserve-currency countries, the United States and Britain." Mr. Heinemann is certainly correct, and that is the core of the problem.

There is considerable European resentment about Washington's new protectionist trade policies since the U. S. has long lectured Europe on the evils of protectionism. Also, Washington has promised its worried allies for over a decade it would end the balance-of-payments deficit, but the deficit has increased. Supposedly, the new SDR's are to be activated only after the U. S. has balanced its international payments, but now the U. S. is pressing for their immediate activation. The U. S. public continues to fulminate against France, which has not added to its gold stock for more than a year, while Italy and Switzerland have recently increased their gold reserves.

Clearly, recent monetary events are momentous, and their impact will be felt for many decades to come. It is difficult for us to believe that the European nations creating SDR's are ignorant of the fact that the elimination of gold and a total shift to SDR's could unleash a colossal paper inflation. Many Europeans have lived through the horrors of a runaway inflation and they fear it in a way Americans cannot grasp. *Can* they eliminate gold? Certainly. There is no limit to man's folly. When the economic illiterates in charge stop blindly cursing gold and begin to realize that gold as such has nothing to do with the basic problem, then they will have come finally to grips with the gut cause of these recurring gold crises—namely, in the long run each nation can spend no more than what it earns abroad. To prevent cheating the paper money used must be limited in some way, preferably by tying it to something "immutable" like gold, silver, mink coats, diamonds, or something similar. In the *New York Times Magazine*, March 31, p. 118, there was an article on gold which completely sums up our position and with which we agree wholeheartedly. It declares bluntly, "There is no escape in simply wanting to do away with gold." It states that the U. S. can end the war in Vietnam, raise taxes, reduce the deficit, control lending and investing abroad, stop all foreign travel, but "The one thing we cannot do is wish away the problem by pretending it is all the fault of the 'barbarous metal' gold."

The recent gold crisis was the last dress rehearsal for the final collapse. When Washington had the gun at its head and was told to either deflate our economy or devalue the dollar, the Europeans were sold on a compromise of a two-tier system. We wonder what the other conditions were which forced the Europeans to accept this compromise. Did it include the dumping of LBJ, McNamara and Westmoreland—combined with an end to the Vietnam war? That a tax increase was involved is obvious. Conceivably there were secret agreements similar to the one reportedly made with Italy last month, whereby we guaranteed any gold losses Italy had when following our lead. At any rate, the gold situation has subsided and in Canada and England the gold-trading ban has ended and the price is stable.

We have long viewed gold shares as *hedges*. If you leave 10% in gold, and gold shares go to zero, the worse you have lost is 10%. But if there is a monetary crisis, many gold shares would sell for *20 times* their present level—this could mean the difference between your being able to meet a margin call or not. Let those who say that gold is unimportant ponder what gold has done here in recent weeks. In the words of an unnamed expert, the *Journal* writes "The European financiers are forcing peace on us. For the first time in American history, our European creditors have forced the resignation of an American president."

April 5, 1968

*   *   *

We think the rally was simply shock reaction when the Europeans obtained Johnson's decision to retire in exchange for their acceptance of Special Drawing Rights. This spared Johnson from deliberately inducing a depression in the United States to support the dollar, much as England is now in the process of doing to itself. If this reasoning is correct then the so-called 'success' of the SDR's is comparable to the safety of a lacerated hemophiliac swimming in shark-infested waters. We think it is only one more postponement of an inevitable dollar collapse. Nothing has been changed or cured; no U. S. leader demands that the U. S. balance its budget and its international balance-of-payments. We now have around $10 billion in gold reserves against staggering claims in excess of $34 billion, whereas in 1949 we owned nearly $25 billion free and clear. If a catastrophe has merely been briefly postponed then golds will become strong again, and the market will collapse.

April 19, 1968

*   *   *

After we went to press last week William McChesney Martin Jr., Chairman of the Federal Reserve Board, rocked the world with the startling statement that "The nation is in the midst of the worst financial crisis since 1931." Apologizing several times "for appearing to be emotional" Martin called for a tax increase, reducing the budgetary deficit, and a correction in the adverse balance of payments "to avoid disaster." According to Alice Widener, publisher of *USA Magazine*, "Over the weekend of April 5–7, West European central bankers held a secret meeting at Basel, Switzerland, and expressed their furious displeasure over U. S. Government failure to take drastic steps to correct the chronic deficit in our balance of payments. Nothing has appeared in the world press about that meeting."

Interest rates across the nation skyrocketed, bonds collapsed, and continental gold prices inched closer to $40 an ounce. Mr. Martin spoke strong words, but he has not yet resigned, and he once said "If I thought we would have to devalue, I'd quit." Martin raised the spectre of worldwide currency devaluation, and pointed out "what procrastination did to Britain." He calculates that this year's productivity gains of 3% compared with wage increases on the order of 7% would lead to uncontrollable inflation, and eventual deflation or depression.

Then, on Monday, April 22, Pierre-Paul Schweitzer, managing director of the International Monetary Fund, expressed confidence there would be no crisis if "adequate measures are taken in the U. S." to reduce the B.O.P. deficit. But without such measures "it is anybody's guess and I won't venture one." The chart of the British pound slipped, stock prices fell sharply, and you once again are getting a preview of what could happen if the gold crisis breaks wide open. The prime rate that commercial banks charge their best clients was increased to a record high of 6½%, and stock brokers are charging between 7 to 8½% on their margin accounts. Chillingly, government bond dealers began discussing the possibility the Treasury might soon sell a 6% note to refinance more than $8 billion of notes and maturing in mid-May. The Federal Government has not offered such an issue since the Civil War. A few years ago we issued the then-widely ignored prediction that Government bonds would yield over 10% before this gold crisis was finished; nothing has occurred to induce us to change this gloomy forecast.

Finally, as we went to press, it was announced the U. S. had its first trade deficit in more than five years (January 1963, due to a longshoreman strike). With March such a disappointment the U. S. trade surplus in the first three months of 1968 was only $73 million. This is gravely serious since the U. S. traditionally runs a heavy trade surplus, the main positive element in the overall balance-of-payments picture. But even this is distorted since we force foreign aid recipients to spend their bounty here. As we have said all along, the picture is far worse than most people realize.

April 26, 1968

*   *   *

We have said all along the market is inextricably linked to the gold crisis, so perhaps recent swings reflect the market's own indecision and confusion about overseas events. That gold shares are not only firm but actually rising to challenge their former all-time highs should hardly comfort long-term bulls. Overseas, the price of gold on the free market is making new highs daily. This represents continued reaction to Martin's pessimism (discussed here last week) concerning the obvious failure of America to achieve a balanced budget and a balance in international payments. The strategy of forcing gold producers to sell gold on the free market ostensibly to suppress the price (and hopefully even drive it below $35 per ounce) is *not working*.

The bad news on how much the gold crisis in the first quarter of this year cost the United States is now in. Around $1.197 *billion* worth of gold was poured down the London gold pool before it was abandoned on March 17. Treasury aides said this was the largest outflow ever, easily surpassing the previous high of $900,000,000 set last December. And the result was futility. An increase above the price of $35 an ounce turned out to be unsuppressible and by March 31 the U. S. Treasury's gold reserve was down to $10.703 billion, lowest in more than three decades, and $2.481 billion less than a year earlier.

May 3, 1968

*   *   *

Another gold crisis is obviously about to break into international headlines. It began in London last Friday when publishing magnate Cecil King charged Britain was "threatened with the greatest financial crisis in our history" and called for Prime Minister Wilson's ouster After we went to press last Friday sterling plunged, along with London stocks. King resigned as Director of the Bank of England after he said the crisis was not to be removed "by lies about our reserves, but only by a fresh start under a fresh leader." The Financial Times Index slumped 14.7 points, sharpest loss since November 23 after sterling was devalued; the record drop was 18.6 points in May 1962. Considering his inside knowledge of the Bank of England, King's comment about "lies about our reserves" alarmed international circles. We have been illustrating such discrepancies for years. It is widely acknowledged that the Bank of England's monthly figures on foreign exchange reserves are a statistical fantasy, with the real position heavily masked because of secret swap arrangements with other central banks.

Yesterday, the price of gold in London went to $40.40 per ounce, a record high since the new two-tier pricing system went into effect April 1. To those who are legally allowed to buy gold, we urge you at all costs not to settle for your present profits. The lowest gold can go is around $37 and the highest is well over $100. And heavy buying at prices well above the official price of $35 could threaten the whole two-tier system (which we think will collapse anyway). The day of reckoning is grinding inexorably closer.

Treasury Secretary Fowler pitched his two cents in at the Senate Foreign Relations Committee when he said he expected the two-tier gold price system to continue for "decades ahead", thereby defying the widespread belief that it's only a temporary expedient. Prodded by generally critical questioning of Senator Mundt (R., S.D.), Fowler again opposed doubling the official price of gold because he could think of nothing "more highly explosive" than the world-wide inflation that would follow. Henry, dear Henry, *what do you think we have now?* Oh well. Back behind the looking-glass, Alice.

From Fowler's other remarks we deduce he is concentrating entirely on the Special Drawing Rights and indicated that "it may well be, for the indefinite future, that the SDR's are the only major step that needs to be taken." We think he is profoundly incorrect, and like most economists he misses the point by concentrating on gold rather than placing the blame where it belongs—the use of the printing press to buy votes for reelection.

May 24, 1968

* * *

On a short-term basis the market is still overbought but becoming less so, and a nice swift crack below 870 could well clean out the weakness we foresaw. On a broader basis we think there is a crash in the wind for 1969, just as the one in 1962 followed the speculative binge of 1961. Crashes inexorably follow excesses.

Monday May 20 the London gold price burst through the so-called psychological barrier of $40. The US is powerless to restrain this price increase, a natural consequence of the lunacy inherent in the so-called "New Economics" which is nothing more than the "Old Chaos."

On Tuesday May 21, French turmoil pushed the gold price to a new high at $42.30, with particularly heavy French demand for gold coins in Zurich—this is second only to the $44.35 quoted on March 15 in Paris at the height of the last gold crisis.

By Wednesday May 22 the new crisis began to pick up steam as gold hit another new high of $42.60 while the Swiss franc was especially strong against the dollar—suggesting the possibility of an outflow of French capital to its traditional Swiss sanctuary. In fact, that's how Switzerland got its French community, as readers of the book "Swiss Banks" remember. Thus, once again, gold has shown it is not a "barbarous relic"; rather it is man's last refuge against the irresponsibles who take over occasionally. The poor pound was driven to a new low of $2.3848 (before the authorities intervened). Foreign exchange in gold markets were unnerved by the startling increase in the U. S. Treasury bill rate to a record high, followed by higher rates for dollar certificates of deposit. Then came a new trial balloon by the Commerce Department.

It proposed to allow American companies to keep the proceeds from their borrowings in foreign countries until the borrowed funds were used in direct investment, provided they pledge they will use proceeds of the foreign borrowings before making any transfers of cash or other liquid assets to their overseas affiliates.

Wednesday Mayy 22 *The Wall Street Journal* broke the news of still another trial balloon—letting Americans trade and own gold, "heard at a top-drawer annual monetary conference being held in Puerto Rico by the American Bankers Association." This is a startling development—a move toward freedom which is in direct contradiction to this country's trend toward totalitarianism. We cannot believe Americans will be allowed to escape inflation by owning precious metals, particularly since the 'reason' for this move is so stunningly illogical we have difficulty grasping it: its advocates feel it will prompt South Africa to start marketing some of its gold! Nonsense—the burst of additional private demand would obviously send the gold price soaring.

May 24, 1968

* * *

Neither paper gold nor France's trouble will save the U. S. dollar, since we are still hemorrhaging more overseas than is being earned there. This net surplus of paper money outflow must wind up *somewhere*, and if it is not France then it will be somewhere else. When dollar bills get to be knee high, and then neck high, foreigners will refuse to accept them, paper gold or not. And that is why there inevitably will be a gold crisis. Nothing will avert it short of balancing America's international payments and really halting inflation here. We don't pin much hope on the tax increase which supposedly will help the balance of payments by slowing U. S. inflation—which in turn, theoretically, will increase U. S. exports while decreasing imports. This remedy is too little too late.

June 7, 1968

* * *

Senator Jacob K. Javits has been advocating an American gold embargo rather than a higher gold price. The president of the Federal Reserve Bank of New York, Alfred Hayes, correctly rejected an embargo because it would not end the U. S. payments deficit, and because it would be "illusory" to expect an American gold embargo to lead somehow to a world-wide rejection of gold and to blind faith in the paper dollar.

Yesterday Jacques Rueff, a man we have long pinpointed as the brains behind de Gaulle's stance on gold, reiterated that the price of gold should be doubled to expand world liquidity. He stressed he wanted the relationships between currencies kept the same. We doubt anyone in Washington is bright enough to see through this scheme, for they have yet to prove to be a match for de Gaulle. Why keep the parity with France, when everybody has been devaluing against the dollar for decades? We still stick to the "Dines Plan" which means totally getting the government out of gold management, permitting all currencies to float freely and find their own level of value from now on. Those who seek salvation in the Treasury's printing press will only find that their paper money buys less gold (or less goods).

An article in June 1 *Fortune* discussed the Unthinkable Topic. It was titled "Ways to Devalue The Dollar—None Of Them Easy." Now that it is finally being discussed in the press, after having been in The Dines Letter for seven years, expect it to become a major topic in the coming months. *Fortune's* article was confused, since it is based on American premises: Americans have been singularly incapable of clearly grasping this topic. The printing-press politicians, so preoccupied with gold, exchange rates, IMF, paper gold, their belly-buttons and what have you, have lost sight of an elementary fact. Forget all the sophisticated devices and simply start with barter. How many U.S. Fords are worth how many tons of French potatoes? Now devise some sort of medium of exchange as their equivalent. And, to save on the salaries of bureaucrats, allow it to operate in a free and open market. Let every currency be expressed in terms of gold, free to fluctuate from minute to minute. But they are not listening to *The Dines Letter* yet. They never have.

On Wall Street, a bust follows every boom; the current binge, which we call the April Fool's Rally, doesn't look as if it will be any exception. We have already begun talking about what will be called the "Crash of '69." The market is now in a blow-off climax, and when the public has the bit in its teeth like this it is hard to say precisely when it will end.

June 14, 1968

*   *   *

Since devaluation depends on emotional politicians and will be sprung on us without forewarning, one has no choice but to keep a permanent position in golds as a sort of insurance. Gold headlines are simmering down a bit, but those who try to sell golds and buy them lower are frequently tricked out of them and need to repurchase at higher prices. So just take your positions in golds and forget about them. If there is a devaluation, and if the market collapses, golds will go up by so much that they alone will enable you to avoid a margin call. To that extent they are certainly worth the peace of mind. Gold is still extremely important, witness the spectacle of politicians forced to press for a tax increase in an election year; it has been made quite clear that a tax increase will be needed to save the dollar. Devaluation will be postponed once again, but nothing will really save the dollar short of an unlikely return to fiscal integrity.

June 21, 1968

JULY 1, 1968: DJI 896.35; ASA - 17

*　　*　　*

It is easy to assume devaluation will cause a DJI collapse, but the English experience is *contra*. Immediately preceding and after their 15% devaluation on November 18, 1967, their stocks soared. The exciting question leaps to mind, is the DJI following suit? Would devaluation be bullish? This topic is emotional, perhaps partially because those to whom money represents love see devaluation as a personal attack, or even a slur on the national manhood.

Economists call devaluation inflationary, despite the fact that when the dollar was devalued early in 1934, it was followed by a deflation which only World War II could rout. But devaluation was bullish for the DJI, which at last turned up from that point. Or is that because prices after the 1929 crash were already unrealistically low? Hard to say.

A small devaluation seems to be bullish, since it stimulates exports without the kind of economic upheaval which disrupts trade. This also applies to slow inflation, because wages and prices adjust, and those who are cheated by it are not informed enough (or organized) to object effectively. But a *large* devaluation, or *overly rapid* inflation, disrupts trade and would be bearish after the initial stimulations were dissipated. The definitions of large and small depend on the size of the nation, the phase of its economic cycle in which devaluation occurred, and other intangibles including mass psychology.

This leaves the investor with a profound dilemma. It is hard to say precisely what a dollar devaluation will do because nobody knows what the reigning politician will do about it. An invigorating inflation might ensue, but when the adrenalin wore off there could be a deflationary collapse. Thus, we can safely rule out the middle and concentrate on the extremes, of which we will have either *or both*. To hedge all bets you should own some reasonably-priced stocks (or land) to protect against inflation, but only if these companies will be able to raise their prices accordingly—*this rules out all companies whose prices are government regulated.* Also, borrow money to pay for them, because our politicians discriminate against lenders by inflating debts away. Avoid holding cash. So much for inflation.

In addition, prepare against a possible deflation by purchasing gold and silver shares, because during devaluationary deflations these shares have skyrocketed. And since they will rise so much, this hedge can be small—10%, or as much as 30% of portfolios—but not more. Don't trade golds, *invest* in them, as if you were buying fire insurance.

July 12, 1968

* * *

The U. S. trade deficit in May was $32.2 million, following March's deficit of $157.7 million. These are the first U. S. deficits in five years, and are in line with our predictions of coming disintegration and then collapse of the U. S. trade position. Inflation makes foreign goods more attractive in the U. S. (increasing our imports) while higher prices make our goods less attractive overseas (decreasing our exports). Result: disaster. So far this year the annual rate of exports is running 6% over last year's, but *imports are up 18%!* Therefore, the U. S. trade surplus in the first 5 months is only at an annual rate of $972 million. L.B.J.'s goal of a surplus of $4.6 billion is so obviously unattainable that even the normally optimistic Andrew F. Brimmer, member of The Federal Reserve Board, threw in the towel last month and said there was "no prospect" of reaching that goal. The trade surplus, last bastion of capitalistic expertise, has been the U. S. ace of trumps for years, offsetting profligacy in foreign aid or other giveaways, tourism and the military. With a smaller trade surplus, the other deficits will have no offset and will look bigger than ever.

July 26, 1968

* * *

Rumors of French devaluation are flying again. The U. S. Treasury has finally retaliated against France for the temporary export subsidies she adopted in the aftermath of the abortive May revolution. The "countervailing duty" of 2.5% will begin in 30 days and end when France abolishes her export subsidies. The inside reason that a higher retaliatory duty was not imposed was to avoid pushing the franc into a major devaluation, which would have to be matched by most European countries, and thereby trigger a chain reaction which even the U. S. Treasury is beginning to understand. The end result almost certainly would have been a devaluation of the U. S. dollar. Thus we are dubious there will be a sizeable French devaluation, partly because there has already been a slight devaluation in the sense that the free market exchange rate on the franc varies from 5.25 to 5.65 to the dollar, and the present rate is at the lower end of the spectrum. Moreover, before a large devaluation of around 30% could be effected, there would have to be a meeting of the IMF. However, we will readily concede that the franc is in trouble.

Meanwhile, across the channel, as we predicted, the British trade deficit increased following the June improvement. We remain most pessimistic about the future of the English pound.

Half a world away in Peking, Japanese and Chinese officials are scheduled to discuss replacing the French franc as the medium for trade between the two countries. The Chinese communists switched from the pound to the franc last April, after the pound was devalued late in 1967. It is widely believed China was badly hurt by that devaluation, since they had large sterling accounts abroad. It's just a question of time how long everyone is going to stumble around before they realize that gold is the only lasting medium of exchange. Where are those who declare gold is useless?

Meanwhile, back at the ranch, the Federal Reserve Board ponders when to reduce the 5.5% rediscount rate set April 19. A faint move toward lower interest rates is partially what is sparking the strength in Savings & Loan shares. Thus the power of gold reaches into many corners of the world. At any rate, fears are that a lower discount rate right now might prompt Europeans to think the U. S. is giving up its anti-inflation fight which began with this summer's tax increase.

Just let those gold and silver shares sit on the back burner in your portfolios, because we have gone through these periods many times during this decade. Each time we heard that "gold is through" it turned out to be the Bottom of another large-scale advance for golds. Some people insist on buying only when gold is in the headlines, and then selling out at the next Bottom when there is a temporary lull.

August 16, 1968

*   *   *

An August 17 *New York Times* headline read "U. S. Balance of Payments at Best Level in 2 Years" while a *Wall Street Journal* headline on August 16 read "U. S. 'Real' Trade Gap Hit $2.2 Billion Rate in 2nd Period, Worst Since World War II." So, is the news good or bad? The *Journal* article discussed a little-noticed statistical series which discounts inflation and analyzes U. S. trade in terms of "constant" 1958 dollars. In the first quarter of 1968 there was a deficit at an annual rate of $100 million—the first deficit since 1959. The real shocker came in the second quarter figures just released, when the U. S. sustained a "real" trade deficit of $2.2 billion.

On paper, the U. S. still had a surplus in the second quarter, but even on this inflated base the trade performance was a bitter disappointment to officials since the $900 million annual rate was sharply lower than the $1.5 billion surplus of the first quarter. Initially, the Commerce Department had reported the second quarter surplus would increase to $2.2 billion.

If Big Brother's inflationary tamperings are subtle, then the outright manipulation of the total payments deficit is blatant. The trade surplus is, after all, only one factor in the balance-of-payments picture. The appearance of disaster was only avoided by outright numerical juggling, when the deficit in the second quarter soared to $780 million from $170 million. These include purchases by the Canadian Government of $500 million worth of special, non-marketable, medium-term U. S. Government Securities—*fictitiously considered as inflow of funds in the balance-of-payments accounting!* Treasury aides admit these dollars are roughly as real a threat to the U. S. gold stock as dollars held in other forms, such as bank deposits. Also helpful was West Germany's regular quarterly purchase of $125 million in similar securities, plus a special $125 million purchase by a group of private West German banks. If you add this $750 million to the admitted deficit of $156 million, the deficit was really $906 million—no improvement at all.

The second factor in offsetting the trade deterioration was foreigners buying $520 million of U. S. common stocks in the second quarter, up from around $280 million in the first. This inflow was "the highest in history" for any quarter, according to Treasury Secretary Fowler, adding it springs mainly from the "confidence of investors the world over in the prospects of this economy." We disagree. The explanation is that foreigners are drowning in their holdings of U. S. dollars which they can't trade in for gold, so they are going to buy America. Literally. In the last year, net foreign purchases of American stock have been nearly $1.5 billion. Adding the $950 million of foreign purchases to the trade deficit of $906 comes out to a grand total deficit of $1.856 billion, more than the crisis level of $1.74 billion in the last quarter of 1967. Thus, deterioration in the trade surplus is accurately reflected.

Some day this elaborate fiscal farce will be understood, and those responsible for it will stand condemned and despised for the damage they did.

August 23, 1968

* * *

The international monetary crisis looks as if it is about to come to a boil again. The U. S. dollar and English pound are plummeting in relation with the German mark. Rumors of an *upward* revaluation of the mark (the opposite of a devaluation) were not dispelled when Karl Blessing, President of the German Central Bank, failed to make a clear denial of the rumor. The really sound currencies have been the mark, Swiss franc and French franc. Of course the French franc was knocked for a loop by the attempted revolution in May.

France came to her senses with the surprise and sudden lifting of all currency controls on September 4. While it was a gamble that the franc could be sustained in a free market, they actually had little choice. The moment controls were imposed, people began to evade the rules and gold poured out. Sometimes it takes a government a while to grasp that controls can do more harm than good.

We stand by our frequently reiterated prediction that the franc will not be devalued although we admit it is a fairly close call. We also expect that the German mark and the Swiss franc will eventually have to be revalued upwards, primarily because their currencies are run in a sound fashion. Sooner or later they will have to gain while the U. S. dollar and the English pound collapse.

As for the U. S., its gold stock fell $5 million, continuing its steady monthly decline this year—interrupted only by the trivial gain of $213 million in June. *But potential claims on the U. S. gold stock were $33.13 billion on May 30, up 11.9% from $29.61 billion of a year earlier.*

And sad but not least is poor old England. Her reserves in August were down by $67.2 million.

The loss was blamed on repayments of previous debts, but who on earth can follow the machinations and journal juggling that has been going on in England since the infamous brain drain began. Our modification of Darwin's Law reads: "Collectivist political systems favor the survival of the unfittest." Those who understand what is going on simply emigrate.

September 6, 1968

*   *   *

[Y]ou are now getting your last chance to get out of the Glamor stocks, or even sell them short. [W]e expect a crash in 1969.

Two top West German officials emphatically denied rumors of an imminent upward revaluation in the Deutschmark and ruled out such a move this year. When Finance Minister Franz Josef Strauss was asked "would the mark be revalued some, or at all?" he replied "my answer is a radical no." This did not stop speculators from extremely aggressive selling of pounds to buy marks and the pound sterling weakened under the backlash.

France seems to have won its gamble by ending currency restrictions. Our prediction that the French franc would not be devalued appears to have been borne out. France's monetary gold and foreign exchange reserves fell by a staggering $249.1 million during August. It was lucky that France had some reserves on which to fall back. That's more than can be said about the U. S.

France's loss was the U. S.'s gain. The U. S. bought $220 million gold from France during the second quarter, a dramatic turnabout from 1962–66 when France bought almost $2.9 billion from the U. S. Query: Since America resented France buying gold from us, is France now justified in hating American gold purchases from her?

Finally, U. S. government statistics in the official Monthly Review of the Federal Reserve Bank of New York showed a substantial improvement in the dollar in the first six months. Mr. Coombs indicated that the improvement in the dollar's position has been a result of the problems of the pound and the franc. Another favorable factor was increased demand for Euro-dollars by major American banks, plus foreigners investing in U. S. common stocks. This Coombs report does not point out that foreigners can sell our stocks just as easily, which could be a sudden negative factor. There are other trouble spots ahead, particularly a major reshuffling of European currency values. If the currencies of West Germany, England and France undergo either a devaluation or an upward revaluation, it would almost certainly trigger an across-the-board reevaluation of all European currencies. The chain reaction effects of this are indescribable now, but one would feel better owing gold shares for financial survival insurance.

September 13, 1968

* * *

The U. S.'s trade surplus slumped to a nominal $100 million in the first half of this year, down precipitously from the previous comparable period's $3.5 billion. Yet, luck still seems to be on the side of those who mismanage America's international balance of payments. Foreigners have bought around $800 million worth of U. S. common stocks in the first half of 1968, which equals all of their purchases for 1967, which in turn was nearly four times that of 1956.

When will it stop? When foreigners own America? There is no way to stop politicians from selling America out, short of that. That was Gold's function, but that has been neatly short-circuited.

Meanwhile, at a symposium on monetary problems, former Secretary of Commerce, John T. Connor, blasted the Administration's BOP policy as "a complete failure," and called for a quick end to direct government controls over private foreign investment. Mr. Connor is now president of Allied Chemical Corporation, and perhaps sees a different point of view than when he was in the Government. Since foreign investment produces a dollar inflow, Mr. Connor was clearly correct when he said controls should be "dismantled and phased out quickly."

We were particularly fascinated to learn that during a closed session of the symposium there was considerable discussion of creating a monetary system with dollars disengaged from gold and all currencies allowed to trade freely. As far as we know, we were the first to put this idea forth a few years ago, predicting ultimately the world would accept it. We are under no illusions that anyone is listening to us but we are glad to see that the idea is finally beginning to appear in print. We like the idea because it accurately measures the value of currency on a daily basis, thus eliminating the disruptions caused by devaluation or anything else. We think you'll hear more of this in coming years.

September 20, 1968

*   *   *

"When war clouds threaten or international financial storms gather, which central banker will sleep more soundly—one whose vaults contain gold or one who holds only Special Drawing Rights?"

South African Finance Minister Dr. Nicholas Diederichs
(Courtesy, *New York Times*)

In a major speech on gold U. S. Treasury Secretary Fowler virtually admitted that he was colossally wrong about speculators. Last year, it was claimed that removal of the $35 official price would immediately cause a collapse in the price of gold, and that was considered desirable because it would punish the speculators. (Notice speculators are evil, primarily because they seek a profit.) Now Fowler says "a sharp drop in price below $35 per ounce in the private market would cause concern about the value of gold held in existing monetary reserves." Continuing in his world of fantasy, Fowler said that "the workability of the two-tier system has dashed the speculator's hopes for a change in the monetary price of gold, and makes his holding more volatile. Many speculators may find it too costly to continue to hold a non-earning asset, such as gold, and recognize they have fought a losing battle. Furthermore they are no longer promised a floor on the market and must consider the risk of loss—even with a market price at or close to $35 per ounce." Fowler repeats every cliche that was mouthed when the floor under gold was removed. Time has proved him wrong in every respect.

September 27, 1968

*   *   *

U. S. import and export statistics continue to deteriorate. Exports in the first seven months are up 7% over 1967, while imports soared 21% above the 1967 figure. The surplus last year was $4.1 billion, whereas this year a surplus of $1.0 billion is hoped for. It is from this surplus, the fruits of capitalism, that items like foreign aid and overseas military expenditures come. When there is no surplus, then the expenditures come out of our gold holdings.

Pierre-Paul Schweitzer, Managing Director of the International Monetary Fund, threw up a trial balloon by defying the U. S. and favoring a "floor" of $35 for the price of gold. Schweitzer emphasized this was a personal view and the IMF had not yet made a decision on the question. After the big fuss made in recent years about removing the floor (which was supposed to punish those nasty speculators by adding an element of risk to their speculations) it is strange to consider its return. Makes you wonder about those who shrieked for its removal.

This week's IMF annual meeting in Washington is causing great speculation. Sharp gold price fluctuations still limited nicely, just under $40 despite massive propaganda about substitutes for gold. These "Special Drawing Rights" will not solve the problem, so ignore them. Nor does it matter whether or not South Africa is forced to sell gold on the open market. There are more people afraid of entrusting their life's savings to paper money, than there are ounces of gold.

October 4, 1968

*   *   *

Last week's IMF meeting has left the world's monetary structure in the following state: 1) Gold will remain the controversial centerpiece of the world's monetary system, with the U. S. still trying to get rid of it, and countries like South Africa and France increasing its importance by tripling the price. No conclusion was reached, as noted above, on what to do with newly-mined gold. 2) The IMF's holdings of $22 billion in members' currencies and gold is incapable of further major strain. The IMF, heart of the present monetary system, holds currencies mainly in excess supply already. There soon will need to be additional promises from countries with "sound currencies" to lend substantial sums to the IMF. This excludes the U. S. dollar. Sad, no? 3) Special drawing rights, or "paper gold," as a substitute for gold, looks more remote than ever. Parliamentary approval will be needed by 67 member nations, only 17 of which have completed action. And smaller members, like Algeria, are beginning to withhold their acceptance in exchange for concessions by "the rich countries." How is that for somebody with *nothing* using that as a bargaining point? 4) Flexible exchange rates, which we favored before anybody in the government admitted there was even a gold problem, continue to gain in popularity. If currency values can fluctuate wider than the IMF allows them to now, disruptive devaluations can be avoided because when a country resorts to printing press inflation, the weakness of its currency will be immediately reflected in day-to-day valuations. Paranoiac U. S. officials are adamantly opposed to flexible exchange rates because they fear it would reflect unfavorably on the dollar's value, and because they are terrified that they will no longer be able to run the printing presses to buy what they cannot afford. 5) Commodity-propping came up as a topic for the already-bankrupt IMF and World Bank (when you are broke, expand your spending) for smaller nations dependent upon extremely cyclical raw materials such as cocoa and rubber. Having learned nothing from U. S. farm policy, commodity-propping will bring over-production, divert consumer countries toward the use of cheaper synthetics, and will delay diversification of these smaller countries. 6) McNamara has urged a change in World Bank policy to raise its annual lending from $1 billion to about $2 billion.

October 11, 1968

* * *

Just as this Letter was among the first, if not the first, to predict easy money this year, we are now thinking in terms of tight money in 1969. *That* could be the cause of a bear market.

October 18, 1968

* * *

We are expecting an important decline to start sometime in November, and we have already begun to let out short sales.

November 1, 1968

* * *

Two top-priority decisions confront Richard Nixon: whether to fulfill his campaign pledge to get rid of Lyndon Johnson's controls over direct foreign investment, and whether to renew the pledge previous Presidents have made to hold the U. S. gold price at $35. Nixon obviously cannot get rid of the overseas investment controls all at once because the dollar is simply too weak to tolerate the additional dollar outflow that would surely follow.

One member of Nixon's economic inner circle is Milton Friedman of the University of Chicago, who is an outspoken advocate of "floating" rates, letting the dollar's price be set by supply and demand in foreign exchange markets. We began advocating this position in *The Dines Letter* in 1962, and still strongly support it. Nixon would have to revalue gold if he is going to dismantle the controls—he can't have one without the other. It appears that the whole gold question is now open for re-evaluation, and a devaluation of the dollar is indeed more likely than ever before. This is an excellent time to hang on to your gold shares and be patient.

Meanwhile, England seems to be slowly bobbing towards the surface, with a net loss in October reserves of $9.6 million. While this is compared with $50.4 million *gain* in September, October did include heavy debt repayments of $124.8 million. Here the results are not too bad, but of course the pound is and has been radically over-extended for a long time, and sooner or later it will cave in. No change in our prediction there.

Across the channel, French reserves in gold and foreign currency fell $108.7 million in October to just under $4.27 billion. This compares with a loss of $227.1 million in September, $249.1 million in August and $666.8 million in July. During the abortive French revolution this spring, we made the rather risky prediction that the franc would not be devalued, and with each passing month it looks as if we will be proven right. As soon as speculators draw the same conclusion, a sharp flow of funds should return to France and the situation will be secure there. We are bullish on the franc.

November 8, 1968

*   *   *

A few weeks ago we commented on the growing strength of the German mark. There are now renewed rumors of an impending deutsche mark revaluation which would increase its value in relation to other countries and thus raise the price of German goods in international trade. The more such talk, the more money pours into Germany and foreigners attempt to acquire marks. The Germans have to figuratively shovel the money out of the country to maintain price stability. Germany's big banks are rapidly becoming titans in international finance.

Germany has kept a sound currency, and its problem now is to restrain it from becoming so strong that it angers the countries whose currencies are in trouble: the dollar, pound and franc. The figures just released show that Germany ran a $3 billion trade surplus in the first nine months of 1968, and it will probably approach the 1967 surplus of $4.3 billion.

Any revaluation of the mark would be a serious disturbance to the increasingly interrelated international monetary structure and might trigger a devaluation in one of the other leading countries. It is already a very shaky structure, and it is hard to tell which little shock will be the straw that breaks the camel's back.

While West Germany gold and foreign currency reserves were advancing by a whopping $206.1 million in the week ended last Thursday, to $6.75 billion, the shaky French franc began to feel the impact of those who were transferring funds to Germany not only to avoid a possible French devaluation but also to cash in on an upward revaluation of the mark.

Fuel was added to the fire when Britain announced that its trade deficit last month soared to $302.4 from $242.4 million in September. Gone with the wind were hopes that England's devaluation a year ago was at last beginning to help. At no point have we wavered in our pessimism about the pound.

The seeds of disaster begin to be sown, with the pound and franc and dollar at their depths, and the mark at its height. The price of gold begins to move up, and suddenly gold shares around the world begin to respond with aggressive upward movement. Hard to say which crisis will at last break the bank, but the world's leaders have been playing fast and loose for a long time. Someday the piper will be paid.

November 18, 1968

*   *   *

On November 1 we wrote "chances are the contracyclical golds will make an important Bottom right around these levels. This is certainly not the time to sell out."

This prophecy bore fruit almost immediately as an international monetary crisis erupted with volcanic swiftness this week. DeGaulle termed a franc devaluation "the worst possible absurdity," apparently referring to the failure of British devaluation. Jacques Rueff, the economist whom we consider the brains behind France's throne, chided the Bank of France for its recent easy-money policy. The $3 billion gold drain from France since June has been offset almost exactly by new credits: the Bank is putting out paper to replace the disappearing gold. Astonishing how an abortive communist revolution by "harmless students" last May has brought France to its knees. Even severe economic restrictions have not slowed the fantastic outflow of money from France into Germany.

The inside story, one you will not see in your newspapers, is that other nations have refused to support France unless certain conditions are met. Some of the demands on France include a return to NATO, canceling its veto of British entry into the Common Market, and abandoning opposition to the system of balance of power between Russia and the United States. *DeGaulle views this as an infringement on France's sovereignty and has rejected it out of hand.*

When a nation's citizens become terrified that the currency they hold will become worthless, there is no stopping them. Thus, it was no surprise when the French government closed all its financial markets on Wednesday, November 20, and the National Assembly approved a $400 million reduction in the 1969 budget. West Germany also passed tax measures which would tend to check the flow of funds into Germany, possibly to reduce Germany's expected trade surplus this year of a staggering $4.3 billion. But at the moment the German currency is safe and the French currency is not, and this is something that the money managers simply do not include on their balance sheets. We cannot be as bullish on the French franc as we were as recently as a few weeks ago, but what a shame the franc had to go this way. This was a sheer speculative attack, not a fundamental imbalance in trade as is true with England. We have been bearish on the pound for a long time, and in London foreign exchange markets were also closed on November 20.

As the crisis exploded onto the front pages of newspapers around the world, the price of gold went soaring. Where are those who only a few short weeks ago were crowing that the price of gold was falling, and induced many silly people to sell their gold shares right at the Bottom. The deadlock of Germany's unwillingness to revalue the mark prompted Great Britain, France and West Germany to close their foreign markets for the rest of the week—this announcement was made yesterday. Pierre Rinfret, an American economist, said he learned on "high authority" that France lost nearly $400 million, or about 10% of its reserves last Thursday and Friday alone, while the Germans' put their dollar intake last Friday alone at $650 million. Obviously, at this rate of loss, the franc could not last another week. And the ramifications are only beginning to spread. Not only are the bond markets and stock markets weak, with typical contracyclical gold action, but companies operating in Europe face delays in plans and new risks due to the currency crisis. Just yesterday Gunther Diehl, the Bonn government spokesman, said that in the last three market days alone $1.77 billion had flooded into Germany, more than a quarter of the West Germany gold and dollar reserves.

The *New York Times* editorialized on November 15 that the only solution would be a "sharp reduction of West Germany's balance-of-payment surplus. . . . Bonn makes a fetish of increasing her gold and foreign exchange reserves year in and year out." This is typical of the dim-witted pronouncements from the *New York Times* editorial board in recent years. However, in an extraordinary flip-flop, their editorial of November 21 suddenly grasped the problem and declared that the fundamental source of turbulence "is the inability of countries under the present rules of the International Financial system, to make smooth adjustments for differences in price levels and productivity growth." They finally conclude that they want fluctuating exchange rates within wider limits. A fantastic reversal of opinion, and of course, in the latest editorial they do not even refer to the earlier outdated one.

GOLDEN DOMINOS

The gold problem we have been talking about since 1961 has burst into the headlines once again. Many do not even know that there is such a thing as a gold problem. Yet, the English pound was devalued on November 18, 1967, and the *New York Times* wrote this Monday that "there is agreement among London economic commentators that the devaluation policy had failed to produce results as quickly as the Labor Government predicted." A year ago we disagreed with the Labor Government and it's good to see that they are coming around to our viewpoint.

To understand the gold problem remember that the United States had $24.6 billion in gold in 1949 which we owned *free and clear*. Now we own $10 billion but we owe $34 billion against it. Thus we went from plus 24.6 to minus 24 in just two decades. The United States dollar was bankrupt in 1961 and has existed on charity ever since. With this backdrop, there must be a gold crisis and a devaluation of the dollar, and this has been our fundamental position for years.

Currencies are all interrelated, and the dollar and the pound should have been devalued long ago. But the fixed exchange rates set up by a monetary conference in 1944 lends an essential inflexibility to the *system, and that is why it must collapse.*

We predict that there will be a devaluation of a major currency at some point, and this will trigger a domino effect which will force a world-wide collapse of currencies. *This is all due to the unreasonable refusal to raise the price of gold from its 1933 depression levels to reflect all of the paper money printed in the last three decades.* When you have an ounce of gold, and assign 35 pieces of paper to it, you can say gold sells at $35 an ounce. But when you print more and more paper, and try to keep the price the same, you are in an impossible situation. You must assign more and more pieces of paper to the unit of gold. Thus, the price of gold must rise. That it has been retarded for so many decades means that when the explosion occurs it will be terrifying in scope. We envision at least $100 an ounce. As we have been telling you for years, leave a permanent percentage of your portfolio in selected precious metal shares, spreading out as much as possible, and ignoring whether these shares go up or down. This is a *permanent* disaster hedge, or an insurance policy.

There will be a domino dollar collapse sooner or later. This is even beginning to creep into some of the world newspapers. For example, this week England's *Daily Telegraph* wrote, "in the not-so-long-run French devaluation could trigger off a general realignment of currencies and even a rise in the price of gold." Thus the unthinkable begins to be *thunk*.

November 22, 1968

* * *

[T]here are too many stocks selling at fantastically high Price/Earnings Ratios, and too many investors who don't know what we mean by that statement. Such is typically the precursor of important bear markets. You will note below that we have cashed in some long-term profits on our Long-Range Lists. Use the proceeds to pay off your margin, or leave it in short-term Treasurys for the moment.

David M. Kennedy, the Treasury Secretary-designate, ignited the gold picture by suggesting that an increase in the price of gold is one of the options open in dealing with the international monetary problem. Normally, this would be an intelligent consideration, since a new man should determine all of the options before setting his own policy. Unfortunately, decay has gone so far, and the structure is so rotten, that even sneezing the wrong way is likely to trigger a renewed crisis. All the presidents since FDR have considered the price of gold "immutable," a posture which became increasingly farcical when the price of gold moved above $40 this year. The U. S. dollar has already been devalued by 12%, and it looks as if unlucky Nixon is going to be left holding the bag.

That this situation should never have been allowed to occur is obvious. If currencies were allowed to fluctuate freely, in opposition to the foolish Bretton Woods Agreements Act, American Presidents would not be forced to make foolish statements in order to save face. Nonetheless, Kennedy's remarks are significant, and without meaning to appear cynical, for all we know, this was just a trial balloon to begin conditioning the American people to a devaluation in 1969.

We are particularly bothered by the fact that the golds are now moving up *with* the stock market, which has never continued for long in the past. Golds are generally contracyclical with the market, and either golds or the market as a whole should soon begin moving in the opposite direction.

December 20, 1968

JANUARY 1, 1969: 947.73; ASA - 15¼

*　　*　　*

We predict, despite almost everyone else's opposite opinion, that easy money will prevail over the very near future. To be sure, many of you remember our long-range prediction of 10% government bonds; we still feel that that will eventually arrive, but not just yet.

Inflation, not religion, is the opiate of the people. In such a climate, no politician could conceivably stop inflation. The withdrawal symptoms would be as real and violent as with a narcotic addict.

In view of the above, the most Nixon might hope for is to slightly lower the rate of inflation, but he will find even this impossible without bringing on a horrendous economic downturn. Inflation will increase every year, until the inevitable collapse, which has followed every inflation in the history of mankind, causes such a social upheaval that people will be willing to endure "cold turkey."

WHAT WILL NIXON DO ABOUT GOLD?

Less irrational than his predecessors, Nixon will probably approach the gold problem realistically. But past mismanagement has been so colossal, it is difficult to see how he can avoid an increase in the price of gold. Back in 1961 we thought the price of silver would rise within five years, and the price of gold within ten. We see nothing to change those predictions, which were so amazing at the time, and the first of which has already come true. That is why we feel that all portfolios should take a position in selected gold stocks immediately.

WILL FRANCE OR ENGLAND DEVALUE THIS YEAR?

We have been bearish on the English pound and are even more so this year. Not to run England down, but several generations of Fabian Socialist writers, and London School of Economics professors have left a legacy of workers who believe in class warfare. They believe they must wring as much as possible from the employer now, whether or not that puts the company out of business.

The English working class fails to grasp what even Jimmy Hoffa knew, that a prosperous company can afford to pay more. Hoffa did everything to encourage the prosperity of the trucking companies, and then cut out a big share for his members. We cannot imagine anything more intelligent but until the English working class goes through the wringer, England will be unable to compete and the pound will have to undergo a massive devaluation. Their top tax-rate is 105%, which is difficult for Americans to grasp since it actually amounts to penalizing a man who is successful. As a result, the capitalist class sits back, avoids risks, and fails to provide jobs for the workers. The English have lost many of their freedoms, even the right to spend their income on a trip abroad, and they are more self-impressed with their ability to "take it on the chin" than in improving their lot.

France provided us with one of the major surprises last year. We could not foresee a student revolution which would bring the franc to its knees. We are no longer as confident about it as we were, and an important devaluation is possible. However, France still retains a large gold position, and we would not feel as comfortable on the short-side here as we would on the pound.

OUR SOLUTION FOR THE GOLD CRISIS

We remain in profound disagreement with what has been done, and what will probably be done. No nation this rich or smart should ever be in a position where it could be knocked off balance by a DeGaulle, or where American tourists are restricted in any way, or American capitalists are hampered by controls. The first thing we should do is go through what most bankrupts suffer, and that is a writing off of all past debts. This can be done by quadrupling the price of gold to $140, and then inviting all those who hold the $33 billion in gold claims against our dollar to be satisfied.

There are some know-nothings who actually have the nerve to say that all other nations would devalue by the same amount, and therefore the net result would be nothing. Superficially, this looks correct, but actually all those who held dollar bills would have them cut down to 25¢ on a dollar. The price of gold would be the same for all countries, but the impact on creditors and debtors would be radically different. If Japan, for example, backed its currency largely with U. S. dollars and these dollars were written down as outlined above, then a yen would be worth less than an urge. The impact this would have on the world is difficult to say, but it would hardly be salubrious.

No solution looks pretty at this point, and we don't want to be in a position where we have to show how to clean up somebody else's mess. We are more interested in preventing this from ever happening again. We have no illusions that Americans will figure out who the dummies were who got us into this mess, so we do not believe they will ever be punished. But, the price of gold can be allowed to fluctuate freely, and "free gold" should be the rallying cry of all economically responsible citizens. Whatever you do, ignore the scapegoatism by ignorant politicians when they speak of the "Gnomes of Zurich." Swiss bankers are amused by that appellation but don't really know what to do about it.

It would be nice to buy gold stocks a month before the actual devaluation, but this is impossible because the price of gold depends on politicians. You can stretch a rubber band, but it is difficult to tell when it will actually break. It could come anytime, and the only thing we are certain of is that there will be no warning. It will follow a denial of devaluation by an important government official, and it will almost certainly be announced over a weekend. It will be followed by a major redistribution of wealth in this world. Gold is power.

HOW CAN YOU PROTECT YOURSELF?

Only in the U. S. and England is it a criminal offense for citizens to own gold, while foreigners can help themselves. Owning heavily-margined gold is the best way to handle it, because debt is wiped out by inflation, while the inherent value of gold increases with every revolution of the printing press. Available at 1934 prices, gold is one of the great bargains of the world.

A really intelligent alternative is the purchase of a strong currency. Last year the French franc looked good, and this year the German mark, but you will have to watch your step, since these things tend to change. Perhaps the only exception is the Swiss franc, which is backed more than a 100% by gold. There is nothing illegal when you convert dollars to Swiss francs, so long as all income is reported. Do this scrupulously. Nothing is worth even an hour in prison.

By far last year's most important development was the creation of the two-tier system, because the price of gold was at last freed from the artificial restraint of the U. S. Treasury. The free price rose yesterday to $42.75 an ounce, exceeding the previous peak reached on May 21, 1968, during the abortive communist revolution in France. This amounts already to a 22% devaluation of the dollar, and contradicts the declaration of Presidents Kennedy and Johnson that the "price of gold is immutable." The higher the free market price goes, the more pressure there will be to raise the dollar price of gold. There are of course wishful thinkers who think that South African sales will depress the price, but we predicted a year ago that this would not happen, and it has not. Predictions of the lower gold price are by the same silly people who predicted that when the U. S. stopped buying and selling gold, that the "useless" metal would drop to $10. Wrong. Wrong. Wrong. How they get re-elected baffles us. On the contrary, we look for radically higher gold prices. Any sharp noise will set off a stampede like the one last November when French peasants stampeded towards the German mark, thereby nearly breaking the franc, or an event such as a war in the Middle East. And, every increase in the gold price flows directly and completely into pre-tax earnings for gold-mining companies, since their production costs and all other costs remain precisely the same. Thus you have the enormous leverage in the gold price.

January 17, 1969

* * *

It should not surprise you that the 1968 trade surplus hit the lowest level since the Great Depression, with imports soaring 23%. The more paper money we print here, the more wages go up, thus pricing U. S. goods out of world markets, and also the more money Americans have to spend on cheaper goods from overseas. Normally, a gold outflow straightens things out in a hurry by bringing on a business slowdown. But our politicians have stuck a penny in the fusebox, so the classic remedy is no longer available.

The new Secretary of the Treasury, David M. Kennedy, has unfortunately joined his predecessors by ruling out a change in the $35 gold price.

January 31, 1969

Our laugh for the day was provided when the Commerce Department showed a surplus of $958 million in the fourth quarter, and $187 million for all last year in the international balance of payments. This nibbling fraud was accomplished despite a fantastic roadblock—the fact that our trade surplus of exports over imports all but disappeared last year. First, Washington twisted the arms of enough American corporations to get them to bring a "large repatriation of funds from abroad" though the actual amount has not yet been revealed.

Then there was the windfall of foreign net purchases of $1.860 billion in U. S. common stocks up from $900 million in 1967 and only $170 million in 1966. They can't get gold for their dollar bills so why not "buy America."

But the fraud came about due to an increase in "special transactions" by inducing foreign governments to switch their dollar holdings into special Treasury securities with a maturity of more than one year. Then, the United States simply does not count it as a liability!

Without this phony juggling of the books, our balance-of-payments would have shown a staggering deficit.

February 20, 1969

* * *

Basically, there is no change in our position. Around a year ago we turned extremely pessimistic on growth, glamor, computer, conglomerate, and other assorted fad stocks, indicating that they would continue moving down until the second or third quarter of 1969, at which point they would culminate in a crash.

February 28, 1969

* * *

The irrational fuzzy-thinkers are still prattling about how vulnerable the gold market would be if South Africa ever starts selling gold. This illustrates their detachment from reality, since South Africa has nothing to do with a runaway printing press inflation in Washington.

It has been our contention all along that the dollar was devalued in March 1968, when the two-tier system was established. The 23% premium that gold sells at above the official dollar price of $35 is the amount of the devaluation. This will be reflected in sharply increasing company earnings in coming months. We are looking for the gold group to be one of the "hottest" groups in the world.

Furthermore, the Japanese are finally waking up to the fact that they are basing their yen too heavily on a very shaky paper U. S. dollar and that if the dollar collapses, their yen will suffer sharp losses. Thus, the Japanese authorities decided to increase the gold content of their monetary reserves by over $3 billion dollars.

March 7, 1969

*   *   *

The French franc survived last weekend, proving that miracles can still happen.

On the demand side of the equation, the new Special Drawing Rights of the International Monetary Fund has been moving through ratification processes in more than a hundred countries, and has already been approved by the U. S. Congress and the British Parliament. Around 40 countries have approved it already. Once the total is over 110, Pierre-Paul Schweitzer, Managing Director of the IMF, will have to determine whether the new reserve assets are needed by the monetary system. It is expected around $2 billion a year would be pumped into international coffers, which in ten years would equal half of the world's monetary gold stocks.

No matter how many SDR's are created, we predict it will not keep up with the massive outpouring of new paper money.

March 14, 1969

*   *   *

The German mark was pushed by buying to its ceiling of 25 American cents, while the franc is pinned to its floor of 20¢. This would never happen if exchange rates were allowed to fluctuate freely, as this Letter, in its pioneering stance, has declared for 8 years. As it stands now, the system will have to collapse due to excessive rigidity. For the moment, it looks as if the franc will be devalued while the mark will be upvalued. The exact amount will determine whether world-wide economic chaos will set in.

May 2, 1969

*   *   *

We bought golds originally in expectation of an increase in the price of gold. The price of gold has indeed gone up, and you can start expecting sharp earnings increases to appear in the next year or two.

An important break came on May 27 when the *New York Times* published an article indicating that the Bank of America had joined Chase Bank and the First National Bank of Chicago, among others, in arguing that international monetary exchange rates should fluctuate more than the tiny bit permitted today. We of course have been advocating completely free fluctuations, but the bankers have not yet gotten that far. The Bank of America wants to widen the fluctuation from the present limits of ¾% either side of par value to new limits of 2½%. As we have been saying for years, the bank indicated that international transactions could be hedged through a futures market (purchase of sale of foreign exchange for future delivery). That is precisely the function of the speculator, to remove the burden of uncertainty from the man interested strictly in commerce. In an era when capitalism is no longer taught, much less lauded, it is no wonder so few people understand the role of gold, or how to solve these problems.

May 29, 1969

*   *   *

William McChesney Martin, Federal Reserve Board Chairman, said that his Board would "consider everything" if the surtax extension fails.

Why is the surtax so important? There will be those who do not like our answer, but it is tied to gold.

[O]n June 9 the Undersecretary of the Treasury for Monetary Affairs said today there was a "developing consensus among the leading nations on the need for activation of the new SDR's or 'paper gold'." Mr. Volcker warned, however, that the SDR plan would be placed in "grave jeopardy" if Congress did not extend the surtax. This is the gut cause. Foreigners are dictating American policy because they know the dollar is bankrupt. It seems everyone knows about it except the Americans.

The price of gold has been declining not only because of talk of this new "paper gold," but also because high interest rates make it difficult to finance speculative gold holdings. Furthermore, the South African Central Bank has lost $119 million in the last four weeks, suggesting that South Africans have been selling some gold in the free market.

There is little doubt the SDR's will be activated, but we really doubt that it will come this September. Furthermore, we think it will be a mere token distribution rather than any massive quantity. West Germany, Italy, Netherlands, Belgium and Japan feel that haste would impede the progress being made by the U. S. and England in getting their international accounts back into equilibrium. These creditor countries, led by West Germany, remind Americans that when the SDR's were approved last year in Stockholm, there was an understanding that the waiting period for initial distribution would be five to ten years. They are also trying to exact a price from Americans by tying in general reforms in other areas. There will be hard bargaining over the summer months in a major confrontation between America and Europe.

On June 9, Dr. Jelle Zijlstra, Chairman of the Bank for International Settlements, which has become a key forum for central bank cooperation, said that the monetary system will continue to operate under "uncertain and unstable" conditions because of the intractable payments problems of U. S., West Germany, Britain and France. In its gloomy 39th annual report, the Bank for International Settlements declared SDR's cannot fully solve the problems of monetary instability. Zijlstra suggested that debtor countries will finance deficits with the new SDR's, and hold back on their gold. Many of you will recognize this as Gresham's Law, with bad money driving the good out of circulation. (Remember when LBJ said that the new copper sandwich silver coins would not drive out the good silver coins?) Unfortunately, if central banks keep their gold and use SDR's to settle accounts, they will freeze official gold reserves and in effect reduce international liquidity! The joke about all this is that all these contortions are supposedly to increase world liquidity! And how will Washington react if they are asked to give up American gold instead of settling in dollars and SDR's. The BIS annual report suggested that an increase in the official price of gold is the proper solution.

June 13, 1969

JULY 1, 1969: DJI 875.90; ASA - 12⅝

*　　*　　*

A behind-the-scenes battle is raging.

The American Bankers Association Annual Monetary Conference met in Copenhagen on June 17. Widespread confusion and conflicting viewpoints were manifest. There was widespread support for wider bands of currency fluctuations. On the other hand, there was considerable discontent with the present system of fixed exchange rates, which are defended from national reserves and changed only in the event of fundamental disequilibrium.

Then on June 29, a group of 38 bankers, businessmen and economists from ten countries met for a week in Burgenstadt, Switzerland, and said that "a majority favored both widening the range (or 'band') within which exchange rates may respond to market forces, and permitting a more continuous and gradual adjustment of parities." This refers to the so-called "crawling peg."

Turning now to individual currencies, massive short-term invest-ments by American corporations and individuals in Eurodollars, where rates are now higher than in New York, are distorting the U. S. balance of payments. The "liquidity" definition of the balance of payments will apparently show another big deficit in the second quarter, although final figures are not yet available. Another element in the second quarter deficit was the outflow of U. S. funds into German marks, at the height of the revaluation crisis in May. It might not look this way in the newspapers because recently there has been "an official settlement" basis. This only counts U. S. liabilities as those dollars held by foreign central banks—completely idiotic in the sense that any foreigner can stick his money in the bank with no warning, and thus suddenly have it counted as a claim on U. S. gold reserves. Since the official settlement basis excludes private dollar holdings, it does not count the dollars in the Eurodollar market, so this might not show up. The U. S. is using nasty, tricky bookkeeping which it certainly wouldn't allow U. S. cor-porations to use when reporting their taxes. Even worse for the U. S., our foreign trade surplus shrank sharply in May. While blamed on longshoreman strikes, William W. Chartner, Assistant Commerce Sec-retary for Economic Affairs, said there were also signs of an "under-lying weakening" in American farm product exports. Excuses, excuses, excuses.

July 11, 1969

*   *   *

Why are golds down? Will the deflation which is obviously ahead kill the golds? Let us make crystal clear that domestic inflation or deflation has had nothing to do with golds in recent years. Normally inflation is very negative for gold shares because their selling price is fixed, while their costs rise. So Nixon's efforts along this score mean nothing. There was something more dramatic going on. What does count for gold shares is whether or not the key reserve currencies, the dollar and the pound, are so over-extended that they will have to col-lapse. Nixon apparently has the courage to bring about a domestic deflation, enough so that our dollars will stop piling up overseas. It is precisely that process which has gone on for two decades which has put us into the mess we are in. We have long speculated on a secret deal behind the scenes to the effect that if Nixon created a recession, the Europeans would still keep on giving us credit. If so, Nixon is risk-ing his political life to carry out the bargain.

Furthermore, the Special Drawing Rights, which Harry Schulz so amusingly calls the Stupid Drawing Rights, seem to be coming. If they are enacted, severing the last link between paper and reality, an international inflationary binge could follow as soon as an expansionist government was elected in the United States. This will probably happen again sooner or later. But, meantime, there could be severe selling of gold and gold shares.

July 25, 1969

*   *   *

Buried way back on page 50, the *New York Times* reported on August 7 that the United States had a $2 billion surplus in the balance of payments during the first six months of this year, according to Paul Volcker, Treasury Undersecretary for Monetary Affairs. That was on the "official settlement" basis. However, Volcker told a House banking and currency subcommittee that on the "alternative liquidity" basis for measuring U. S. transactions "we have recorded a very large deficit." We have nothing but scorn for the so-called "official settlements," which is a deliberate distortion of facts.

August 8, 1969

*   *   *

ANOTHER MONETARY SURPRISE—JACQUES-IN-THE-BOX

Throughout this decade we predicted change in the international monetary structure, and that it would come by surprise on a Friday just after the market's close. As a hint of what it will be like, on August 8th, at 2:45 P.M., 15 minutes after the market had closed, Finance Minister Valery Giscard d'Estaing reduced the value of the franc to 18.004 U. S. cents from 20.255, a devaluation of 11.1% in terms of gold or the dollar (it is 12.5% to the Frenchman who wants to buy foreign currency).

The secrecy was astonishing. The *New York Times* reported "the preparations for it were carried out in a secrecy unmatched in anyone's memory here. Giscard d'Estaing said that only eight persons knew about it. A large part of the Cabinet itself was excluded."

The devaluation De Gaulle characterized only a few months ago as "the worst possible absurdity" has occurred.

August 11, 1969

*   *   *

Something we have been expecting for a long time happened this week. A story in the *Wall Street Journal* on September 3 looked like a "trial balloon" if we have ever seen one. It looks as if the Nixon Administration is at last moving towards the so-called "crawling peg."

This refers to the fact that exchange rates would make small but frequent changes up or down against a U. S. dollar which is fixed in terms of gold at $35 an ounce. It seems our leaders understand that the post-World War II monetary system, based entirely on fixed exchange rates and supported by government intervention in foreign exchange markets, is unworkably rigid. Until now, the United States has been adamantly against any altering of that system. But, even as they tiptoe towards the crawling peg, the U. S. is understandably cautious about saying anything which might feed speculative pressures against other currencies. Furthermore, it would be unseemly for the U. S. to display great zeal in asking other countries to do something we don't do ourselves, since the U. S. dollar will be fixed. So, it appears that the study which a group of high officials have been making since the Nixon Administration took office has at last made its decision.

The U. S. will not openly initiate early adoption of a crawling peg, but in a few weeks when the IMF meets in Washington, there will be a ritual. By that we mean, as the *Wall Street Journal* puts it, "Strategists hope that Treasury Secretary Kennedy will be in a position to respond favorably to suggestions they anticipate will come from other nations." Translated into English, some other country will be picked to say, "Hey, why don't we have a crawling peg?" And then David Kennedy can say, "Gee, we hadn't thought about that, let's study the idea."

This crawling peg idea was outlined last April in a speech on Germany by a member of the President's Council of Economic Advisors, Hendrik Houthakker: The only danger here is that strong currencies must be allowed to "crawl" upward, because if all currency crawling were down, the U. S. dollar would be left high and dry and America's exports would become less and less competitive in the world markets. In practice, devaluations and upvaluations have been politically difficult, so this is a clumsy effort to go to the free market, a position we have been taking since 1961.

But if this fixes the U. S. dollar at $35 gold, that means the price of gold is not going to be able to get very much higher for a long time, at least until the U. S. dollar folds. Nobody seems to confront the question as to what would happen if suddenly Washington began to print more and more paper currency. How about 10% more in a year, 20% in a month, 100% in a week? At which point does the system cave in? That is pushing it to an extremity, but the principle remains the same. The only real solution we see, and have seen, is to allow a completely free exchange market, fueled by speculators, as is done in the commodity market for ordinary potatoes. Money is just a commodity, like potatoes, or anything else.

September 5, 1969

*   *   *

This January 1 the new Special Drawing Rights, SDRs, so-called "paper gold," will be allocated. Of the $3.5 billion total, the United States will receive more than $800 million, largest single share. Over the next three years a total of $9.5 billion more will have been distributed, a staggering 68% increase over total IMF borrowings of $19 billion created over its 25 years of existence. There can hardly be any doubt that the world is in for a staggering dose of inflation, because the more paper money you place alongside the same amount of goods, the more pieces of paper must be assigned to each piece of goods.

Nobody will ever see these SDRs, which were ironically designed to foster trade growth. Its proponents hope it will perform gold's function. If, for example, a country next year is running short of gold or dollars because its balance of payments is in deficit it can pay with SDRs, and another country must accept limited quantities. Naturally, Gresham's Law will apply in that "bad money will drive out the good." In other words, nations will pay first with SDRs, then with dollars and hold gold for last. Officials hope all three currencies will coexist. Remember when Lyndon Johnson said that the new silverless coinage would coexist with the silver coinage? L.B.J. did not repeal Gresham's Law— pure silver coinage now sells at a premium. Similarly, it is difficult to see how the price of gold can be stopped from going up.

The German mark closed at a premium of 6.38% to the U. S. dollar which, translated into ordinary English, and not stated as such in American newspapers, amounts to a devaluation of the U. S. dollar by 6.38%. An American going to Germany would find his money was worth just that much less, and considering the staggering U. S. payments deficit, this will hardly be the last devaluation. *Look for an announcement around mid-November that the overall U. S. payments deficit in the quarter ended September 30 rivaled the record $3.71 billion total of the June quarter.*

More and more the world is coming over to our oft-stated position that money rates should be allowed to fluctuate freely; dangerous pressures will not be allowed to build up. The sad thing is not that we are smarter than anybody else, but that a relatively obscure market letter on Wall Street could predict what has been openly denied by the heads of important governments for this entire decade. A year ago, the "consensus of experts" refused to even consider the possibility of flexibility in the monetary system. Now they are sidling into "crawling peg" discussions and other variations of the flexibility concept. Money has not been allowed to stray more than 1% up or down from par. With Germany and France junking any pretense at following this IMF rule, more and more nations are beginning to ignore this completely unrealistic approach, even though it was founded by Lord Keynes.

The reason fixed rates cannot possibly work is that inflation proceeds at different rates. Consumer prices last year showed the United States having risen 6%, England 5%, France 7%, Germany and Italy 3%, El Salvador 1% and Iceland 22%. On top of that, add the different quantities of paper money floated by each country, and you can see how preposterous is the idea that money rates can be held in rigid relationships.

October 11, 1969

*   *   *

One of our long-standing predictions came true this week with the announcement by the New York Produce Exchange that it inaugurated trading of futures contracts for six foreign currencies—to be traded on a newly formed exchange called The International Commercial Exchange. We hope this will herald the final death of the Bretton Woods Agreement, which fixed currencies. We have long favored permitting them to fluctuate freely, despite the outcry of those who are involved in international commerce. We think even they will come around when they realize they will be much better off than paying huge insurance premiums to forestall possible devaluations. They can still hedge themselves by buying futures on this exchange, and letting speculators take the risk —along with profit or loss on such speculations.

Gold shares continue to decline, indicating it is not propitious to initiate new commitments just yet.

October 17, 1969

*   *   *

Now that the West German mark has been revalued 9.29% currency speculators are moving in to reap their profits. Since margin requirements for currencies are extremely low, a gain of 9% means many times the amount of the actual quantity risked. In this way, currency speculation is much like any other speculation in commodity futures.

With marks being sold, buying attention shifted to British pounds, Belgian francs, Dutch guilders, and even French francs. Thus, other currencies were bolstered by the German action. This is how a free market establishes stronger currencies, and criticizing "evil speculators" does nothing but create imbalances.

October 31, 1969

* * *

We have been deemphasizing gold shares in recent months because we felt they would calm down. Indeed, they are now approaching the levels where they are capable of becoming attractive again. The price of gold in London and Zurich is now in a climactic type of final plunge towards its theoretical floor at $35 an ounce. The French devaluation and the German upvaluation, along with developments on Special Drawing Rights, have all combined to postpone an official gold price increase, thus disappointing short-term speculators.

It is interesting as to why the influential Swiss bankers are allowing such a sharp decline, this week to as low as $36.65 an ounce. Some European Central Bankers believe Russia is selling gold for the first time since 1965—denied by other Central Bankers.

However, allowing the price of gold to fall increases the chances South Africa and the U. S. might finally reach agreement on how to handle sales of their newly-mined gold. The U. S. wants South Africa to sell only on the official market, but South Africa would like to be able to sell either to that official market or the free market. The latter could be at a higher price.

It is during such present moments of relative monetary calm that have proved to be the best gold buying opportunities in the past. We don't mean for traders. We have always looked on these stocks as a form of "fire insurance." They protect against an international monetary collapse.

November 21, 1969

* * *

Speculators this week in the gold market did their worst by putting the gold price down to $35.30 anounce. This in effect eliminates the free market premium over the official level. With gold down to its floor, perhaps gold stocks will at last begin the Base Formation that would lead us to upgrade them from "hold" to "buy."

December 5, 1969

JANUARY 1, 1970: DJI 809.20; ASA - 7¼

We are turning bullish on gold shares again, and they should all be regraded "buy."

The agreement signed between the IMF and South Africa was much better for the golds than we had feared. In effect, as we read the agreement, the IMF will support the price of gold at $35 per ounce, which puts a floor under it. But no ceiling has been put on top other than that South Africa must sell its gold on the open market. South Africa has lost its right to withhold gold from the market, but South Africa never really wanted this anyway.

On January 5 the *New York Times* editorialized that "Gold's power to disrupt the international monetary mechanism has now been greatly reduced and possibly ended. Under Secretary of the Treasury Paul A. Volcker has voiced the hope that this latest move will dispose of gold as a 'contentious monetary problem.' "

Gold, of course, never disrupted the international monetary mechanism. It was the wild inflation of paper money which disrupted gold. Indeed, we think it was Nixon's anti-inflationary moves which knocked the price of gold down, by reducing the fear of a runaway inflation. The *New York Times* continued: "But complacency about this country's position would be premature; inflation, which the new tax bill threatens to aggravate, is undermining America's payments position. Hardheaded and deft negotiations by the U. S. Treasury abroad can still be ruined by weak fiscal and monetary performance at home."

We have not yet obtained an actual copy of the agreement, but what we see of it so far suggests to us that gold now has a floor, but no ceiling. It will be interesting to watch the price of gold, to see if it can get much above its floor in coming weeks and months.

January 9, 1970

* * *

The U. S. payments balance was at last announced, showing a record deficit in 1969 of $6.99 billion! This compares with 1968's slight surplus of $168 million and the previous record deficit of $3.9 billion in 1960. Thus, they can play with Special Drawing Rights, or any other games they like, but there are still more dollars flowing overseas in the hands of foreigners than we are earning there, and this flood of dollars overseas is going to have to be resolved somehow. A raise in the price of gold would solve the problem, although this solution will undoubtedly be bitterly resisted to the end.

Much hullabaloo was made of the sharp improvement in our fourth quarter balance-of-payments, with a surplus of $1.1 billion. However, when you get down to the fine print, about a billion dollars of this could be attributed to a combined reduction of purchases of foreign securities by Americans, and an increase in purchases of American securities by foreigners. This is the type of temporary thing which can easily switch directions. It's like leverage being a two-edged sword, as conglomerates are now learning to their sorrow.

For all of 1969, the crucial trade surplus was only $674 million, barely higher than the record low of 1968. *We are not earning our way in the world.*

Also, corporations overcomplied tremendously in bringing dollars home in the last quarter of 1969, to meet their restrictions on foreign direct investments, just as they did a year earlier.

So, when all the dust has settled, another $7 billion worth of paper money has been spewed overseas, as a potential claim on our tiny remaining hoard of gold. There is no way all of this could ever be repaid at the present price of gold. Therefore, investors have no choice but to retain a permanent representation in selected gold and silver shares as a permanent hedge against devaluation or deflation.

February 20, 1970

We are dimly beginning to grasp that the real reason gold shares went down is that foreigners would rather collect 10% interest in the U. S. than take gold which yields no interest. Or, worse, margin their gold holdings paying 10% interest. There is a 20% spread involved, which is too stiff for speculators to condone. However, if this is true, then when U. S. interest rates fall, there should be a rush of foreigners' dollars cashed in for gold.

February 27, 1970

*　　*　　*

Here come the golds again!

Gold stocks soared this week. From the beginning of the year, the following leading golds have risen the following percentages since their low point reached in January this year:

American-South African	42%
Giant Yellowknife	29%
Homestake Mining	34%
Anglo American	36%
International Mining	42%

We think it is increasingly clear the golds have at last resumed their contracyclical function of moving in a direction opposite from the market as a whole. If so, the market must be graded bearish.

We think the reason they lost their contracyclical function temporarily, and this was a new phenomenon to everybody, was tight money.

However, news that interest rates abroad might soon decline suggested money would flow into gold bullion while awaiting other investment opportunities. Particularly at these depressed prices. We think there is an element of truth in that.

March 13, 1970

*   *   *

We pioneered the idea on Wall Street that golds move contracyclically with the market as a whole, but we are going to have to modify that. In recent Letters we have been stumbling towards the idea that there is an exception to this rule. During periods of exceptionally tight money, golds will move down, as in 1966 and 1969. So, the rule can be restated that golds will move up in bear markets, *except when the bear market is caused by exceptionally tight money.*

March 26, 1970

*   *   *

If history books are correct about the 1929 crash being caused largely by excessively thin margins (as little as 10%) then perhaps the conglomerate craze caused the present crash because effective margins were as little as zero.

Most of us had to put up 80% in cash to buy stocks in 1968, but a conglomerate could buy an entire company simply by printing up some convertible paper!

Conglomerates also soaked up staggering quantities of money when borrowing to buy other companies—extremely comparable with the 1929 binge atmosphere of people selling everything to buy stocks. Except here, instead of the masses it was a group of people who ran conglomerate corporations. This money was drawn away from other areas of the economy, such as construction, and was put essentially to a non-productive use, namely buying out other corporations.

It is a fascinating idea. A rich individual instead of buying stock on 10% margin illegally goes out and gets cheap control of a company. The company issues paper, and acquires a half-dozen other companies, issuing paper as it goes along, and borrowing money to the hilt. But instead of buying a certain number of shares on 80% margin like us, the conglomerate buys *all* the company's stock at zero margin!

At the same moment, stock market excitement about these conglomerates induced widespread investor movement of their funds from banks to stocks. Just as banks were making more and more loans to the conglomerators, savers were withdrawing money to buy stock. So the banks ran out of money to lend and money got "tight," or scarce.

At the time of the conglomerate binge, we pointed out these stocks were not "growth" stocks despite Harvard Business School garbage about Synergism, but were, rather, cyclicals. And, arch-cyclicals at that, because conglomerators were concentrating heavily on the most cyclical type of stocks.

As money was siphoned off from every other area of the economy, interest rates began to rise, naturally, but still the conglomerators' appetite was unsated. The higher interest rates and indeed, absence of money, began to actually hurt construction, and other more fundamental areas of the economy. Thus, the stock market *caused* a decline in the economic fundamentals, Charles H. Dow notwithstanding.

As stocks began to decline, people felt poorer, an emotional reaction, but they nonetheless began to cut back their spending, and thus was born a chain reaction: the more they cut back their spending, the more stocks dropped, and the vicious cycle à la 1929 was triggered. We have never heard in this area of New York (admittedly a small part of America) so many complaints about how bad business is— from every small businessman we encounter. Big companies are paying late, forfeiting small discounts for prompt payment. Perhaps overall national profits are not off by more than a few percent, but many individual companies are showing declines of 50% or more. We won't even mention Wall Street as a disaster area.

Nixon better pay attention, or this thing will get out of hand.

May 22, 1970

*   *   *

In another development, Canada released Canadian dollars to float on the open market. We have been recommending floating rates for many years. This allows currencies to adjust to each other immediately, without any false and arbitrary tensions developing.

Canada might have been influenced by West Germany's decision last fall to experiment with a floating mark. That worked out quite well. It's amazing how the free market can replace overpriced bureaucrats.

June 5, 1970

*   *   *

Just why are the golds moving? In group analysis there are only three groups with positive momentum. Aside from cigar and coal, you have golds. We think this latest crisis was triggered primarily by Nixon's prediction that despite all efforts to the contrary the United States will show a deficit in the next year. This is inflationary and shows the United States has become incapable of balancing its budget.

This might be difficult to grasp, but remember that those who advocated the Special Drawing Rights had hoped to remove American deficits as a factor in providing liquidity for the world. SDR's theoretically were supposed to be a deliberately managed reserve growth rather than the haphazard U. S. deficits. Without meaning to be cynical, we have always rejected these Special Drawing Rights as a cheap excuse for our federal deficits, and that because of the social structure of our society the United States was incapable of curing the problem.

June 12, 1970

JULY 1, 1970: DJI 687.84; ASA - 11

*   *   *

In previous bear markets, losses ranged in the 10 to 20 percent level. But, here, many people are sporting capital losses of 80 to 100 percent. Just what is going on, and how permanent will it be?

The economy is in a transitional period from inflation to deflation. This is the result, whether you approve or not, of the unlikely election of a president who really dared to try to halt inflation. A whole bull market based on inflationary expectations suddenly had to adjust to an entirely new set of ground rules reminiscent of the pre-FDR era. Whereas two years ago there was a flight from cash to property, today there is a flight from property to cash—in both cases an unreasonable stampede.

July 17, 1970

*   *   *

Chances are some day it will be reported the Europeans accepted the "two-tier" gold system along with the SDR's, only on the condition the United States deliberately create a recession and diminish its inflationary balance of payments deficits. That is why we had the "Crash of '69" and why the stock market has dropped so far. We think these agreements are secret, and it will be many years before they are revealed. That is our position.

July 24, 1970

*   *   *

Having caused the Crash of '69, the gold crisis continues to build.

July 31, 1970

*   *   *

Incidentally, there is still an incredible absence of awareness about how severe this crash has been. Except for some scattered references here and there in the financial press, we alone have been referring to this decline as the "Crash of '69" since before it began. When this decline comes to be widely referred to as a "crash," and we predict it will, is when we should be close to Bottom.

August 7, 1970

*   *   *

Gold prices jumped to a new high for the year on September 30, to $36.40. This appears to re-confirm our January prediction that the low in the price of gold was in the process of being seen. Looking back, our hunch proved to have hit the result smack on the head.

Part of the reason for the gold price rise, assuming it is not merely the typical contracyclical function of gold with the stock market, was due to Middle East tension.

Furthermore, U. S. gold stock fell $117 million last month, with five out of the last six months showing declines. The U. S. gold stock is now $11.82 billion.

The situation was hardly helped when the U. S. foreign trade surplus shrank last month in the teeth of record high imports. It is from such surpluses we can pay for foreign aid, war, and other items.

Most serious of all is the flood of foreign holdings of U. S. dollars accumulated from all of our past overseas deficits. This has been no problem so far because in the late 1960's Americans borrowed all the loose dollars they could find. High interest rates paid for these Euro-dollars kept them in short supply.

But now interest rates are dropping, and so is loan demand, so foreigners are turning these dollars in to their banks for their own currencies. It is from these banks it might flow to Fort Knox in exchange for gold. This could turn into a run on the dollar, and it would imply a gold crisis dwarfing anything imagined today.

October 2, 1970

*   *   *

The South African Mining Index lunged up to a new high above 101 this week, while on October 6 the price of gold soared to a new recovery high at $36.825. This is surprising in view of the fact that the Middle East has simmered down on the heels of Nasser's sudden heart attack.

Strength in gold on no particular bad news has always preceded the most severe monetary crises in recent years. It is hard to say what is brewing now behind closed doors, but it is surprising that gold shares are not responding more strongly than they already have to strength in the gold price.

Britain's gold reserves declined in September, ending a long period of recovery. Part of September's heavy reserve loss was due to heavy speculation against the pound early this month, before the Copenhagen meetings of the World Bank and International Monetary Fund. The Bank of England apparently dipped into reserves to drive the pound back from its low of $2.3812 up to $2.3901.

October 9, 1970

* * *

We've been howling about gold shares since January, when we flatly predicted the Bottom of the price of gold was being reached. Now suddenly it is in the headlines, and newspapers are actually contacting us for our opinion on why the price of gold is rising. Gold reached a new high at $37.45 in London this week. Now that interest *rates are moving down speculators can* afford to load up on gold again. With a floor at $35 it is difficult to lose, considering the enormous upside potential.

We do not believe it is too late to buy gold shares, and strongly urge that all portfolios contain several percent of our long-favored gold shares like *American-South African, Dome Mines*, and others on our Long-Range List #5.

The gold crisis that should have broken out in 1968 was postponed for two years because of tight money. With the present general easing of interest rates, and another large balance of payments deficit in our International Balance of Payments another gold crisis could well be in the wind.

October 16, 1970

* * *

In line with our predictions, the United States continued to run a negative balance of payments in the September quarter, and dollars piled up in foreign central banks at a terrifying rate. These banks can cash such dollars in, if they so choose, for gold at Fort Knox, and they own far more dollars than Fort Knox could ever possibly hope to supply. Thus is the bankruptcy of the U. S. dollar becoming increasingly visible to all who would see.

The cumulative balance of payments deficit so far this year is $3.3

billion. While lower than last year's corresponding deficit of $7.4 billion, unfortunately if you strip away the phony treasury dealings, Eurodollar flows and allocation of Special Drawing Rights from the International Monetary Fund, the current deficit would be around $4.5 billion for the first nine months.

Meanwhile, England's October trade surplus increased slightly, which might just postpone the next devaluation of the pound, an event we deem inevitable. Unfortunate timing, because England finally seems to have a government bent on lowering taxes and reducing government interference with a free economy, events which if pursued years ago might have eliminated the need for the financial crisis we see ahead.

November 20, 1970

*   *   *

A number of gold stocks have begun to move down in classic contracyclical fashion with the market as a whole. This is interesting because the price of gold holds stubbornly above $37, and has not weakened along with gold stocks.

This suggests the next gold crisis has been postponed. U. S. gold stock edged slightly higher in October to $11.495 billion but this is less important than the fact that dollars in foreign hands continued to climb in August to a total of $43.97 billion. That comes out to more than the gold we have with which to redeem those dollars.

December 4, 1970

JANUARY 1, 1971: DJI 830.57; ASA - 10¾

*   *   *

While the near-term fundamentals look good, on a very long-range basis there are still serious problems which will return to haunt us. You can expect a new wave of inflation, particularly since recent wage boosts have been in excess of productivity. This will in turn reawaken the chronic gold problem.

On a long-range basis America is in a profitless inflation with diminishing markets increasingly vulnerable to foreign competition at lower prices. More American corporations are manufacturing in Taiwan for importation into the United States, but this will show up in our loss of gold. Lower corporate profits mean lower tax collections, leading to higher government borrowing and eventually higher interest rates again.

There is no change in our warnings in recent years of a coming gold crisis. The U. S. will run another huge deficit in its balance-of-payments this year, and more dollars will pile up overseas. With more than $40 billion of our paper in foreign vaults, backed by only around $11 billion in gold in Fort Knox, eventual bankruptcy of the dollar is inevitable.

January 15, 1971

* * *

The gold, silver and platinum groups continue to act poorly, and this is all in line with our expectations over the near-term.

Long-term there is no change in the coming gold crisis. The U. S. gold stock fell $406 million in December to its lowest level in 20 months, down to $11.07 billion.

On another front, U. S. ability to earn gold suffered a blow in that the 1970 trade surplus, while the biggest in three years, was well below Nixon's expectations. And officials say the outlook is for a poor showing this year. Last year's surplus was $2.7 billion, but this is not very impressive when you remember a lot of the foreign aid we give away forces recipients to spend this money in the U. S.

January 29, 1971

* * *

It has long been our feeling that the present calm is completely illusory. Perhaps more people are beginning to at last realize this. Last year Western Europe, Canada and Japan added a staggering sum of $10.8 billion to their holdings of foreign exchange (almost all U. S. dollars) a sum nearly equal to our entire gold holdings!

Our balance-of-payments deficit accounted for the massive hemorrhage which alone bankrupts the dollar, and came in the teeth of the Special Drawing Rights or "paper gold" which were supposed to take the place of dollars. The first distribution of $3.4 billion SDR's did nothing to solve the world's problems. We trust readers of this Letter are not surprised.

February 12, 1971

* * *

Your editor has just returned from a trip to Europe snooping around as to the cause of why the price of gold is pushing up against its highs for the year. The Europeans are terrified about the huge dollar amounts piling up in their vaults, as we continue to run deficit after deficit each year, but there is no easy solution for them. They try to stick as many

U. S. dollars on to West Germany as they can, who is forced to take them because their political life depends on American military might, but there are even limits there. Europeans resent the fact that U. S. dollar outflows are causing inflationary pressures in their countries, but if they blow the whistle and demand gold they will bankrupt us and make all their *remaining* dollars worthless. The Europeans have painted themselves into an astonishingly stupid corner and will remain baffled until the inevitable stampede out of that corner begins.

We predict that the U. S., out of sheer embarrassment will soon refuse to even issue its true deficit and will instead harp on what it calls "the official payments balance." The U. S. will only count those dollars held by foreign bankers, since only bankers can get gold from us. But there are many individual foreigners who are holding U. S. dollars, and who can turn them in to a bank on a moment's notice.

Turning overseas, Japan has boosted its gold reserves to 11.7% of its $4.53 billion reserves. This is nowhere nearly enough to guarantee a stable currency, but at least they are beginning to get the point.

Britain's gold and convertible currency reserves rose more in February since February 1966, now standing at $3.19 billion. This is the highest level since July 1951, and the fifth consecutive month in which reserves have gained. The pound remains strong because England's interest rates are higher than other countries, thus attracting "hot money."

March 5, 1971

*   *   *

If the Europeans had any brains, which they clearly don't, judging from their international monetary performance in the last decade, they would do something with all these dollars. Since they can't buy gold with them (because we don't have enough gold to pay them off) they should throw it in our faces and just try buying our key industrial stocks like *General Motors* or *IBM*. They should just wind up owning American industry via the stockholders' route.

March 19, 1971

*   *   *

Germany, Switzerland, the Netherlands, Austria and Belgium refused to accept U. S. dollars because of an inability to turn those dollars into the U. S. for gold, thus shattering the system set up in 1944 at Bretton Woods which put the dollar in the center of the international monetary universe and defined it as $35 per ounce of gold.

So bankrupt is the dollar, and so dismal and sorry its prospects, a mere casual remark last Monday night started it all. In a dinner discussion in Hamburg, French Finance Minister Valery Giscard d'Estaing suggested the price of gold should be raised as a means of stemming the dollar outflow to Europe; in other words the dollar should be devalued. This upset the Germans and Italians because they are big dollar holders and are thus reluctant to see a dollar devaluation. France meanwhile opposes greater currency flexibility because of fears a general upward realignment of currencies would force the franc higher and place it at a competitive disadvantage.

More and more you hear suggestions *The Dines Letter* pioneered a decade ago, strongly recommending all currencies be allowed to trade freely at all times, with all currencies freely interchangeable into gold. Alas, nobody has ever listened to us. Representative Henry Reuss, head of the international exchange unit of the Congressional Joint Economic Committee, expressed open support of the idea of European currencies being allowed to float upward, and included Japan in the approval.

Meanwhile, the bland, dumb complacency continues. George P. Shultz, Director of the Office of Management and Budget, said while the Nixon Administration is not ignoring the Balance-of-Payments problems, he clearly indicated "domestic and economic considerations are considered more important." Secretary of the Treasury, John B. Connally, Jr., emphasized "that no change in the structure of exchange parities is necessary or anticipated." And on May 3 the Chairman of President Nixon's Council of Economic Advisors said "there were heartening signs that the worst may be over." Paul W. McCracken added "both logic and the facts of experience" suggest the overall U. S. balance-of-international-payments will get better, not worse, with an expanding domestic economy." (We suggest to Mr. McCracken the U. S. balance-of-payments has never gotten better, despite years of bland assurances).

In effect, all that has happened so far is the Europeans have hardened their hearts, and recognize the dollar as the major threat to international monetary stability.

The price of gold meanwhile reacted also, and in Paris leaped to $40.02 an ounce, highest since the monetary upheavals of 1969. In London, there was an 18-month high of $40.10, and in Zurich the major Swiss banks stopped quoting gold prices at all because of the chaotic foreign exchange conditions.

May 7, 1971

*　　*　　*

The U. S. dollar was devalued this week as market rates were allowed to float freely for the Belgian and Swiss francs, the German mark and the Dutch guilder. So far they have only risen a few percent, but by this amount the U. S. dollar has been devalued, as tourists will discover to their pain when they exchange their money for foreign paper. The world will never be the same, and Bretton Woods is more dead than it realizes. In October 1969 the German mark was upvalued by 9.29%, devaluing the dollar by that amount, but the present crisis brought four currencies along with it, with perhaps more to come. Will Japan be next? The next tremor could send the whole house tumbling downwards. It will be our primary task in coming weeks, months and perhaps even years, to solve the problem of timing in switching you from golds back into stocks. That will be a challenge and a half.

European leaders still have not yet dimly begun to understand how to stop the runaway printing presses in America; instead of upvaluing their own currency, they should downgrade the dollar, and they have not quite yet thought of this. Until they do, the gold crisis in its full fury will most likely be postponed. After all, Europeans need dollars there, since they have become used to them and even addicted to the situation, as an individual would be with heroin.

This is the fifth monetary crisis in 3½ years, and they will continue to occur until basic reforms are made—this is a basic *Dines Letter* prediction. In November 1967 the pound was devalued; in March 1968 gold prices soared; in August 1969 France devalued the franc; in October 1969 West Germany upvalued the mark.

The only solution that will work is to understand and recognize that the dollar has been bankrupt since 1961, and remove it as a reserve currency. At the same time, *the price of gold should be allowed to float freely against all other freely-floating currencies.* When more liquidity is needed (America's excuse to print more paper) then the price of gold will rise and automatically create new reserves—and this will be a splendidly self-regulating mechanism. Second, America should make balancing our international payments of paramount priority, with an aggressive attitude towards increasing our exports—no less than Japan's definitively pro-business attitude.

If the Europeans don't demand our gold, we can be thankful it has not dawned on them to ram those dollars back down our throats in tender offers for IBM, General Motors, GE, Standard Oil of NJ, and a few other corporation. *That* would be one way to stop us.

Just as things were simmering down yesterday, it was announced since May 3rd the Netherlands, Belgium and France have taken $422 million of American gold for their surplus dollars, dropping Fort Knox' contents to the lowest levels since World War II. And Jacques Rueff made the headlines, whose latest book "The Monetary Sin of the West" is a best-seller (although you would certainly never know it from the American press).

As Rueff so correctly points out "The evil began when the world departed from the gold standard in 1922 and again after the war." We have never seen this in print before, but *The Dines Letter* believes the 1929 Crash can be blamed on that 1922 decision.

Rueff warned ten years ago in *Fortune* the gold exchange standard would lead inevitably to a permanent dollar deficit, worldwide inflation and U. S. foreign debt that could not be honored. He recommended the price of gold be doubled, with a return of the gold standard among Europeans immediately, while working out a freeze of existing U. S. dollar debt pending an arrangement to eliminate it by easy stages. Rueff pointed out tartly "You Americans are not angry, like us. You have no foreign money in your reserves."

And what were the golds and gold stocks doing all this time? Astonishingly enough, they were not moving together, and we have not seen this phenomenon before. The price of gold moved up all right, but gold shares remained definitively retrograde. The tension is being stored up like an underground pre-earthquake, and sooner or later something is going to break loose. The situation wasn't helped when good Senator Jacob Javits declared our pledge to exchange dollars for gold as "obsolete" and suggested the Treasury might "close officially the gold window." This in itself could trigger a rash of gold demands from smaller nations anxious to get gold before it is too late, bringing about the long-dreaded "gold run," akin to a run on the bank, which could well knock the Dow-Jones Industrials down by several hundred points.

A truly fascinating situation unfolding now.

May 14, 1971

*   *   *

The last crisis has shown decisively the world doesn't disintegrate when some major currencies float in violation of the IMF's rules, so you can expect to see additional floatings in the period ahead. The big question is when the dollar itself will be allowed to float in relation to gold, because then we envision a colossal advance in the price of gold which will drive gold stocks much higher than anyone dreams possible today. When will this happen? There are undoubtedly Top Secret plans somewhere along these lines, because we cannot allow our gold to dribble out indefinitely. Chances are when our reserves get down to $10 billion, or some other round number, something will happen.

The U. S. lost $35 million in gold in April, the 6th consecutive monthly fall, even before the big outpouring in May. We now have $10.93 billion in gold at Fort Knox. At the rate we are losing gold, the $10 billion figure could be reached within the next few months, after which chaos could ensue.

May 28, 1971

* * *

Over the week-end, at the American Bankers Association in Munich, Treasury Secretary John B. Connally took rather a firm stand, describing it as without arrogance or defiance. Connally said "We aren't going to devalue the dollar; we aren't going to change the price of gold; we have the right to expect more equitable trading arrangements to be initiated by others; and we also expect Europe to accept the responsibility to share more fully in the cost of defending the Free World."

In this tense situation it was announced the U. S. trade balance deteriorated sharply in April. This was the first monthly surplus since February 1969. Actually, the figures make the picture look better than it really is because imports are recorded without including insurance and freight, and exports include shipments financed by government foreign aid! Sounds like the Penn Central before it went broke, doesn't it?

June 4, 1971

* * *

[R]epresentative Henry S. Reuss, Chairman of the Joint Economic Subcommittee on International Exchange and Payments, introduced a "sense of Congress" resolution in the House of Representatives which would formally end the United States commitment to sell gold to foreign banks at $35 an ounce, letting the dollar "float." The Treasury quickly denied we are going to devalue the dollar, but Reuss' statement still stands.

Also this week a new study showed floating exchange rates were good for Canada, and is perhaps the best way to avoid the problem of a world dominated by the U. S. dollar. This study is published and endorsed by the Canadian-American Committee, of which the U. S. sponsor is the National Printing Association.

And thus continues to rage the debate on the question "Do Floating rates hamper trade and investment?" Our answer for years has been a most emphatic *no*.

June 11, 1971

JULY 1, 1971: DJI 893.19; ASA - 11¼₆

* * *

As expected, the $357 million drop in the U. S. gold stock in May was a shocker. This seventh successive decline leaves only $10.57 billion in Fort Knox.

We can't prove it, but we have the strangest feeling something is going to happen when the $10 billion level is reached. If the U. S. keeps losing gold month after month, it is going to be reached soon, and that might just trigger a run on the bank. We think there must be Top Secret contingency plans in the White House for this.

After years of lies, misguided optimism, distortions, misinformation, and unbelievably ingenious devices like SDR's to keep the U. S. dollar out of bankruptcy, yet a new trick has been uncovered. The Commerce Department just announced a new set of measures in the United States balance of international payments. The *New York Times* said "The new measures are designed to be more meaningful than the old, which had become increasingly difficult to interpret." Translated, that means it was increasingly difficult to hide how bankrupt we were. But even with all the tricks, the figures still look terrible.

July 9, 1971

* * *

The big news here was David Callahan, president of the West Coast Commodity Exchange, who decided to probe the loophole against Americans owning gold. It is legal for Americans to own gold coins minted before 1934, so now gold coins will be traded as a commodity! We are watching this development with intense interest because it will be the first time in the 300-year history of commodities exchanges in this country and abroad in which open bidding will determine the price of gold.

Meanwhile, back at the ranch, gold shares basically continue to amaze us by showing no reaction whatever to the continuing Uptrend in the price of gold. And, worse, a new financial international monetary crisis is brewing, this time in France.

The Bank of France has been buying dollars to prevent the franc rate from breaching its upper limit of 19.5¢ amid rumors the French government has been considering an upvaluation of around 3%. With the franc in high demand, France's Finance Minister, Valery Giscard d'Estaing, said in a television interview France would not raise the value of the franc. But as he spoke, the price of the dollar kept dropping. This is the bloodlessly cynical judgment of the market place.

And, frugal Germany continues to watch the spectacle of its mark floating to a new high at 28.69¢.

The West German Central Bank is determined not to show a monthly increase in its dollar holdings during this present period of a floating mark. The German Central Bank has been selling some of its massive hoard of dollars accumulated before the May 9 decision to float the mark.

When we turn bearish on the market next, we will most likely go into gold stocks as a hedge. Right now the golds are near their lows, and there has been no error in postponing our switch. We are watching the situation closely because important timing will be involved.

July 16, 1971

*   *   *

FRACAS OVER NEW GOLD TRADING IN CALIFORNIA

Last week we mentioned the big gold news David Callahan was going to institute gold futures trading on the West Coast Commodity Exchange. The situation has been in the financial headlines ever since. The Treasury declared on July 16 "the law does not permit the speculative trading in gold or gold futures as proposed by the West Coast Commodity Exchange." Callahan ignored the Treasury ruling and insisted he would begin trading in gold futures at 7 A.M., July 20. The next day, Callahan sued in the federal courts charging the Treasury with inhibiting the Exchange's business, and requesting a declaration the gold trading plan does not violate the Gold Reserve Act of 1934.

The situation is important not only because it is the first time in the 300-year-old history of commodities trading in the U. S. and abroad in which gold prices were set by open bidding, but it would also break the monopoly of five London companies which establish the price of gold each day and also provide U. S. speculators with a hedge against deflation.

On July 20 the Treasury began a formal investigation of gold futures that could lead to prosecution by the United States Attorney.

The fact the Treasury did not take any direct action to halt the trading, and yet is suing, suggests it is taking a hedged position. It wants to see whether the sky caves in now that gold is freely traded, and is thus keeping all its options open. Later, it can drop its suit if necessary, or prosecute it, depending on what happens.

Because interest rates got too low in the first and second quarter of this year there was a tremendous outflow of short-term money seeking higher interest rates elsewhere. Morgan Guaranty Trust Company leaked the unbelievably bad news this week the U. S. deficit in its international balance of payments soared to a record $7 billion in the second quarter. This compares with the $9.8 billion deficit in all of last year! Adding on the first quarter's deficit of $5.5 billion, the $7 billion estimate shows that in the first half of this year alone we ran at a deficit of more than all the gold we have in Fort Knox! We're so bankrupt that with every passing year we are bankrupting ourselves over and over again.

The international impact of this news stirred up nervousness about the dollar overseas again. The dollar continues to sink to new lows while the price of gold creeps higher. There has been consistent speculation over a possible upward revaluation of the French franc. The price of gold in Paris last week reached the highest point in 21 months.

July 23, 1971

*   *   *

Secretary of Commerce Stans warned Congress this week the U. S. might have a negative balance-of-trade this year for the first time since 1893. Stans based this on the trade deficits in April or May, the first time in 20 years imports exceeded exports in two successive months.

The balance-of-trade is not the same as the balance-of-payments, which has been in deficit for years. The balance-of-payments is much broader, including not only trade, but all payments to and received from foreigners.

Little wonder then the U. S. gold stock dropped $61 million in June, carrying U. S. reserves to the lowest level since before World War II. This 8th consecutive monthly drop in our gold holdings brought our total down to $10.6 billion, and because more than $548 million in gold in the U. S. stock represents double-counted deposits and investments of gold owned by the 119-Country International Monetary Fund, *the gold owned by the U. S. edged below the $10 billion level for the first time.* Many regard $10 billion level as the danger point at which the U. S. might abandon its pledge to pay out gold for surplus dollars held by other countries, of course, denied by officials. So rumors are beginning to fly all around the world the dollar will be devalued soon.

We have been predicting a dollar devaluation since 1961, and we think it gets closer with every passing year. It is a miracle the dollar has survived this long.

Meanwhile, the price of gold has skyrocketed to $42.05, its highest level since July 1969.

The U. S. dollar has already been devalued in relation to the German currency, by 5.35% to be exact, and France could be next. Even now the Bank of France is desperately making massive purchases of dollars to hold the franc within its permitted range (under International Monetary Fund rules of 5.5125 francs to the dollar, or 18.14 cents for the franc).

This week the Treasury Department banned all speculative trading in gold coins effective immediately, ending the experiment by the West Coast Commodity Exchange reported here in recent weeks. The outrageous law was passed by anonymous people, so we don't even know against whom to vote. The Treasury regulation now has been revised "to reaffirm and clarify the long-standing gold policy of the U. S. that speculation in gold in any form is prohibited." The Treasury said the amendments "explicitly prohibit the trading of gold in any form on commodity exchanges and the acquisition of American or foreign gold coins of any description for speculative purposes." This outrageous violation of your rights as an American is now backed up by criminal penalties. There is no reason in the world Americans can't buy and sell gold coins for speculative purposes, any more than they can with stocks or wheat, or anything else.

July 30, 1971

*   *   *

[T]he price of gold has skyrocketed to another new high of $42.50 on August 4, and rumors of a devaluation of the dollar are racing around the world.

Meanwhile, French Finance Minister Valery Giscard d'Estaing served notice on July 30 France would oppose a new distribution of Special Drawing Rights, the "paper gold" used to supplement reserves of the International monetary system, and will try to convince her Common Market partners to do likewise. Giscard d'Estaing said with American deficits creating ample dollars overseas there was no need for creation of additional reserves. We have long said nothing can replace gold, and were pooh-poohed by experts who said gold was obsolete, a barbarous relic, and that paper gold was certainly as good as the real thing. There is no change in our prediction SDR's are doomed, like all the previous artifices to conceal the bankruptcy of the U. S. dollar.

Giscard d'Estaing also stressed France's intention to maintain the parity of the franc at its present 18¢ level even if the mark were upvalued.

Clyde H. Farnsworth in the August 1 *New York Times* wrote "it's not whether it would happen but when. This was the way many European authorities were talking about the devaluation of the dollar." Rather an alarming way to begin an authoritative article in the *New York Times*.

As we go to press a flight from the dollar continues throughout Europe, according to Swiss bankers. The central banks of England, France, Belgium and Switzerland are all supporting the dollar while the floating mark and guilder moved to record highs against the dollar. The Deutsche mark was around a 6% premium from its official 3.66 mark-to-the-dollar parity while the guilder was at a 2½% premium. In case you don't understand what that means, the U. S. dollar has been devalued by that amount in those two countries. Swiss banks temporarily suspended dealing of dollars but in this kind of atmosphere it's hard to say what will happen next.

Our 1961 prediction that the dollar would eventually be devalued still stands.

August 6, 1971

LATEST ON GOLD—WHAT WE WOULD DO ABOUT DEVALUATION

The price of gold skyrocketed to a new all time high at 43.94 on August 9. Yet, we think the gold crisis will temporarily pass for the moment, but the next crisis will once again start a general bear market. But that will be later.

With the dollar selling at record low prices, Britain's gold and dollar reserves climbed for the tenth successive month to a new high of $3.87 billion in July. The pound is very strong right now.

France is likewise in excellent financial health, this week virtually eliminating exchange controls today for French tourists traveling abroad. Just as the United States is beginning to think about imposing them, the French are at last benefiting from their relative financial prudence. Finance Minister Valery Giscard d'Estaing said gold and foreign currency reserves had jumped by almost half a billion dollars in July and stood at $5.802 billion. Part of the inflow of course was speculative money expecting an up-valuation of the franc.

France acted to stem the speculative inflow somewhat, asking bankers to differentiate between speculation and normal business needs.

Germany's floating mark was 7% higher since it was "floated" on May 10, which means the U. S. dollar has been devalued by 7% since then. Swiss banks also temporarily halted trading in dollars to try to stem the rising international monetary panic.

Part of the reason for the currency panic was publication on August 7 by a Congressional sub-committee of a conclusion that the United States dollar is "overvalued" in relation to other currencies and urged in one way or another it be devalued. The sub-committee declared only by altering the international exchange value of the dollar could the chronic deficit in the U. S. balance-of-payments be cured, it would help create jobs at home, stimulate exports and raise the cost of imports to Americans, retard U. S. investment abroad, and attract foreign invest-ment in the stock and bond markets and in American firms. The report was issued by the Joint Economic Sub-Committee on International Exchange in Payments of which Representative Henry S. Reuss, Demo-crat of Wisconsin, is Chairman. Hubert H. Humphrey of Minnesota was a signer, and three Republican members dissented.

The Treasury Department promptly dismissed the sub-committee report as not being representative of any wide body of Congressional opinion. The subcommittee's report came like gasoline on an already incendiary situation; just this week the Financial Times of London said devaluation was "not whether, but when."

Economist Milton Friedman rejected the sub-committee conclu-sion because unilateral dollar devaluation would not work without cooperation of other countries. Robert V. Roosa, whom we discussed at length when he was a Kennedy Administration official, and who did an incredible job of bungling the silver situation, agreed with Mr. Friedman.

Finally, just as the general market rally was beginning, the U. S. finally took action by formally proposing to the Executive Directors of the International Monetary Fund the ban within which currencies are allowed to fluctuate be widened to 3% from their present 1% levels. In Washington a spokesman for the Treasury Department would neither confirm nor deny the report. Should this proposal be adopted by the 118 nation IMF, economists have theorized this would dampen some of the massive currency flows which have been disrupting international mone-tary markets in recent years.

In the opinion of *The Dines Letter*, this ploy will fail like all the previous hare-brained schemes before it. Lest you think we are com-pletely negative, we reiterate our long-standing position we should aban-don trying to define all other currencies in terms of the dollar, and pick an abstract absolute such as gold. It can be platinum, diamonds, or anything else which is rare and has the other time-honored qualities of gold. That doesn't matter. Then, the price of this commodity, be it gold or something else, should be permitted to float freely at all times, permitting an international commodity exchange to thrive so international businessmen would have definite prices on which to base their business decisions.

August 13, 1971

*   *   *

As we long predicted, the devaluation occurred over a weekend, and it came with *wehrmacht-blitzkrieg* speed. There was no warning, again as we long predicted it would come. To see how governments devalue, for the future, here is what appeared on the Friday before devaluation in the August 13 *Wall Street Journal:* "The currency situation, however, remained unsettled. Some analysts viewed a report the U. S. has proposed wider fluctuations for currencies from their par values as a sure sign the U. S. doesn't plan any devaluation of the dollar. One rumor popping up yesterday was that the White House has plans for a sizeable tax cut to help spur the economy. However, the rumor was denied by the administration."

On August 15, 1971, President Nixon declared a Chapter XI Bankruptcy on the U. S. dollar. In his speech he unilaterally changed the 25-year-old International Monetary System and declared henceforth the U. S. would cease to convert foreign-held dollars into gold. The highlight of his Day of Reckoning Speech was a 90-day freeze on prices, wages and dividends. That this was only a 90-day measure ties in with our theory the other points in Nixon's speech were simply a smoke screen. Nixon is an avowed "free marketer," a *laissez-faire* type.

Finally, Nixon slapped a 10% surcharge* on all imports except those such as coffee which are not subject to duty and those such as sugar which are limited by quota. We consider this a colossal blunder, which can only antagonize other nations into equivalent retaliatory actions. The specter of an all-out trade war looms large unless Nixon gets rid of this one quickly. We think he will, as it is only part of his overall smoke-screen technique to devalue the dollar.

So there you have it. A staggeringly complex force has been unleashed in the world, with effects which will not be fully understood for a long time. Stock exchanges around the world tumbled as the U. S. stock market skyrocketed 32 points on Monday, the largest one-day number of points gained and the highest volume in history.

Clearly, the dollar is now a fiat currency, worthless, like the Russian ruble and the Lower Slobovnik. This is the first stage of a classic

---

* The courts ruled this act illegal in 1974 and ordered refunds made.

gold crisis. For the moment, the world currency situation is leaderless. Commerce will first turn to the mark, the yen or perhaps even the Swiss franc, but eventually everybody will have to turn to gold. All currencies will be defined in ounces of gold. So, the message of the future is— *back to gold!*

This typical first stage, of closing the gold window, has occurred in every important devaluation of a major currency in the history of the world. For the moment, the dollar will not be valueless because people can still spend those dollars here. But, if allowed to float freely, the dollar will continue to decline in relation to other currencies because we are running at a big trade deficit every year. If we continue to pile up $10 billion or so more outside our shores each year, obviously people will no longer accept these dollars for anything. We will then have the classic runaway inflation panic which has been seen so many times before in this world. The dollar might well become inconvertible, which we predict will be the next stage, but events can move with alarming alacrity once they get started, so investors have no choice but to retain selected positions in certain gold shares.

If Nixon gets rid of the 10% surcharge, and allows the dollar to find its own level and yet be exposed to the full brunt of foreign competition, thus coming to the position we have been advocating and predicting would eventually occur, then there is a whole new ball game and the U. S. might just yet pull out of the coming depression. With a floating dollar the surcharge is completely irrelevant because with a cheaper dollar our goods suddenly become all that more competitive. The surcharge is a blunder because it can only provoke other nations into realizing the trick we are playing on them by devaluing the dollar. Germany, which has more dollars than we have gold in Fort Knox, will lose by an equivalent amount of the devaluation, which will in turn undermine her own currency—so additional financial turmoil seems certain in the near future. But the U. S. must cure the basic reasons for the devaluation such as outlay for foreign wars, maintenance of U. S. troops overseas without compensation, foreign aid, excessive tourism, unfair trade restrictions by Japan and other countries, and above all a halt to U. S. inflation which prices our goods out of world markets. Until these problems are tackled head-on by Nixon, there is no way to avert the coming gold crisis. By throwing the book at the economy as Nixon has just done, and in recent weeks we have been predicting Nixon would do *something*, the day has been postponed. Nixon has shot his wad. What can be his encore?

And, there is no way domestic American inflation can end as long as Nixon keeps on pumping up the money supply in excess of the size of our population. The root of inflation is not with the people, except when they are unthinking enough to elect governments which insist on over-inflating the money supply.

August 20, 1971

*   *   *

Well, here we are in a wage-price freeze with a floating dollar, and the world has not ended. The impact and ripples are still spreading across the nation, and it is way too soon to tell what is happening yet. Of the many significant aspects, a way has actually been found to stop organized labor. We have in the past complained a small segment of the American economy keeps getting 10% a year wage increases, while the rest of labor does not get them, so in effect organized labor is profiting at the expense of the rest of labor. The average working man doesn't see this, but the resulting inflation makes him understandably furious. Obviously, organized labor cannot be stopped forever this way, but at least Nixon has found a temporary solution towards making our products more competitive on world markets, and might just turn our deteriorated trade balance into a more favorable posture.

Well, this is the first week with the dollar only paper, a fiat currency for the first time since the American Revolution. This is the typical first stage of a gold crisis, where the bankrupt country slams its gold window shut, nobly declaring gold is not relevant to its own particular currency.

On the first day of trading, the floating dollar was surprisingly strong in its debut on Europe's reopened market, but in following days, it began to weaken. Such weakness is part of our continuing predictions and we predict the dollar itself will eventually become inconvertible until we allow our gold to be sold again.

That is a shocking prediction, and by inconvertible we mean foreigners will not accept it at all. The only reason it is being accepted now is because we still do have $10 billion in gold, so that the dollar does have *potential* value. Ten billion dollars might not sound like much compared to our liabilities overseas of at least $40 billion (assuming our government has told us the truth) but if we allow a radical increase in the price of gold then our gold holdings will be worth that much more and we can once again afford to sell gold at a much higher price. For example, if the price of gold were quadrupled, then we would suddenly have *$40* billion in gold.

The U. S. Government will be the biggest gainer from a gold price increase because we still have more gold than any other country. Even domestically pressure will begin to mount for floating gold price on industrial gold as manufacturers of gold products are squeezed by high world prices and a fixed selling price. For so long as the price of gold hangs around the $43 an ounce level, don't look for too much pressure from these quarters. But should the price of gold go higher, there will be a hue and cry from the squeezees.

August 27, 1971

* * *

[T]here is no letup in the growth of money supply. So long as printing presses are running wild, without even the moribund restraining influence of gold, it is preposterous to think inflation can possibly end.

Well, the U. S. dollar is floating and in general there has been a devaluation of a few percent around the world. Whether this is the calm before the storm, the eye of the hurricane, remains to be seen. But for years we have been urging whoever was listening to favor floating exchange rates. Apparently recollections of the chaos which accompanied erratically floating rates in the world depression of the 1930's led most governments to cling to the certainty of a fixed-rate system. This depression psychosis continues to haunt current money managers. Unfortunately, the fixed rate syndrome is so ingrained governments continue to hamstring current floating rates right now. Japan for example, which only allowed the yen to float after a week of expensive resistance, wants to keep their currency values deliberately low to continue its export boom. England, a long-time deficit country, is now enjoying a trade surplus for the first time and is afraid an upvaluation would change things. Another important reason is for negotiating. With all major currency rates up for realignment, nobody wants to be in a position where they must accept a higher rate because the market pushed it there. So, rates are now floating, but on rather short moorings. Switzerland has created a maze of regulations, the Netherlands came up with new controls, and France has a two-tier market system which keeps a fixed rate for trade but channels capital movements on to a free market where the franc floats. German banks are also constricted.

But what about the price of gold, and gold shares? The general feeling now among most market letters is extremely negative, as you can see from this quote from the *Professional Investor* of August 26, 1971.

> ### IN MEMORIAM
> Here lies a goldbug. He passes into history as he joins the Tasmanian Wolf, the Dodo Bird, and the Passenger Pigeon. Alas poor goldbug, we knew him well.

The price of gold is edging lower as speculators cut their commitments. Still, it is above $40 an ounce, not that much below its all-time highs and we think it will rest here for awhile until the world realizes that no paper can be trusted and they will have to return to gold. At that point we are looking for explosive advances in gold shares.

We predict pressure for an outright American devaluation of the dollar against gold will begin to build up among the major western nations. It was precisely to squelch such rumors Paul A. Volcker, U. S. Undersecretary for the Treasury for Monetary Affairs, said on September 4 "We are not looking for any change in the dollar price, up or down." You know the moment it is denied it is coming! Remember when our government swore it would never stop selling gold?

September 10, 1971

*   *   *

One piece of good news was the rate of growth in the nation's money supply slowed sharply during August to an annual rate of around 3½%, according to the Federal Reserve Bank of New York. This is the smallest percentage gain in the money supply in the last 7 months. Whether or not this was a side effect of the international financial crisis is unclear, but without a sound currency all of Nixon's policies, in our opinion, are eventually doomed to failure, and we insist the growth in the money supply be strictly tied to the price of gold.

September 17, 1971

*   *   *

The price of gold continued to rise this week, and is now actually challenging the all-time high at $43.80, reached on March 10, 1969. There continues the astonishing divergence between the enormous strength in the price of gold, and the glaring weakness in gold mining shares. The situation cannot continue for long.

On September 21, Representative Henry S. Reuss (D., Wis.) long a foe of higher gold prices, saw the light and recommended the dollar be devalued in relation to gold! We have quoted Reuss many times over the years, sometimes unfavorably. In Reuss' speech was revealed the concept he still thinks fixed currency relationships are desirable, but at least he has begun to move away from the silly insistence the price of gold remain unchanged. Reuss said "Now is no time for false pride, for quibbling that we will never 'devalue' when all the world knows that a devaluation is in fact occurring."

The Dines Letter would like to make clear its position a small devaluation in the order of 9-10% as Reuss suggested would in no way solve the problems of the world. You have a $50 billion pile of dollars sitting overseas which is hanging over our heads like the Sword of Damocles. This must be wiped out through an official bankruptcy. The price of gold should be raised to $200, and made freely convertible into gold again. Which would write down the $50 billion to a figure we can cope with. Thus, the Europeans who are seeking a higher gold price are working for their own bankruptcy! And the United States is resisting this! Do they know what they are doing?

The present attitude of seeking a $13 billion swing in the balance-of-payments surplus is folly, because this will have to come out of the hides of other countries who are already teetering on the edge of recession. There is no way to pay off the debt. We should declare bankruptcy and write the dollar down to where it belongs.

Then, on September 25 the managing director of the International Monetary Fund, Pierre Paul Schweitzer, came out in favor of a small devaluation of the dollar by an increase in the price of gold, again opposed by Nixon.

Amidst all this maneuvering came the announcement the U. S. had a trade deficit in August, the fifth month in a row, although this was quickly shrugged off saying Nixon's controls imposed on August 15 probably had not yet had any real impact as of that time. If there is not a dramatic improvement in September, watch the fan.

October 1, 1971

*   *   *

Again the spectacle of the price of gold right up near its all-time highs at nearly $43 an ounce, while gold shares plummeted. Gold stock weakness was attributed to a statement by a London banker to the effect that at a recent International Monetary Fund meeting there was a considerable body of opinion holding gold would cease to be of monetary significance within this decade.

October 15, 1971

*   *   *

Until last week Britain and the United States were the only bullish stock exchanges, and now with Britain alone in the world, it is difficult to see how they can long resist the grip of the worldwide depression we envision somewhere in the next few years. There is no change in our position that because of the incredibly inept bungling of the gold crisis the DJI will eventually wind up below 400. That is because the U. S. refuses to solve the gold problem in the only way which will solve the problem, namely by a stratospheric increase in the price of gold.

October 29, 1971

*   *   *

The unwise handling of the international gold crisis was perhaps best typified when Connally, who is supposed to be solving this mess, said a higher gold price would be "exactly what we don't want to do. We want to demonetize gold. It can't be the standard of the future. We're not going back to convertibility in terms of gold. There's no economic reason for changing the price of gold. If you want to do it just to help someone's pride, we'll be glad to consider that—we'll talk about anything when we know what they are willing to do."

Meanwhile, the bankrupt dollar is becoming even more bankrupt. Our balance-of-payments deficit in the quarter, as we hinted a few weeks ago, reached the staggering sum of $12.1 billion. While the September quarter deficit occurred mainly before Nixon's August 15 actions, and the 10% import surcharge plus closing off our gold sales will moderate the results for the final quarter, we strongly doubt the problem has been solved. The Treasury only had $10.2 billion in gold left to back convertibility of the dollar at a time when the official basic balance-of-payments deficit was running at an overwhelming annual rate of some $48 billion!

To put this deficit of $12 billion in perspective, the full deficit of all 1970 was $9.82 billion.

November 19, 1971

*   *   *

We suppose the only reason the United States is resisting a higher price of gold is it might interfere with our ability to issue an unlimited amount of dollar deficits overseas, with the Europeans forced to accept them in any quantity. The discipline of gold would, true, eliminate world inflation, but it would hamper our dollar imperialism.

It is no surprise at all to us that exchange rates among the leading countries have really been quite stable because it is not the relationship of currencies which is wrong, *it is the price of gold.*

Meanwhile, we really don't know what to do with our gold stocks. Their charts are terrible, and their fundamentals couldn't look better.

November 26, 1971

* * *

What's basically rotten about the whole state of negotiations is that everybody is trying to get back to the same bankrupt system of fixed exchange rates which has just collapsed. For years *The Dines Letter* has been a strong proponent of freely-floating exchange rates, and the establishment of a commodity market by which businessmen can buy futures in money the same way Hershey Foods Corp. can buy cocoa futures. The major advantage of a floating system would be no unwanted accumulation of currency could conceivably occur because rates would respond immediately to the forces of supply and demand. It would also allow all governments to pursue independent domestic economic policies, irrespective of some trading partner which holds to an unrealistically high or low rate of exchange. If fixed exchange rates come back, then we foresee yet another monetary crisis on top of the one we have been predicting for a decade.

As we go to press in a highly volatile situation, an announcement came out of Rome the U. S. was prepared to devalue the dollar, a major reversal of U. S. policy! It was a hypothetical discussion in which the U. S. asked the other nations what they would do if the U. S. devalued by 10%. There was an astonished silence, and then a hastily-called recess of an hour and forty minutes.

The U. S. dollar is collapsing on foreign markets as we go to press, suggesting something might happen this week-end.

December 3, 1971

* * *

Our predictions since 1961 were meant to withstand countless denials and statements by presidents, Treasury officials, and Official Spokesmen to the effect the dollar would "never" be devalued.

This week the dollar was devalued.

How come the little *Dines Letter* was right and so many other great experts wrong? Simply because the numbers pointed to the fact there *had* to be a major dollar devaluation.

The world reacted with astonishing euphoria, particularly Congressional leaders. Representative Henry A. Reuss of Wisconsin, Chairman of the joint committee's Subcommittee on International Exchange and Payments, said the action was "glorious" even though a few years ago he said the dollar would "never" be devalued. The unfettered joy even spread to Senator William Proxmire of Wisconsin who called Nixon's devaluation "most welcome news." We think the tune of these gentlemen will change later when harsh realities reassert themselves.

At the moment the majority opinion is looking for a dollar devaluation in the order of 5–8%, combined with currency upvaluations of foreigners by 3–5%. We predict even this swing will not help the U. S. balance-of-payments that much. The basic problem is that domestically unions are looking for 40% increases in their wages and this is simply untenable in a competitive world. There is already ample evidence the 10–15% change in parity rates has not drastically altered the current steel import picture. Foreign steel is still pouring into this country despite the 10% import surcharge imposed in August and despite the upward float of other currencies against the dollar already!

December 17, 1971

JANUARY 1, 1972: DJI 889.30; ASA - 9

* * *

[W]e had to know the dollar would be devalued (or, the price of gold raised) which we did. And then we logically concluded gold stocks would rise. They didn't. Why? *Because our prediction remains unfulfilled!* We were looking for a massive rise in the price of gold, which we unwaveringly predict lies yet ahead. In other words we take the shocking position *the dollar will be devalued again.* In other words, the 8.57% devaluation of the dollar on December 19 from $35 to $38 is not the last devaluation of the U. S. dollar!

As if to add credence to our predictions, gold this week reached an all-time high at $44.525 for a Major Upside Breakout. This shows a *persisting* monetary uncertainty, which is in vivid contrast to the belief by almost every commentator in the world that at last the monetary issue has been "solved." How many times with us in the last ten years have you seen the monetary solved only to see it once again raise its ugly head some months later?

The South African Gold Mining Average made a new recovery high this week at 86.9 and is in a solid Uptrend. Strength in golds was apparently prompted by C. Cordon Tether writing for London's *The Financial Times* that Washington might consider a gold price increase to "say, between $70 and $100 an ounce, if not $140" to make the dollar again convertible into gold. The conjecture drew the comment "ridiculous" from Washington Treasury sources. The U. S. Treasury spokesman said "The program is to phase gold out of the system." The free world gold market does not agree. Furthermore, lower interest rates have made it cheaper to speculate in gold on margin, a force which is spurring demand at the moment.

January 7, 1972

* * *

THE COMING GOLD CRISIS

That heading might surprise you, since we have been writing similar headlines for almost a decade now. Didn't the last dollar devaluation finally satisfy us? After all, we got what we wanted, didn't we? The answer is an emphatic *"no."*

We stand practically alone in the world, as we have so many times on this subject, brushing aside as completely false all the current euphoria about how the international monetary crisis has been "settled." The engineering of public opinion has been so proficient we find it terrifying. From a position of "we will never devalue" to Congressional leaders demanding devaluation, is an amazing mass psychological reversal. A *New York Times* headline said "Why It Feels So Good to Devalue."

We write this despite the fact that Nixon hailed the results as "the most significant monetary agreement in the history of the world." The so-called "Smithsonian Agreement," signed in the Smithsonian Institute in Washington, will mark just one more milestone in the long, downhill degradation in the U. S. dollar.

Not only is the agreement irrelevant, but it is preposterous. To declare gold is now worth $38 when the free price is $45 and making a new high almost daily, is an exercise in totalitarian doublethink.

All those billions of dollars on the books of foreign central banks are sitting there, or else being used to buy U. S. Treasury paper and thus financing an unbelievably large Federal deficit, for what is one of the most incredible public relations gambits since the South Sea Bubble.

It has emerged from the recent gold crisis *there is no practical alternative to gold.* The very fact the U. S. refuses to sell its remaining hoard strongly indicates we are preparing for the next phase of dollar devaluation which will be a much larger one. That is because there is no way to get the dollar convertible into gold again. If it is made convertible into gold at current prices, our gold hoard would be cleaned out overnight. You see, we owed, according to Federal Reserve and Treasury figures (including dollars or securities readily convertible into dollars) approximately $60.7 billion at the end of the last September.

Considering recent turbulence it would hardly be shocking if we did not owe around $65 billion.

Now, when you have only $10 billion in gold to back that up, obviously you're going to need a massive increase in the price of gold, to at least $70 or even $140 an oz., which would make our $10 billion suddenly worth $20 or even $40 billion. Then, the remainder of the $65 billion could be pinned down by long-term loans to the U. S. to make the dollar solvent again. *Until this happens, the dollar is as bankrupt as we have been describing it in recent years.* This is is the monetary iceberg. *[T]here is only one price for gold, and that is the free price for gold on the open market. Nothing else counts.*

In conclusion, the 8.57% devaluation of the dollar to $38 was a publicity stunt which glossed over all the real problems, making the dollar *still* uncompetitive and inconvertible, and forcing Europe into a slump to pay for it. Europe, with a long proclivity for falling for schemes such as tulipomania and the like, is now deeply involved in refusing to get out of the current joyride. The U. S. devalued out of weakness, not strength, and although it improves the ability of America to compete, we are confident 40% increases for labor unions will soon erase this edge in short order. The devaluation will not stop imports because our wages are spiraling.

January 14, 1972

*　　*　　*

This week saw the dollar falling to a new low trading against the German mark, and the price of gold skyrocketing to a new all-time high at $46.17 on January 17 in the free gold market, as confidence in the dollar plunged to a new low. Promptly they dusted off Roosa and Volcker to make reassuring cooing noises but nothing seemed to help. That's quite a bit of commotion for a monetary crisis which was supposed "solved," particularly since gold stocks began to move this week after our switch in the last *Dines Letter* to a buy signal for them, and many golds are in strong Uptrends—an ominous sign for the stock market as a whole.

January 21, 1972

*　　*　　*

Foreigners are not rushing to send their dollars back to the United States as had been widely predicted; such expectations were behind the recent DJI Uptrend which assumed Europeans would now stampede to buy U. S. stocks at the devalued dollar's discount price. Not only do Europeans not want a weak currency which might be devalued again but American budget deficits at astronomically high levels presage a probable return to inflation! Furthermore, interest rates are too low now to encourage dollar repatriation. Finally, Nixon has not yet actually gotten the dollar devaluation bill through Congress. With the price of gold reaching a new all-time high at $49.25 on February 2, efforts to get the official price raised to $38 are revealed as a purely asinine fantasy in futility. No gold is being transacted at the official price, and the free market price insures there will be no transactions at the lower official price.

Before gold shares began their recent rise, we flashed a buy signal on them, and quite a bit is happening among the South African gold mining shares as a result of recent currency developments. Since the South African currency unit, the rand, was devalued against the dollar in December, with the new exchange rate being 1 rand equal to $1.33, it is time for a new look at the gold mining shares. Expressed in rands (gold at $38 is equivalent to 28.50 rands) the earnings for example of East Driefontein would *triple* from 10¢ to 31¢ if the price of gold were doubled.

February 4, 1972

*   *   *

One of the reasons for this week's market rally was Nixon's formal presentation to Congress of the devaluation of the dollar and an increase in the official price of gold from $35 to $38. The whole gold syndrome cooled off, the free price of gold backed off slightly from its highs, gold stocks dropped out of the headlines, and even the dollar recovered slightly on foreign currency exchanges. Thus, the ludicrous charade continues. Washington is gassing about raising a price from $35 to $38 when the free price is near $50. How irrelevant can anybody get? There is no change in our feeling the price of gold will rest and consolidate here for a while, before lunging upward to new highs yet again.

In other words, for the umpteenth time, we brush aside those who say the gold problem is solved. It is not solved, and has not been solved, and nothing we see in the headlines indicates the problem is even close to being solved. The 8.57% devaluation of the dollar against gold will be a brief fillip to the economy, but compared to the staggering deficits being run up these days, the deficit will soon be used up to the detriment of our foreign trading partners.

Meanwhile, with the free price of gold in a long and well-established Uptrend, there has come a development which for years we have predicted would eventually come, and which we warmly welcome. The Chicago Mercantile Exchange announced it would provide futures trading in 7 foreign currencies: the Japanese yen, Mexican peso, Italian lira, British pound, Canadian dollar, German mark and Swiss franc. Computer programs should be ready this summer at which time the Exchanges will open for trading. Quite an addition to its normal fare of commodities!

The Exchange's Chairman, Leo Melamed, said the purpose is to permit importers, exporters, international business concerns and financial institutions to use the current futures contract as a hedge against the risks inherent in potential price swings in a specific currency between the time a transaction is agreed upon and payment falls due. This eliminates the need for everybody to have fixed currency rates, and throws the cost of international business on the international businessman, where it belongs, instead of the tourist.

February 11, 1972

* * *

Ominously, the dollar is being buffeted on international markets again, dropping to the floor rates set by last year's so-called "Smithsonian Agreement."

Cause for the doubts were the Uptrend in the price of gold, doubts about the ability of the U. S. to ever solve its balance-of-payments deficit (occurring when foreigners obtain more dollars than they send to us), continued inconvertibility of the dollar into gold, low interest rates domestically which causes money to leave the United States in search of higher interest rates elsewhere, and the astronomical budget deficit which seems sure to set off another inflationary holocaust.

On February 16, Secretary of the Treasury John B. Connally told Congress the United States "can't afford" to return to the convertibility of the dollar this year. Connally said it would be "sheer folly" to resume convertibility before there was convincing evidence of improvement in the nation's underlying balance-of-payments situation. Connally, along with everybody else in Washington in recent decades, continues to ignore our solution to the problem, which would be a radical increase in the price of gold and the declaration of immediate convertibility, along with the right of anybody in the world to buy gold from the U. S. Government, *particularly* U. S. citizens. This would be a Chapter XI on the U. S. dollar, long overdue and desperately needed to stave off a horrendous depression.

February 18, 1972

* * *

The fact that interest rates are higher overseas encouraged multi-national companies and speculators to leave dollars overseas to get the higher return rather than bring them home for the lower interest rates here. This is why we think the Fed will nudge rates upward here to encourage the reverse process. Meanwhile, however, Europeans are drowning in a flood of fiat U. S. dollars, a process we warned would occur, and we continue to warn it must lead to a new international monetary crisis.

Arthur F. Burns, Chairman of the Federal Reserve Board, returned from a regular monthly meeting in Basel, Switzerland, March 12 declaring "I am a very calm man" after a week-end of talks with central bankers from western Europe, Japan and Canada. He really must be a calm man because it is our opinion anyone who is not panicky right now doesn't understand the situation! Undoubtedly Burns was pressed during the secret meeting on how to stop the continued hemorrhage of U. S. dollars every time we run a new balance-of-payments deficit, and when the dollar would be made convertible into gold again. Yet, he remains calm.

Meanwhile, Britain's increased overseas buying last month put the nation's trade balance into deficit by $83.2 millon. This is a topsy-turvy world, remember? The deficit is considered "good" because it shows consumer demand is strengthening. Fantastic. Meanwhile, the Japanese balance-of-payments surplus widened last month to $790 million from $742 million the month before, and they say that's good. Now, one of them has to be wrong.

March 17, 1972

*   *   *

The dollar firmed in overseas trading on news of higher U. S. interest rates and at President Pompidou's signal, French monetary attacks on the dollar are being called off at least for the time being. For years we have been predicting the dollar was in for a major collapse, and this turn of events in no way deters us from our conviction the dollar is going to have to go through a Chapter XI bankruptcy before all this is over.

Meanwhile, with the free price of gold holding staunchly near its all-time high around $48 an ounce, the laughable charade about whether the U. S. will raise the price of gold from $35 to $38 continues, amid the nervous palm-wringing normally associated with incipiently terminated virginity. The Senate passed the devaluation measure earlier this month and the House has just approved the bill while ruling out-of-order a proposal to allow private trading in gold. It figures. This is the same bunch, pretty much, who got the dollar into trouble in the first place. They remember nothing and they have learned nothing.

Instead, Treasury Secretary John B. Connally declared there ought to be new international monetary rules to include an "automatic discipline" against countries which run chronic surpluses in their balance of payments! How *dare* anyone beat the United States at its own game of capitalism! Let's *punish* them, right, Connally? The nerve of anyone working harder and saving more and balancing their budget more than we do. There should be a law.

Meanwhile, while gold stocks have not done very much in this country, the South African Mining Index has just made a new high above 93. Furthermore, gold *momentum*, has just turned defiantly positive. The Paflibe group chart of gold showing the Upside Breakout of momentum above 100 is very bullish and suggests an important rise in gold shares could be just ahead.

March 24, 1972

*   *   *

The big news was that the February trade gap between exports and imports widened to the second largest reading for a month on record, coming to a staggering $597.6 million. True, the dock strike distorts figures but with the dollar maintaining only one-half of one nostril above water, the bad news could not have come at a worse time.

The international monetary scene was then rocked by some rumors the U. S. Treasury was considering sale of some gold to depress its price and punish the speculators. A more asinine idea has not been heard since someone in the U. S. Mint decided to mint an extra few hundred million silver dollars to punish silver speculators, who then bought all those dollars several times over. It would be like trying to drown the Sahara Desert.

March 30, 1972

*   *   *

[M]onetary authorities from 10 industrial western nations agreed after their parley the dollar has successfully weathered its early tests. Dr. Otmar Emminger, Vice Chairman of the West German Central Bank, said there had been no net increase in his institution's dollar holdings so far in March. We agree this is an encouraging sign.

The experts concurred American trading figures would worsen for awhile before getting better, but Emminger felt the average 12% dollar devaluation was "realistic" and would endure "for quite some time—a matter of years."

April 7, 1972

*   *   *

The London gold price on April 18 made a new all-time high at $49.50 per ounce, but it did not make the headlines. The South African Mining Index, however, noticed it and made a new high at 95.1, continuing its flawless Uptrend. One of these days, golds are going to burst into the headlines again and frankly, we can't understand why they have not already done so.

Puzzlingly, the dollar is now comfortably off the floor of its lower intervention limits so big money is apparently not betting on an immediate resumption of the ongoing international monetary crisis.

Particularly so when the Morgan Guaranty Trust Company, in its highly regarded monthly publication World Financial Markets, has suggested our first quarter balance-of-payments deficit might have even exceeded the $1.5 billion deficit recorded in the fourth quarter of '71, perhaps as high as $2 billion. On an "official settlements" basis perhaps the deficit got up to about $3.5 billion, even excluding the $710 million in January for Special Drawing Rights and before seasonal adjustment. It takes nerve to ignore all this bad news, but international speculators are apparently doing so.

April 21, 1972

*   *   *

In recent weeks we have been wondering why golds were not attracting attention, and this week they at last broke into the headlines as the price of gold in London soared to a new all-time high at $54. The South African Mining Index made a new high within a flawless Uptrend at 99.2, challenging the old high of 110.9 reached April 29, 1969. So far as gold is concerned, 1972 is chillingly reminiscent of the way 1971 developed, with a full-fledged gold crisis yet ahead.

If the price of gold were tripled, then instead of $10 billion in gold, we would suddenly, with the stroke of a pen, have $30 billion. This would be enough to manage the debt we have created. For some reason, totally unfathomable to us, this solution has been vigorously resisted by politicians of every stripe. Perhaps because it will enforce restraint in the issuance of paper money.

Thus, our solution, unlike most everybody else, is not so much concerned with the parity ratios between the U. S. dollar and other currencies, but the relationship of all currencies to the property it represents (gold). In other words, we have to write off the inflation of the past 30 years, and those who are holding dollars now will be forced to pay the piper. Some ignorants suggest such a devaluation followed by everybody, would put us back right where we started from. The parity of all currencies would remain the same, but the relationship of all paper to all property would be changed and *that* is the root cause of the chronic currency crises which have occurred since we began writing about golds.

Some say there isn't enough gold to back currencies in this day and age of a greatly expanded economy. This is ridiculous. If you raise the price of gold high enough you can have a value high enough to reach any amount. *It is not the quantity of gold which is relevant, but the price at which it is sold.* And, it is an outrage, which the American people have somehow never been bright enough to demand, to allow foreigners to buy our gold without permitting *Americans* to buy it! We think any American should be allowed to own either gold or paper as he so chooses to protect himself from the ravages of inflation. As it stands now, there is precious little place to hide, except to buy selected gold shares.

May 12, 1972

*   *   *

It does not appear to be a monetary crisis but rather simple bullishness on the price of gold which is going on for the moment. For one thing, South Africa announced they were going to slow down their gold sales. We were wondering when they would wise up and hold their commodity off the market for higher prices. Also, Connally, in a surprise shift, resigned as Secretary of the Treasury, yielding his position to George Shultz, who is well-known for his more classically Republican, free-market orientation. Shultz is known to be against wage and price controls, and is generally more conservative in his economic thinking. This could presage an important shift in U. S. policy, particularly in that it comes right before Nixon's trip to Moscow, and we are picking up consistent rumors Nixon will offer a higher gold price to Russia as part of the package to help them buy American goods.

This is the sixth monetary crisis in four and a half years, and they will continue to occur until basic reforms are made, as has often been predicted by *The Dines Letter*. In November 1967 the pound was devalued; in March 1968 gold prices soared; in August 1969 France devalued the franc; in October 1969 West Germany upvalued the mark; and of course you know about the 1971 crisis which ended in a devalued dollar.

The Dines Letter still feels the only lasting solution is to remove the dollar as a reserve currency (because it is bankrupt) at the same time recognizing the free price of gold is the *real* price of gold and that its 1932 level of $35 an ounce is preposterous. All the excessive paper printed since then will have to be wiped out by a massive increase in the price of gold, which is another way of saying the same thing. Then, the price of gold should be allowed to float freely, probably to at least $150 an ounce, against all other freely floating currencies so that liquidity is automatically either expanded or contracted depending on how crazy governments are with their printing presses. This would be a splendidly self-regulating mechanism.

In reference to man's trust in paper money rather than good old gold, the price of gold got up to an unbelievable $57 to $58 an ounce in London on May 17 continuing an Uptrend which has flabbergasted even those who still disagree with our predictions of a higher gold price.

May 26, 1972

* * *

The South African Mining Index hit a new high at 101.8 on May 31 while the price of gold reached a new all-time high at $59.45 on May 31. Yet, despite our numerous predictions the price of gold would rise are at last coming true, there are those who still indulge in the fantasy the price of gold is rising because of "industrial demand." It is not. It is rising because of the numerous fiscal blunders made in recent decades.

Perhaps one of the causes can be shown by the April U. S. Trade Deficit being the second largest on record. Clearly, the effects of the dollar devaluation last December have not yet taken hold.

June 2, 1972

* * *

Having been so anti-gold for so many years, Wall Street continues to be flabbergasted by the price of gold being reported in the headlines every day as making new highs. They have tried every excuse, whether it be South Africa withholding gold from the market, speculators and hoarders at work, ominous gnomes of Zurich, industrial demand, South Africa finally achieving a healthy balance of payments, and so on. Everything except the truth.

The South African Mining Index meanwhile skyrocketed to its

new high at 103.2 on June 7 this week, challenging the all-time high reached on April 30, 1969 at 110.9. The price of gold itself is in a runaway Uptrend and made a new high at 64.85 on June 7. The accompanying chart is already obsolete!

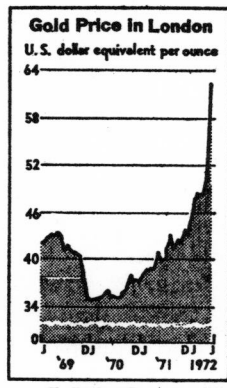

Gold Price in London
U.S. dollar equivalent per ounce

The New York Times/June 7, 1972

There is an uncomfortable reminiscence of a similar gold crisis last year at this time just before the dollar became inconvertible and well before the dollar was devalued. Yet, there are significant differences. For one thing, the price of gold is skyrocketing this time without any serious pressure on the dollar itself. This ties in with our theory the price of gold is rising to wipe out the excess of *all* paper money in the world.

When Fort Knox has $10 billion of gold and our international trade deficits over the decades have left a mountain of $50 billion overseas, clearly the dollar is bankrupt because if foreigners asked for their money in gold, not everybody could be satisfied. Therefore, we have long held the position the price of gold would have to rise radically to wipe out that mountain of paper in the sense one would need more and more dollars to buy the same ounce of gold; in other words the price of gold had to rise.

June 9, 1972

JULY 1, 1972: DJI 928.66; ASA - 13$^{15}\!/_{16}$

* * *

Meanwhile with stock exchanges around the world plummeting, the South African Gold Index made a new high at 109, highest since May 12, 1969. Gold's relative strength is in a roaring uptrend, and, most important, in the 112 Group Momentum ranking gold is now in 7th place. This is very good for gold shares.

Meanwhile, England raised its bank rates for the first time in three years from 5% to 6%, which makes it more attractive to hold pounds and to some extent compensates against fears of devaluation. Japan cut its bank rate from 4.75% to 4.25% to stave off capital inflows and head off pressure to upvalue the yen again—although if all European currencies float the yen probably would also. Switzerland took Draconian measures to halt new capital inflows including an 8% *penalty* for new deposits! That is a "negative interest rate." All you need is a currency backed 125% gold like the Swiss have. The French franc also is strong, because they have so much gold also. See the pattern? Since August 15, 1971, when the dollar was made inconvertible bankers have squirmed to avoid taking in more U. S. dollars.

U. S. interest rates will have to rise to make holding dollars more attractive, which will hurt our business boom. That's why the wild monetary turmoil was construed negatively on Wall Street. The dollar must be made convertible into gold again, since the U. S. owns almost one-quarter of all the world's gold. But the price of gold will need to be much higher or the U. S. will lose all its gold at $38. That is why the free price of gold is now $66. Until then the chronic long-term weakness of the international monetary system will continue, we predict, and it could choke off trade and capital flows if inconvertibility continues.

July 6, 1972

*   *   *

It is our profound conviction the high price of gold will lead to amazing earnings increases by gold mining companies in coming weeks and months, and the only reason gold shares haven't really begun to rise yet is because of the antagonistic attitude of most analysts towards gold. This will change as great earnings reports begin to glitter in.

Silver and platinum are also turning very bullish along with gold, and those who play the commodity markets should take aggressive positions here also in our opinion.

July 14, 1972

*   *   *

Ominously, on July 18, with the DJI plunging and the number of new lows of 210 vastly outnumbering the 8 common new highs, no less than 2 of those were precious metals (American-South African and

New York & Honduras Rosario). Golds and silvers typically make new highs during bear markets.

The dollar rallied strongly when the Europeans failed to let the dollar float against their currencies, and then the U. S. shifted its policy and actually acted to prop up the value of the dollar. It took the form of heavy selling by the New York Federal Reserve Bank of the German mark at declining prices. This show of cooperation is a reversal of former Secretary of the Treasury John B. Connally's policies. At the same time Federal Reserve Board Chairman Burns indicated the U. S. would reactivate its reciprocal borrowing arrangements with foreign central banks which have been suspended for almost a year. The sale of German marks is the first significant intervention since last August 15 when Nixon formally severed the link between the dollar and gold, events which led to devaluation of the dollar. But, in our opinion, this is all too little, too late. We won't have enough marks to keep this up indefinitely, because the dollars are accumulating overseas at a far faster rate than we can earn German marks.

July 21, 1972

* * *

It has been our theory all along, aside from looking for a selective summer rally, we were late into a bull market although it was not yet quite over.

We also have pointed out the amazing similarity between this bull market and the one which began in October 1966. If the similarity holds true, then there should be one final speculative binge starting quite soon, which should lead into the Top.

The price of gold screeched to a new all-time high at $70 on August 2, 1972 at last fulfilling our numerous predictions in recent years the price of gold would eventually double from the then "immutable" price of $35 an ounce. The South African Gold Mining Average made a new all-time high at 109.4 on August 2 but the really big jump in gold mining shares we have been long awaiting still has not yet happened. We are supremely confident it will and again urge all accounts large enough to afford a position in gold to take some. We are particularly interested in *American-South African* (ASA - 54¼) which manages a portfolio of many individual South African gold mines.

From what we gather, the reason for the latest outburst of gold activity was French President George Pompidou's proposal at last week's Franco-Italian summit meeting for a possible upvaluation in the gold price for certain central bank transactions between Common Market countries. This would make it easier for Italy to reaccept strict integration with the European Economic Community Agreement reached in Basel, Switzerland in March on the repayment of debts arising out of joint Common Market support operations. Thus the Europeans have slowly stumbled towards the position we have long predicted they would take: raise the gold price to bring it in line with the reality of the open market price now around $70.

The price came in for Britain's attempt to support the pound before it was allowed to float freely a few weeks ago, and it comes to a staggering $2.61 billion. There are now only 16 countries still pegging their currencies to the British pound and thus currently floating with the pound: 18 other countries and Hong Kong, once a part of the sterling area, now peg their currencies to the U. S. dollar or gold (which amounts to the same thing). Still sticking with the pound and hence floating are Ireland, India, Jamaica and South Africa. Bermuda just shifted the Bermudan dollar to the U. S. dollar from the British pound as a result of England's latest float.

The price of gold is going to continue to have to rise because governments persist in running the printing presses to get themselves re-elected. While it can be seen inflation has tempered moderately in a chart of the Consumer Price Index the long-term picture of consumer prices shows a runaway Uptrend for anyone who would see.

Treasury Secretary George P. Shultz said on August 2 the U. S. had no present plans to seek changes in the two-tier system of gold prices. Asked if that system was still credible in view of the recent high of $70 an ounce, Shultz said "It is under a strain . . . but it is holding." We would rather use "rout" than "strain," but that is a matter of personal taste.

August 4, 1972

* * *

[I]gnoring the Technical picture, the Fundamentals for gold shares have never in their entire history looked better. The price of gold has doubled and it has not yet been reflected in earnings or in the price of gold shares yet, and they both must obviously come. *We continue to make the prediction, which will startle some, perhaps many, the price of gold will eventually reach $200.* And to those who say gold is worthless and so unimportant, we ask why does Washington refuse to sell what it has. The U. S. is hanging on for dear life because it knows *gold is the ultimate currency.*

The older generation of money managers, remembering how the general breakdown of the international monetary system played a key role in the worldwide depression, are stridently insisting on some form of fixed exchange rates. We think that is the root cause of part of the recurrent monetary crises. Let all money float freely against one another, which reduces political chicanery, just as they are floating right now! We think one of these days they're going to wake up in Washington and say "Hey, these floating rates work fine the way they are, so let's leave them alone!"

September 8, 1972

* * *

The South African Mining Index finally smashed to a new all-time high at 112.4 on September 11, while the price of gold holds near its all-time highs around $67 per fine ounce.

The situation was not aided by the nonetheless truthful remarks of Wilbur D. Mills, Chairman of the House Ways and Means Committee in an interview with the German news magazine *Der Spiegel*, that the United States dollar is still grossly overvalued in terms of the Japanese yen, the West German mark and the French franc, and should be adjusted. This prediction of another dollar devaluation by a leading member of Congress sent jitters through the international currency markets. What none of these gentlemen seem to understand is *all currencies are overvalued in relation to gold, and not so much in relation to each other*. A massive increase in the price of gold would solve their problems, but *The Dines Letter,* despite weekly commentary on this subject for almost a decade, still has not gotten its point through to whomever has been in Washington all those years.

Then, the IMF called on the U. S. to raise short-term interest rates to ease its balance-of-payments problem. This could lure back much of the capital which has moved abroad and strengthen the dollar's market value, the IMF said in its January report, along with a prediction American foreign trade will not improve much during 1972. The IMF is not expecting very many benefits from last December's Smithsonian Agreement until 1973, or even later.

We think here is the guts of why the prime rate has been moving up lately, and do not think it has anything at all to do with the tight money normally associated with the end of bull markets. We think this is strictly an international monetary problem, and the solution has escaped most analysts because they do not really think much about the international scene.

Also on September 11 the IMF suggested further distribution of Special Drawing Rights, so-called "paper gold" (which is not gold at all, of course) remained in doubt. While they were proving readily acceptable after three years of use (distributed January 1, 1970, 1971, and 1972 under a 1969 agreement calling for total creation of $9.5 billion worth of the reserves unit) before any further distribution can be made, "further discussions" were considered necessary. Between the lines we think this means Europeans are becoming increasingly impressed with real gold rather than paper gold.

September 15, 1972

* * *

Also worth noting here is the world's first gold futures market scheduled to open in Winnipeg in late October. The U. S. Treasury has warned Americans they will be breaking the law if they buy such futures, however we hope some investors will challenge the Treasury in the courts by claiming they will be holding pieces of paper rather than actual gold bars. The Treasury is probably distressed an American could easily give a Canadian bank a Swiss address affording no way for the Treasury to check on who is actually breaking the law. We understand the margin requirements will be only 10% of the price, enabling the speculator to borrow up to 90% of the gold contract.

Meanwhile West Germany's Economics Minister Helmut Schmidt on September 14 formally rejected raising the official price of gold as a solution to international monetary problems. Behind this is his pique at the prospect of those with the largest amount of gold benefiting the most from such an increase, while countries like his which have accepted so many U. S. dollars would be left holding the bag (so to speak).

September 22, 1972

* * *

Last year at this time when the price of silver was $1.35 we wrote, "The price of silver appears to be in the same position as was the price of gold last year when it dropped to $35 an ounce, its floor. There is no way to go but up. We are waiting for more positive signs before becoming bullish on silver for the first time in six years."

It looks as if silver has started a long bull move. Our favorite in the group is *Hecla Mining* (HL - 15⅛).

For those with a more speculative bent we recommend *Fresnillo* (FRE - 15) which we have written up many times in the past, and fortunately took profits at higher prices.

October 13, 1972

* * *

The South African Mining Index is in a Downtrend as is the price of gold, and we would expect in the near future the price of gold will get down to the $55-$60 level. This will be part of a normal and healthy Correction of the enormous advance from $35 to $70.

Another reason gold stocks have been soft lately is they are contra-cyclical: golds move opposite to the general market trend, and the stock market has been moving higher lately.

On November 15, 1972, gold futures trading began in Winnipeg, Canada at a brisk pace estimated around 70 contracts. The trading unit is a 400 oz gold bar which U. S. citizens cannot trade under the same outrageous provisions which prohibit their owning gold itself. Closing prices were $62.55 for gold delivered in January 1973, $63.65 for April, $64.75 for July, $65.70 for October, and $66.51 in January 1974. The minimum amount of cash needed for the 400 Troy ounce unit graded 995 parts per one thousand of fine gold will be $2,800. The contract is worth around $25,000 at the current London price around $62.80 an ounce. The broker's commission will be $50 a contract for the public. This is an historic move and one which we predict is the fore-runner of a similar exchange somewhere in the United States sometime in the future, hard as that is to believe now.

November 17, 1972

* * *

A unanimous sub-Committee of the Congressional Joint Economic Committee recommended leading financial nations permit the sale of monetary gold on the free market if they want to, although not to buy it. The aim is to depress the price of gold, another in a long list of desperate attempts by governments to do so since we first became bullish on gold in 1961.

The question is, who is going to be selling gold in exchange for paper money these days? Even the United States, which postures about how it wants to phase gold out as a monetary reserve asset, is clutching onto its gold with all its might. As it should. Then, who is going to be the seller?

November 24, 1972

* * *

SHOULD YOU LEAVE MONEY IN SWISS BANKS?

If you mistrust dollars and want to leave them in a strong currency, why not Swiss francs? They are backed *over* 100% in gold, so even though as an American you cannot legally buy gold itself, here is a loophole which also earns you interest! This is an interesting alternative to U. S. government bonds.

Now that fanatical American devotion to a $35 gold price has been smashed, and the phrase "barbarous relic" is heard less often, we think sooner or later governments will say "Why not?" to a much higher gold price. Perhaps a deal has already been set to raise the price of gold, which would help Russia buy U. S. goods. If so, it would solve American monetary problems for years to come, and solidly reestablish the dollar as a prime mover in the world.

How high do we expect to see the price of gold go ultimately? We would not be shocked by figures close to $200 an ounce. Mark that down as a definite prediction.

[M]any mines are taking advantage of the higher gold price to work marginal ore with a view to extending the lives of the mines. There is little evidence as yet previously uneconomic areas are being brought into production, and when that happens the demand/supply equation could be shifted in the other direction. In fact, gold might even be mined in the United States again and this would be of enormous help to America. We do not expect any politician to agree with us on this, but that's the way it's going to be some day, mark our words.

December 1, 1972

JANUARY 1, 1973: DJI 1031.68; ASA - 11⅝₁₆

* * *

The South African Mining Average is quietly sneaking up towards the challenge of its all-time highs, and many gold shares are reversing to the upside. Apparently our major study on gold shares on December 1, 1972, caught the recent Bottom quite nicely. Gold action suggests there will be another monetary crisis late in the first quarter. We still recommend all large portfolios take at least a small position in the gold shares on our recommended lists.

January 12, 1973

* * *

The Smithsonian Accord reached on December 18, 1971, as we predicted has not worked in its intended purpose of bringing trade and payments among major trading partners into approximate balance. In fact, the large industrial nations have never been more out of line in their trade and balance-of-payments, aside from probably contributing significantly to Europe's skyrocketing inflation rate. Conceivably, the

Smithsonian Accord is beginning to distort trade patterns and is maintained only through rigorous foreign exchange controls. For all of 1972 the U. S. deficit reached a staggering level of at least $8-$10 billion, an amount nearly equivalent to our entire gold reserves!

The Dines Letter still believes the only lasting solution is for currencies to float freely against each other.

There is no change in our long-standing opinion Europe is financing colossal trade deficits by the United States and that it must eventually reach the breaking point. It is only the fact the U. S. has around $10 billion in gold bullion that people accept the U. S. dollar at all, for otherwise it would be backed by paper only and would be ranked with the Russian ruble. We still think the official rate of $38 is ridiculous and only the free price of gold on the open market counts. At that level the U. S. already has $20 billion in gold against overseas debts of $65 billion. Therefore, if the price of gold tripled from around these levels to $210 an ounce the dollar would be solvent and it could be made convertible once again. It sounds like a radical prediction, but so did our predictions the price of gold would go up at all, when we first made them a decade ago.

January 19, 1973

* * *

On September 13 Rueff spoke before the Joint Economic Committee in Washington and suggested a "Marshall Plan" for the U. S. consisting of a worldwide upward revaluation of gold, with all windfall profits to be loaned to the U. S. to sop up the huge $65 billion deficit overhanging the monetary world like the sword of Damocles. The official gold price he envisions is $65–$70.

We objected his plan made no provision to force the U. S. to end its deficits, and therefore the official price of gold would be whatever the free market price is, even up to $200 an ounce. Rueff is known to us as favoring fixed exchange rates between currencies, so we asked why not let all currencies float against each other? He appeared most unhappy with these pointed questions, and after commenting there would be too much uncertainty if "The Dines Plan" were effected, rather abruptly ended the interview. Considered a radical by some in his country, we found him right but too cautious and conservative.

The Dines Letter believes in total reform by sweeping away the old rules and letting the free market determine the price of gold and all currencies, with gold treated as the *common denominator* of all currencies. We've been a voice in the wilderness for years on this subject, but the future probably will agree with us. This week Switzerland joined Canada and Britain in illegally floating their currencies under present international rules.

January 26, 1973

* * *

[T]he U. S. dollar is again the buffeted, unwanted currency on world markets. Even West Germany raised its barriers to the U. S. currency over last week-end, with speculation Japan might be forced to upvalue its yen against the dollar once again. The root of this is West Germany had the largest trade surplus in history while the United States experienced its widest trade deficit. Also unwritten is that the U. S. deficit will decline this year but will still be enormous, perhaps in the $8 billion range as against last year's $10 billion according to the Morgan Guaranty Trust Company.

Fuel was tossed on the flames when Herbert Stein, Chairman of the President's Council of Economic Advisors, suggested the yen and mark be allowed to float upwards which would amount to another devaluation of the dollar, and he also suggested there be a special import surcharge much like the one which was in effect briefly in 1971.

How amazing nobody agrees with us that all currencies should be allowed to float against each other, particularly in relationship to a free gold price, and yet *that is what the world is stumbling towards*!

February 9, 1973

* * *

[T]he U. S. dollar was devalued this week. There is no change in our long-standing prediction these crises will continue to recur until all currencies are floating against each other.

This week's action corrects some relationship between the dollar and other currencies, but it does not do anything substantial for the relationship of all paper currencies with the ultimate store of true value, gold.

The price of $35 an ounce was set in the Gold Reserve Act of 1934, and lasted until the December 1971 devaluation up to $38 an ounce. Now the price is $42.22 an ounce, which is just as ridiculous as the previous price was because the free market price is now at an all-time high of $75.93 in Hong Kong! Can you imagine grown men playing games like this with a commodity at completely unrealistic prices while the free market price is much higher?

Higher prices because of devaluation will sneak into the United States in subtle ways in some cases. For example, if there are imported electronic components, or foreign oil, such raw materials will radiate out in a subtle way throughout the economy. It's also going to hurt American tourists who want to travel overseas, and as usual the tourist bore the brunt of it with many foreigners refusing to accept dollars at any price.

On December 18, 1971, President Nixon hailed the Smithsonian Agreement as "the most significant monetary achievement in the history of the world." At the time, we scoffed at that assertion, and this dollar devaluation proves we were correct.

France, Austria, Australia and Hong Kong reflected the full 10% devaluation sought by the United States. However, Italy decided to let the lire float which angered the French who are opposed to floating currencies. Portugal set a new dollar parity 6.4% higher than her old one, not giving the U. S. the full 10% trade benefit it was seeking. Spain, Sweden, Norway and Denmark are expected to make similar moves soon, with varying rates. Greece, Israel, and Mexico are determined to devalue in terms of gold just as has the U. S., therefore their trade with each other will not change.

Since around one-third of our oil comes from overseas petroleum, electricity rates will eventually rise because they depend to a large extent on petroleum imports. The same applies to users of gasoline, home heating fuels and asphalt shingles. This is the time to buy a foreign car, foreign art, antiques, coins, and stamps. Watch for American tourists to buy less, stay at cheaper hotels, eat in less expensive restaurants, particularly if airline fares are adjusted for this devaluation.

Our recent bullishness on silver came home with a vengeance as silver prices soared in heavy trading up to the $2.24 level.

February 16, 1973

* * *

What a fantastic week! The price of gold soared above $90 an ounce yesterday, just about bringing home our long-standing prediction that the price of gold would soar well over $100 an ounce. Gold mining companies are going to start reporting some really fantastic earnings and we are astounded the contumacious gold shares have not really exploded on the upside. We think this is one of those rare opportunities of a century where killings can be made even by Johnny-come-latelies getting into gold shares at this time. See our basic gold Letter of December 1, 1972, which caught the gold shares right at the Bottom!

Inflation raised its ugly head again, although we still think it is the gold crisis rather than inflation which is bothering the market. Not only did the cost of food leap between 2% and 3% in January (the largest monthly rise in 20 or 25 years) but Arthur Burns, Federal Reserve Board Chairman, commented the economic expansion has entered a "more sensitive phase" and that expanding demand for goods and services could begin to pull prices upward to reinforce cost push pressures. Burns added there is need for greater moderation during 1973 in the provision of new supplies of money and credit, which is not good for those who are looking for an expansionary policy.

February 23, 1973

*   *   *

Anyone reading the newspapers these days is aware another gold crisis is under way. As we go to press, the dollar is being pounded heavily on overseas markets while the price of gold soars, and we are confident this is no surprise to long-time readers of *The Dines Letter*.

In the greatest rip-off since the Mississippi Bubble, the tripling of the price of gold in the last few years has reduced by 66% the real value to all the people who hold U. S. dollars. In terms of the amount of gold those dollars buy, it has been reduced by two-thirds since 1970.

Some feel this rise in gold is a flash-in-the-pan. That is what they have been saying for years, and we continue to disagree quite vehemently. Instead, we urge you stock up on any items containing gold you were planning to acquire in the near future because they are going to be a lot higher-priced next year. The price of gold is up to stay. It has been held at the preposterously low level of 1932 prices for decades, and now it has blown its top. Get out of its way.

There are two problems. First, the overhang of $82 billion in Europe, representing our past deficits, needs to be wiped out in terms of gold, and the current rise in gold prices is doing just that. When gold gets up to $200, Nixon can then make the dollar convertible into gold again and tell everyone to go whistle for their value.

The second facet is the halting of further U. S. deficits. Nixon is moving aggressively in this area, of course too little too late, but there are signs of tightening in international trade in that those who wish to bid against American firms must state what kind of government aid they are receiving.

We do not think the dollar needs to be devalued against other currencies any further. That is official Dines Letter *policy. However, we think all currencies should be upvalued in relation to gold itself.* There has just been too much paper printed since 1932, and it all needs to be wiped out *à la fois* by a massive increase in the price of gold. The party's over. If this is not done soon and skillfully by Nixon, the two facets of the problem we have outlined above will become confused, and competitive devaluations will cause such international chaos there could be a breakdown in international communications and a real depression. Isn't *anybody* up there listening to us?

We think the government's move to raise interest rates this week was a belated and feeble effort to make holding dollars more attractive. In the midst of panic, however, people don't look for an extra half-percent interest.

Some advisors are falsely recommending you take profits (even though they have been recommending that you sell short gold stocks for years) on the grounds that this has been discounted. Nothing could be further from the truth. *We think the really big gains are still ahead.*

On the February 22 Most Active stock list, there were three precious metal stocks: Western Deep Levels, Welkom Gold and Agnico-Eagle Mines. We have never seen golds on the OTC Most Active List before. It's historic!

March 2, 1973

* * *

Many advisors are now recommending you take profits on gold shares. No! We are still as much against it as we have been all along, because golds have not yet really fully discounted the tremendous earnings increases ahead. We show the same ruthless devotion to facts about the gold shares as we have the last ten years, so simply avoid the crowd which has been wrong so often on this particular group. Some are wondering how high gold can actually get, which is the same nonsense we heard at $40, $60, and $80. We *still* stand our ground predicting gold at $200. The world is now about to get much more competitive, and these competitive devaluations are simply a beginning. It's high time since competition will bring lower prices and tend to counteract inflation, which the former idiotic system of fixed exchange rates tended to encourage. So, this is the end of an era in as yet unforeseeable ways. By 1985 United States will need to import 50% of its oil, and at this level of the dollar's value it could cost over $30 billion annually. That means we will have to earn a surplus of $30 billion to pay for our oil imports, an obvious impossibility. This is a very serious challenge to Detroit, not to mention the pollution disruptions, and it suggests big changes are ahead.

On March 5 European finance ministers decided to keep their official currency-exchange markets closed all this week to help push the U. S. to act in this crisis. This means major currencies continued their uncoordinated "floating" this week by refusing to buy dollars to hold down the prices of their currencies, or to hold up the dollar's value. Thus, *the entire world is floating right now, an action we have been strongly recommending for over a decade, despite the fact everyone of political importance has disagreed with us.* France put heavy pressure on the United States to raise its official price of gold and make its dollar convertible, another prediction we have been making for some time will come true.

On Wednesday March 7 Representative Henry Reuss (D., Wis) said, referring to speculation the U. S. might start massive intervention with its cash reserves or by borrowing to support the parity of the dollar, was "an excellent way of squandering millions of taxpayers' dollars." Instead the U. S. should allow the currency markets to determine the real value of the dollar. Hurray for Representative Reuss!

On the same day Paul A. Volcker, Under-Secretary of the Treasury for Monetary Affairs, made the stunning slip that dollar holdings of foreign central banks and other official agencies are estimated at $70–$80 billion with another $20 billion or so in the hands of private parties abroad. With nearly $100 billion outstanding and only $10 billion in gold at Fort Knox the price of gold is going to have to go a lot higher to make these two numbers equal each other. Perhaps $200 gold will not be enough!

March 9, 1973

* * *

Busy meetings are now going on in important financial centers, so the crisis is not really quite over yet. If it is mishandled from here, the whole thing could break out again and wash away all the hastily constructed dikes. The Europeans seem to be holding out for something, although it is not quite clear what America can do to help at this point. They have ruled out the only solution which will really work, namely a radical increase in the price of gold and a free float of all currencies, which is why they are all going crazy trying to figure out a solution. Meanwhile, they are awash in a growing sea of dollars which take away their jobs on one hand and buy up their industry on the other. With the ogres of a disastrous trade war and depression lurking in the background, the people in charge are running out of time, and they had better wise up.

On March 12 West Germany announced the mark would be upvalued by 3% when foreign exchange markets reopen on March 19, if they do open on schedule. Then, West Germany, France, Netherlands, Belgium, Luxembourg and Denmark will link their currencies in a common float as a defense against the influx of dollars. Britain and Italy will continue to float independently and Ireland will follow Britain.

For the first time this century foreign exchange markets remain closed for the second straight week, a really unbelievable admission the people in power really don't know what the dickens to do. What could possibly be going on behind closed doors in such long secrecy? We hope they are considering returning to gold as the ultimate measure of paper money, but we really doubt they will stumble on this solution, since the gentlemen running the international monetary scene have been amazingly silent on this subject for a decade.

The Common Market countries will float their currencies jointly because if they are floated individually the entire work of a generation to develop a Common Market might just break up the whole shaky bundle. Nor did it help on March 8 when the Senate Banking Committee recommended sales of our reserve gold in the free market by the United States to "make it clear that there was an important downside risk to gold speculation." The same silly concentration on speculators, rather than the cause of all these gold crises, shows the abysmal ignorance of the people in charge towards the whole concept of international monetary finance.

On March 13 the Bank of Japan announced it would support the joint float of six Common Market currencies against the U. S. dollar and will keep its intervention to support the floating yen to a minimum when foreign exchange markets reopen. It seems the unwise policies of our monetary leaders are driving the world together as communism and war never did.

Then, on March 14 West Germany dropped a blockbuster, announcing the mark would from now on be defined no longer in U. S. dollars, but in Special Drawing Rights. (Actually this follows the lead of United States in declaring the new par value of the dollar in terms of Special Drawing Rights as well as in terms of a fraction of an ounce of gold.) When Congress officially devalues the dollar it will equal 0.828948 Special Drawing Rights. Since the Drawing Right is defined in gold, this is a long step towards moving back to a gold standard, another one of our long-standing predictions.

March 16, 1973

* * *

A new era dawned this week, and our prediction of freely-floating exchange rates has at last taken a long step towards reality.

The system of fixed exchange rates born in Bretton Woods 1944 is finished, as we have long predicted would happen. Perhaps now the world can follow the example of a country like Canada, which for year after year allowed floating rates and nothing terrible happened. Of course, central bankers and governments are unhappy about floating currencies because it diminishes their ability to control their domestic economies. This is of course correct, although there is something to be said for the beneficial long-term impact of the discipline of the market place on fiscal and monetary sanity.

On March 6, 1973, the *New York Times* wrote: "On the limited evidence of yesterday's performance of the foreign-exchange market . . . the private foreign-exchange market appeared to be able to function well without governmental intervention." The discovery of the free market is always an exhilarating phenomenon.

Then, on March 16 the *Wall Street Journal* had a lead editorial likewise signaling a shift in official thinking towards what *The Dines Letter* has been advocating for years: "Paradoxical though it may seem, the adoption of floating rates by all the major trading countries is likely to bring a greater degree of tranquillity to the international monetary system than it has seen for several years."

Then, last week-end, the sharply higher short-term interest rates in the United States began to attract some dollars back to America, causing the dollar to firm up in the free market. We would guess that behind closed doors, even though denied by the United States, America had agreed to foster higher interest rates in this country as part of an international financial package. Then, with Sweden and Norway announcing they would join the float of the six common market currencies, people like Sam I. Nakagama, Chief Economist of Kidder Peabody, said flatly: "The monetary crisis appears to be over." We don't think it is quite over yet, although the worst part has been surmounted, and the really big boom in gold mining shares is, we believe, still ahead. *Even this floating system will not work unless the dollar becomes convertible into gold.*

March 23, 1973

*   *   *

One market letter this week wrote: "The days of the goldbugs are numbered and if you have profits as the result of the 'gold bubble' you had better start salting away your rewards from speculating on man's irrationality." Thus, the same hostility to our bullishness on gold as we have endured for the last decade.

While money managers wring their hands and piously hope for some kind of settlement in time for the annual meeting of the International Monetary Fund in Nairobi, Kenya this September, nobody is allowed to use the nasty, filthy, rotten four-letter word called "gold." Unfortunately, that is the only solution, and it is the only one excluded from their discussions.

As if to celebrate this new confidence the price of gold soared to a record high in London on March 26 at $90 after Samuel Montague & Co. predicted gold "may easily" top $100 an ounce because of monetary uncertainties and other factors like Japanese restrictions on gold imports are about to be eased. We think the Japanese will be fools if they don't get rid of some of their dollars by buying gold with them.

Nonetheless, we hew to our position the price of gold will settle down at around $80 an ounce for awhile before the big push above $100, which we predict will leave flabbergasted those who have been putting gold down in recent years.

March 30, 1973

* * *

Also significant this week is the gold rush in Tokyo when it became legal at the beginning of April for any Japanese to trade, use or hoard gold. We have not the faintest hope in the world monetary authorities will take this as an indication people really want sound money, and will continue to blithely ignore such examples.

April 6, 1973

* * *

Gold shares soared this week after a *Wall Street Journal* article predicted sharply higher earnings.

Meanwhile, back at the ranch, the unbelievable news of a $10 billion loss in the first quarter was the second highest quarterly deficit on record, according to a preliminary estimate of economists at Morgan Guaranty Trust. One has to go back to the $12 billion deficit in the third quarter of 1971 to find a similarly shocking loss. To put it into perspective, the first quarter loss this year equals the entire gold holdings of the United States at the official price! At that rate of loss, who can believe people will still be accepting U. S. dollar bills overseas much longer?

So far, Nixon has handled the gold situation fairly well, considering what he inherited. However, unless the dollar is made convertible promptly there will be another crisis which could bring down the whole house of cards. We have been predicting the current phase of the international monetary crisis would calm down temporarily, and it appears to be coming true, but convertibility is urgent. We don't really expect the dollar to be made convertible, since nobody in government has been listening to us for the last ten years and we certainly don't see any reason for them to change their ways now.

If the United States could only stop losing money internationally on their balance-of-payments there is a chance a rise in gold price could yet bail the dollar out. After all, we only have $11 billion at the official rate, but at the free market rate our gold holdings are closer to $30 billion. It will have to triple again to get anywhere near our overseas liabilities of $100 billion, which is why we are looking for a much higher gold price. Don't ask anyone if they agree with us, because we don't expect any more agreement than we have gotten in the last decade.

Meanwhile, the West German Central Bank said on April 23 "major world currencies are likely to continue their float against the American dollar until the United States balance-of-payments approaches equilibrium." The Germans said this was unlikely before 1974.

April 27, 1973

*   *   *

[T]his is a bear market of unknown depth, length or cause.

Gold shares shook off their correctionary tendencies and spurted to new all-time highs when the price of gold skyrocketed to an eye-popping $124 an ounce in Paris while the dollar was falling to record lows. We now envision gold prices over $400 an ounce. Gold and silvers are the top two groups in in our group analysis (see your *Paflibe* and *Crosschart* chart books) for typical general bear market action. With Arabs dumping dollars and fleeing into gold, they're beginning to develop the real power the U. S. lost long ago due to its unwise policies.

Meanwhile, gold shares are virtually alone on the new high list.

We envision disaster unless Nixon immediately raises the price of gold to $400, makes the dollar freely convertible into gold, ends foreign aid, and makes a surplus in our trade balance a matter of top national priority.

Until then, keep buying gold shares.

May 18, 1973

*   *   *

All they are talking about in Europe is Watergate and the rising gold price, which soared to a mind-boggling $128.50 an ounce in Paris and $110 in the much larger London bullion market. Several times your editor overheard people in bars trying to connect the plunging dollar with Watergate. We reject such thinking out of hand because it does not comprehend the real nature of this gold crisis as we have been describing it in recent years. The *real* problem with the dollar lies in the fact that it is not convertible into gold, which means the U. S. can continue to pile up dollar surpluses overseas. Foreigners can do little with these dollars because the United States' products are not competitive, so the surplus dollars are going into gold. This dumping of dollars is causing a continuing series of devaluations, which is not the real problem. We think the United States' dollar is fairly valued in relation to other currencies, but that all currencies are radically out of whack with the price of gold. The dollar is now getting *underpriced* in relation to other currencies, which is bound to cause severe international strains, and which in turn could lead to a depression if gold is mishandled by ignorants who do not really understand the situation.

We have been afraid to use the $400 figure for gold in print, but now that the price of gold is over $100, we can be more candid.

To make the dollar convertible again, there must be enough gold to meet all demands from those who hold U. S. dollars, which comes out to $100 billion right now. Therefore our current hoard of gold, worth around $10 billion at $42.22, would have to go up in price by around ten times to match our liabilities. That is how we get a figure of over $400. There is most emphatically no change in our prediction that until the dollar is made fully and freely convertible into gold again these monetary crises will continue, and what annoys us in the current solution is not solving the problem but instead, will create others. The U. S. dollar is not overvalued in relation to other currencies, and refusal to make it convertible will force the dollar even lower and could trigger some rather ugly international economic consequences.

May 25, 1973

*   *   *

Since prehistoric times only 100,000 tons of gold have ever been recovered from the earth. If all the gold that was ever mined was melted into a single cube, it would fit inside the baselines of one baseball diamond to a height of 90 feet. The total annual production of gold averages 1% to 2% of the existing stock of gold. Precisely why it is so valuable. Gold doesn't have to be used as a monetary medium. Anything scarce and indestructible could be used. Platinum, diamonds or even neon. But people accept gold, and it's pretty difficult getting hold of new gold, which forces the printing presses to bear some relationship to a tangible reality.

For this and many other reasons we think the gold share boom is *about to begin*, and if there is a slight profit-taking dip which begins around June 6, exploit it by buying golds to the hilt. Or to the gilt. Most advisors, after long pessimism towards gold shares, now admit they are good but that the stocks are "too high." This is typical of all new bull markets in any commodity or stock. We do not believe golds have fully discounted the price increases we see ahead.

We think gold shares will catch up and have a furious rise which will devastate short sellers and leave most onlookers breathless.

June 1, 1973

*   *   *

There have been a number of speculative manias recently, such as in Japan, Singapore, and Hong Kong. Looking at individual charts in those markets shows they are just about where Wall Street was before the Crash of '69; they don't know about charts, but we suspect they are going to get a rude lesson in the next few years. Commodities are another area which has gone wild.

The future will someday explain the most puzzling phenomenon of all, how there could be such a boom going on, replete with a complete sellout of farmers' commodities, and yet transportation earnings are *lower!* How come the earnings of container companies collapse in the middle of a boom? Is this some sort of weird, invisible depression?

LATEST ON GOLD
THE SINKING OF THE TITANIC

Be stubborn and tenacious. Those of you who bought *ASA Ltd* at 5 in 1961 and have so far made 16 times your money didn't do it by being easily frightened. Around the time of our gold seminar some spectacular second quarter gold mining company earnings will begin to be reported, *so we are looking for a big gold move this summer.*

Events have been unleashed by this gold crisis no man can foresee, but we predict 1973 will be looked back upon as their watershed. It could mean, for example, the beginning of the end of the era of inflation, what with food prices rising climactically and suggesting the beginning of the final phase of inflation—runaway inflation. Such final phases are always followed by a bust. The implications of this are overwhelming, since our whole economy is based on giving labor increases which are promptly erased via inflation. If workers can't get raises any more, there are going to be some fairly distressed people. Make that a topic at your next cocktail party, and just sit back and listen.

Reading *Le Monde* on May 17, 1973, we came across an article by Jacques Rueff. He was still looking for gold at $70 because "since President Roosevelt in 1934 U. S. prices have more than doubled." Presumably his reasoning is that since gold was $35 in 1934, it should be doubled now. He also thinks Europe should have a reverse Marshall Plan tying up the remainder of that $100 billion sloshing around Europe into long-term low-interest loans to the United States. He's desperate. He's trying to bail Europe out of its biggest ripoff since the Tulipomania and Mississippi Bubbles.

As we told Rueff in January, $70 is unrealistic. Time has borne us out, what with gold reaching an all-time high at $127 in London this week, which is a long way from $35, and nowhere near where we think it's going to wind up. With the price of gold now at $121.50, we think gold looks cheaper than it did a year ago at less than half its present price. This is because rising gold prices are now "respectable," and the Arabian sheiks are getting into the habit of switching the dollars they are getting for oil into gold. The Arabs have been stuck by two dollar devaluations, and now all they want to listen to is that solid yellow metal clanging on their Caddy's fenders. They fear for our runaway food inflation, increasing currency restrictions, and the Watergate-induced lessening of confidence in U. S. government.

Gold is the point of reference for all monies, its common denominator, as we have pointed out on many occasions. The final high price of gold depends on when, where, and how convertibility of the dollar is restored. Observing Washington over recent decades has given us pitiful little confidence they know what to do next. We even have our doubts they understand what convertibility is, many of them. But, gold is the real money, and it is now immobilized in the banks around the world at the preposterous "official" price of $42.22, at which price no gold is changing hands. It's the *real* money, which is why nobody is letting go. The trouble is, such gold held by governments is sterile and is not functioning as money because it is not changing hands. This must lead to chaos because gold is vitally necessary to the monetary functioning of the free world. Meanwhile, Europeans continue to panic over our $100 billion debt. What can be done with it, moan the Europeans wringing their hands? Will it be voided and a new U. S. currency backed by gold come into play? Something *radical* must happen fairly soon to solve this crisis. The U. S. dollar is now moving too low in relation to other currencies, because the price of gold is too low. After all, France will not permit the United States to sell Cadillacs in France at $1,000 in terms of French money. No way. Thus, *look for rapidly-rising trade barriers all over the world, and eventual unacceptability of the dollar overseas.* Hard to believe? So was our prediction the price of gold would move to $100. Let others scoff, but if you are a tourist pay for your trip in advance, right now. Also, *get some money out of the United States.* Those who listened to our advice and switched into German marks have made almost 50% on their money (because the mark was upvalued and the dollar was devalued) since then. It is not too late to get money out.

Meanwhile, back at the Winnipeg Commodity Exchange this week, gold contracts for delivery in July 1974 went up to $139.40 an ounce. Nonetheless, Peter M. Flanigan, Assistant to the President for International Economic Affairs, said on June 5 he was puzzled by the recent performance of the dollar. When Chairman J. William Fulbright asked about the dollar's decline abroad, Flanigan replied, "Frankly I don't understand why that's so." And, Professor Henry C. Wallich, of Yale, said in Washington before the International Monetary Fund last September, "If efforts to negotiate a new international monetary system should fail, if in some crisis national or international credit instruments should cease to be universally acceptable, world-wide belief in the 'intrinsic value' of gold, now buttressed by mounting industrial demands, might again restore gold as the basic world money."
Hallelujah.
Silver tends to run at approximately $\frac{1}{15}$ or $\frac{1}{30}$ the price of gold. Even if the price of gold falls now to the $90-$100 an ounce level, that's a heck of a lot higher than the price of silver right now. You can buy a bag of silver dollars now for under $2,000. Why don't you put a bag or two in a safe place, just in case.

June 8, 1973

*   *   *

Now that gold is beginning to force Americans to earn their way in the world, just like other people, perhaps a sign of the times was a notice we spotted on the way to work one morning which read "Why not bring a friend from Europe to the United States?" under which some enterprising American wrote "Because he eats too much."
The collapse of the dollar began to accelerate at such a rate, and the gold stocks were climbing so quickly, that even the incompetents who are running the international monetary system got frightened last week-end and came up with their latest "solution" to solve the gold crisis. Of course, it will fail like all the rest before it and there is absolutely no change in any of our opinions along these lines.
Just to give a slight example of how intertwined the gold crisis is here, we must keep interest rates high to attract dollars back to the

United States. If we keep interest rates low, dollars will rush out overseas for the higher yields and the dollar will collapse even further. Thus, the United States will be forced to keep its interest rates high for a long period of time, and it will incidentally create a credit crunch and a recession, or even possibly a depression. That is a function of high interest rates, and rates are high now not only because of the boom but because of the gold crisis overseas. Events are conspiring to gang up on the dollar all at once.

In recent *Letters* we have been asking, when the government runs at a deficit, who pays it? We don't just mean a budget deficit internally, but a balance-of-payments deficit where we import and spend more overseas than we are earning there. Who pays for it? The answer is, the government simply prints more paper to pay for it, and since there is more paper chasing the same amount of goods, prices must go up. This is a source of inflation, and every time you pay more for any item you get, you are paying a tax which is meeting the federal deficit! The collapse of the dollar, and the primacy of gold is forcing the United States to earn its way in the world. But, instead of becoming the breadbasket of the world, Nixon is limiting the amount of soybeans which can be exported (thereby additionally antagonizing Japan—what has Nixon *got* against Japan?). Meanwhile these asinine price freezes are solving nothing except that in addition to the recent wholesale destruction of baby chicks, it is bringing a like state to unborn piglets. Recently 39% of one corporation's sows were pregnant compared to a normal average of 5% to 7%. It will take well over a year to rebuild livestock breeding herds, which means we are in for some real food shortages. This is gold's way of punishing America for living beyond its means, and it's going to mean more hungry bellies here and fuller bellies for countries whose monetary managers knew what they were doing.

LATEST ON GOLD: THE ENDLESS EMERGENCY

Events are marching inexorably from crisis towards catastrophe as the monetary authorities lose control over the international monetary system at an accelerated pace. Nothing is being done to put a complete stop to U. S. deficit spending or to restore the convertibility of the U. S. dollar into gold; these are not even stated objectives of our leaders, so there is obviously no hope for a lasting solution.

Washington in the past said Europeans should use their accumulated dollars to buy more American goods. When they started doing it, America slapped on export restrictions because European demand was driving our own prices up too high. Food was coming out of American mouths to feed foreigners. Now the dollar is not only inconvertible in terms of gold, it is also apparently inconvertible in terms of soybeans.

Some feel gold is too valuable to use. To which we shout "Never!" Gold is the only thing valuable enough to back currencies. If it becomes too expensive for jewelry, that's too bad.

On July 7 Paul A. Volcker, Under-secretary of the Treasury for Monetary Affairs, ruled out any new formal devaluation of the dollar. "Washington's belief remains that action is not necessary and that the situation will correct itself in the market."

The U. S. ban on soybean exports is distressing the Japanese, and making enemies, which could undo all the good of foreign aid and the policy of helping our former enemies. The ramifications of the current international monetary crisis will be felt for decades, in our opinion, hard as that is to believe now. *The Dines Letter* remains among the very few really alarmed by the situation. While Volcker was saying "I see a turnaround coming" the *New York Times* of July 7 ran a headline "Money Exchanges in Uproar Abroad as Dollar Slides. U. S. currency declines to record lows for ninth day in succession." That is why over the week-end they decided to support the American currency, but no one knows how, when or why or with what. This is an effort apparently to keep the "speculators off balance." Meanwhile, on the same day the *New York Times* had an important article about a looming shortage of lox in the United States. An official of the Washington Fish & Oyster Company in Seattle confirmed foreign markets were forcing prices up and said the Europeans, especially, "take almost all we have." Of course, what else do you expect them to do with all those dollars they're drowning in? And nobody has even begun to consider the impact on our economy of that flood of dollars coming back into the United States, and its inflationary impact.

No paper in the world anywhere is a satisfactory substitute for gold, and no government official can make it so. How else could anyone illustrate the power of gold: despite the official policy of the entire world, every government, every army, almost every leading economist, with an official U. S. price of $42.22 per fine ounce, people wanting protection from inflation are willing to pay 282% higher for the free market gold price of $119.00. Threats to sell gold on the open market have always been a complete hoax. With runaway inflation, central bankers will certainly not exchange a safe asset like gold for rapidly-depreciating paper money.

Isn't it sad to watch a group of determined men shoveling excrement against the tide? If only we could sit down and talk to the international bankers of the world for about two hours. But they have never heard of us and certainly wouldn't be interested in what we have to say.

Governments are now hoarding gold. *Why shouldn't you?*

We award this week's booby prize to former Deputy Treasury Secretary Charles E. Walker for suggesting in Washington the U. S. should sell $1 billion of gold in the free market to help bring down the soaring price of precious metal. Gold is all the dollar has left! Selling gold will cause a temporary dip in the price of gold, and send it skyrocketing even higher later because the dollar will be worth that much less. We heard the same foolish advice when we were right on top of the silver crisis a decade ago as people like Roosa (who is partially responsible for the absence of silver in your coinage) recommended selling enough silver to overwhelm the speculators. Every time they tried that, the silver disappeared into a bottomless pit.

June 13, 1973

Well, here it comes. The British-owned Brown & Williamson Tobacco Corporation offered to pay $23 a share for all the common stock of Gimbel Bros. in a proposal involving more than $200 million.

This will help our balance-of-payments picture to some extent, but most of that $100 billion we owe overseas is owned by governments, and of course they cannot use IBM or Gimbels stock to back their own currencies. They must either have gold or a satisfactory substitute, of which there are none. Therefore, look for a new wave of foreigners buying American industry, just as we have been doing to them in previous decades. We suspect there is going to be a lot of nationalistic hullaballoo before this is all over in response to this new wave of foreigners finally getting something for those dollars they earned from us in recent decades. Less than one out of 100 financial institutions own a single share of a gold mining company. Can you imagine what will happen once they start taking a hedge position on the gold? Yet, Americans are so uninformed they still have not thought of asking their government precisely why U. S. citizens are *prohibited* from owning gold, if it is so unimportant.

This might sound like idle, theoretical speculation, but it isn't. Edson Gould's excellent *Finding & Forecasts* points out that since the beginning of the year the dollar has been devalued by almost 50%, and by 72.36% since the end of 1969! You don't believe it, you say? Okay, here are the figures.

End of:	$ U.S. Decline in Terms of Gold		
	London Gold Price P/oz.	$1 U.S. Would Buy (oz.)	Depreciation $ U.S. in Terms of Gold
1969	$ 35.200	.02840	$1.00000
1970	37.375	.02675	.94190
1971	43.625	.02292	.80704
1972	64.900	.01540	.54225
6/5/73	127.250	.00785	.27640

At least, if you had held gold, you could have been protected, but holding paper money you were absolutely at the mercy of the lunatics who have been running the gold show the last few decades.

The press continues to display unbelievable ignorance about the gold situation. For example, on May 27, 1973, the *New York Times'* John M. Lee wrote, "The price of gold, even above $100 an ounce, is an irrelevant curiosity, interesting, but unimportant."

June 15, 1973

*   *   *

Most sources of advice remain quite negative on gold, so obviously the Top of the gold move has not yet been seen. We still believe the coverage on golds we have been giving you in recent years is simply unavailable anywhere else in the world. Many have been flabbergasted by the soaring gold price, except us. Our message loud and clear is still that the prices of gold and silver are going to go *much* higher. This is the time to buy gold jewelry, and as we said on June 16, 1972, if you use gold or silver in your business, *stock up*. Those who did last year should certainly be gratified now. The rest of us American slobs who are barred from owning gold because of some stupid laws designed to cheat us, can't do much more than buy gold shares.

Otmar Emminger, Deputy Governor of the West German Central Bank, and a man at whom we have scoffed many times, is at last beginning to see the light. He said in a speech on June 17 that the world might have to live with floating currency exchange rates indefinitely to avoid the flaws of the old system, at least until the "key currency of the system (the dollar) is firmly established as the stabilizing anchor of the system." He said the old system of fixed exchange rates had been perverted "from being an instrument of discipline on deficit countries to one forcing monetary debauchery on surplus countries." Welcome to bright sunlight, Otmar.

June 22, 1973

We haven't seen this anywhere else in the world in print, but we think the widely-discussed energy crisis will benefit gold more than anybody else. After all, the international oil companies are at the mercy of foreign governments. Yet, when these foreign governments get dollars for their oil, particularly the Arabs, the first thing they do is throw it into gold, which wends its way back to the gold mining companies! Therefore, *gold could well be the primary beneficiary of the oil crisis.*

Minora Sagawa, Chairman of Minora Securities Co., one of Japan's leading brokerage houses, predicted Japanese investors would put $4 billion to $5 billion in American stocks in the next few years, and a continued growth of direct Japanese corporate investment in the United States. Japanese stock investors now hold some $1 billion in American companies, despite the loss from two devaluations of the dollar in combination with upward revaluation of the yen. Thus the wave of foreigners buying America has commenced. The race between Europe and Japan buying us out, and gold moving above $400 and dollar convertibility to stop them, is on!

When the U. S. April trade balance moved up, we pooh-poohed it as an aberration. Sure enough, the U. S. trade balance retreated deeply into deficit in May. Imagine the spectacle of even two devaluations not enabling U. S. trade balance to move into the black! It looks like another devaluation of the dollar is beginning to lurk in the wings. Protect yourselves.

June 29, 1973

JULY 1, 1973: DJI 880.57; ASA - 24½

LATEST ON GOLD: KEEP THE FAITH, GOLDY

Two weekends ago there was a meeting in Basel, Switzerland at a moment of great panic, when the bottom was falling out from under the U. S. dollar. *Our sources inform us the French had threatened to press for a common market gold-backed currency, which would in effect cut the Gordian Knot and raise the official price for gold radically higher.* It's the only way Europeans can stop this insane game Washington is playing, but once again the Europeans backed away from the only solution which could work, probably due to incredible U. S. pressure. Instead, the group came out with a "compromise" in which Western monetary authorities would intervene in exchange markets to halt the collapse of the dollar. Since Washington is trying to phase gold out of the monetary system, it panicked at the prospect of a European gold currency, which would surely replace the dollar as the official unit of all international transactions, and which should have been done years ago considering Washington's dogged deficit spending.

According to our informants, the big bone of contention was who was to take the risk of attempting to support the dollar. It was decided United States would bear half the risk, and the lender would pick up the other half. This enables the United States for example to borrow marks, and sell them on the open market in an effort to lower the mark and bring it into closer relationship with the "undervalued" U. S. dollar. The amount of short-term credits was raised from $12 billion to $18 billion, which is a pretty potent weapon.

Our answer is that nothing will help. If Washington has really taken the desperate step of committing any of our shrunken gold reserve in an effort to placate the Europeans, when that is lost *there will be an even weaker dollar.* Their policy is so shortsighted as to be beyond credibility. They should throw in the towel, and raise the price of gold radically, to at least $400 an ounce, and remove all restrictions of any kind from gold. A few months of supply and demand at that rate, and currencies will once again go back to very narrow daily changes. Until then, efforts to cure the disparity between individual currencies fail to grasp that all currencies are overvalued in relation to gold. The gold price has been held at artificially low levels for decades, while the amount of paper money has ballooned beyond belief. We have predicted continued currency crises until this gap is closed.

Then on July 15 Sabena Airlines of Belgium ordered 10 American airplanes from Boeing instead of 10 French ones because U. S. devaluations have made our planes 30% lower than the European ones! This comes on top of the French industry's disappointment over cancellation of the Concorde. Back in 1969 when the dollar was worth 5.5 French francs instead of the present 4.1, the prices of the two planes would have been roughly equivalent. Yet, the French, as is true of most Europeans, have not yet even dimly begun to grasp that the way to outsmart America is to buy out their stock market. *Forbes Magazine* points out in an editorial this week that Britain and France are spending around $2.5 billion developing the supersonic Concorde, which aside from their own captive airlines has as yet no customers and in terms of economics will probably never fly. *Yet, Forbes pointed out for less than $2.5 billion, they could buy 51% or absolute control of every American major aerospace company!*

If the people in Washington knew what they were doing, they would raise the price of gold immediately before the Europeans get smart enough to do precisely what we have long been warning of, and which now *Forbes Magazine* has begun to pick up on. There's got to be somebody in Europe in a high enough position who is going to read one of these things one of these days, and that's going to be it.

July 20, 1973

*   *   *

For years we have been suggesting this decade would be called the Sickening Seventies, and we still think it is not worth your selling out your golds and silvers yet. Sit tight, be patient, and continue to have faith in us as you have over these many years.

With the prime rate now at 8½%, which is the price banks charge their best loan prospects, 9% is now being discussed widely. If 9% is the minimum loan charge, you can just imagine what ordinary folks have to pay to get a loan these days. We still predict, as we did at our recent Seminar, a prime rate between 10% and 15%, and if that doesn't cause a Crash we can't imagine what on earth could. If you were planning on selling some property, accelerate your plans.

Now, you are undoubtedly going to ask us whether it pays to put your money in any of these instruments. We are as bearish and pessimistic and negative on bonds as we have been for we don't know how long. In our opinion, inflation is going to wipe these fixed-income instruments off the face of the earth. In terms of gold, it has almost happened in the last two years alone. The more money they print, the less your fixed income is worth.

The Federal Reserve Treasury confirmed Fed intervention in foreign currency markets since July 10, and it was speculated central banks soon will sell gold in the open market. This did not help the dollar, which continued to lose ground, and soon must lead to a large balance-of-trade surplus, which can only antagonize foreigners using the dollar even more. Meanwhile, the price of gold is digesting its tremendous runup and would certainly be entitled to a Technical Correction back towards the $100 an ounce level. This would in no way change any of our opinions, and in fact would make us more bullish on gold than ever. We know few people agree with us, but that was true when the price of gold was $35, and we turned out to have been right.

*   *   *

After we went to press last week an 8¾% prime rate was initiated by New York's First National City Bank, making our predictions of a 10–15% prime rate not so ridiculous as they first sounded. And anyone who can envision prime rates that high (which means most other interest rates are going to range in the 20–30% area) should wonder how any boom could possibly survive the withering blast of such a hostile environment to profit.

This is one of the reasons we think there is a tremendous disaster ahead for the banking system, because we don't believe people can pay these kinds of interest rates and yet stay solvent.

August 3, 1973

*   *   *

Few people understand how the depression on Wall Street, skyrocketing food prices and interest rates, all kinds of scarcities throughout the economy, rising gold prices, are all interrelated. Here is why they are.

The ballooning quantity of paper monies in recent decades in relation to a stable gold price has created a tension, a torque, which has left the price of gold grotesquely underpriced. The United States, seeking to discard gold, has tried artificially to force the price down to zero for decades, despite gold's obvious triumph when it soared above $100. These efforts have by no means been terminated. Since the United States is running its printing presses faster than anybody else, and running bigger international deficits than anyone else, there are more dollars piling up overseas than can possibly be spent by foreigners here. Therefore, with supply and demand of paper being what it is, the excessive quantity of paper dollars overseas has been forcing the dollar lower in relation to other currencies. That the dollar has plunged too low and is therefore "undervalued" is irrelevant. So, foreigners begin erecting all kinds of barriers, such as two-tier currencies (France will take dollars from tourists, but not from banks) and tariff walls. Since the United States must thus take some of those overseas dollars off the market, the best way is to raise U. S. interest rates high enough so that foreigners are tempted to move dollars to the U. S. despite the possibility of another dollar devaluation. That is why we have been predicting a prime rate between 10% and 15%, and you will note the new 9% prime rate is already spreading. Skyrocketing interest rates (a desperate and vain gambit to retain dollars here) and the resulting "credit crunch" will cause a bear market just as surely as you are breathing. That is why Wall Street firm after Wall Street firm is going out of business, as people simply cannot afford astronomical interest rates to finance their margin accounts, and must therefore liquidate. That is also why many high-quality bonds are yielding 9%, and Treasury bills were sold at a record average yield of 9.802% this week, "by far the highest yield of any Treasury security ever sold" according to a Treasury spokesman. With such high yields available, money is further attracted away from Wall

Street's common stocks, and a wave of bankruptcies on Wall Street appears inevitable.

According to the *Holt Investment Advisory*, 277 Park Avenue, New York, the cost of short-term money exceeded that of long-term capital only five times: 1920–21, 1928–29, briefly in late 1959, 1966, 1969–70, and now. On each of the previous occasions, without exception, there was a substantial drop in stock prices.

In addition, because of the stupid refusal of the United States to raise the price of gold radically, the collapse of the dollar overseas has made U. S. raw materials so cheap in relation to other sources of supply that the entire international trade pattern is out of whack.

That is why, despite record grain crops this year, farm bins are emptying.

Corn traders bid prices to an historic high of $3.0125 for a bushel of corn. This has to be translated into much higher meat prices, despite the government's belated removal of controls on beef prices. Actually, this might be a blessing in disguise for those who like their meat rare, because in a few weeks getting *any* kind of beef will be rare! And, aside from the widely discussed gasoline shortage, there is now a shortage in cotton denim.

Also behind this inflationary surge is the feeling of desperation by those who own dollars overseas (since they have not been able to get gold for those dollars since August 15, 1971) that they had damn well better spend those dollars really fast or they'll wind up using them for wallpaper. Thus they throw these dollars back in our faces as quickly as they can, only adding additional fuel to inflationary commodity pressures. Do you see how all these facets are interrelated, and how we have tied the whole package together for you?

With all those dollars flooding back from overseas, the control of the money supply is not even in Washington's hands any longer! The mischief this will distribute throughout our economy is incalculable. This is another reason Washington is completely out of control of the situation, and particularly because they do not understand it all goes back to gold. Instead, Washington fumbles with half-measures such as price and wage controls which completely miss the point.

August 10, 1973

*　　*　　*

We have not changed our opinion about the long-term bear market, and we are still awaiting evidence to the contrary. The Crash of '69 squeezed the water out, second phases of bear markets are typified by amazingly "underpriced" stocks, and final stages of bear markets feature panic liquidation of debts, some of which has already been seen in Wall Street margin accounts.

[T]here are more problems coming. The oil in the United States, which will be paid for with ever-depreciating dollars, will cause an incredible balance-of-payments deficit which is now being concealed by skyrocketing agricultural prices; rising wheat prices are as much a windfall for us as oil is to the Arabs. The fiat dollar could lead to tariff wars or even shooting wars. Once again we are told the gold crisis is over. *Nonsense.* Nothing has changed.

We had been looking for a dip in the price of gold below $90, and this week it actually got down to $89.50.

At any rate, the $463 million balance-of-payments surplus in the June quarter, the first quarter in the black since 1969, is not enough to begin earning back the unbelievable sum of $200 billion accumulated overseas from past Washingtonian squanderings.

We hope we will be forgiven if we do not go along with these government figures. Considering the long record of incorrect predictions and duplicity in the ways these deficits are measured, we have little confidence there was actually a small surplus. We predict some day the truth will come out the government often did not tell the truth about these figures, and the picture was much worse than it appeared.

There is no change in our long-standing predictions the U. S. dollar will be toppled from its throne and the currencies heavily backed by gold will take over; this includes the Swiss franc, the French franc, a convertible Russian ruble and the South African Rand!

We are still standing by our daring prediction of an all-time high in the prime rate above 10%, and possibly near the 15% level. This is why we are looking for a massive bust in the economy as the impact of these chillingly high rates begins to permeate the cost structure of American business. In July, the nation's mutual savings banks had a net deposit outflow of $600 million, the largest "disintermediation" in the history of the industry.

You are living through interesting times.

August 17, 1973

*   *   *

On August 18 a key Congressional subcommittee concluded that in current circumstances, floating exchange rates are "the best available alternative and are clearly superior to fixed parities." The Joint Economic Committee, headed by Representative Henry S. Reuss (Wis., D), unanimously concluded that it would be "a mistake" for the U. S. to try to establish fixed exchange rates again, at least until a reformed international monetary system is in operation "in two or three years." The last comment reveals a fatal weakness in their attitudes. These gentlemen think they have lots of time, whereas actually we think the latest swap agreements did not buy more than a few months. We are looking for another gold crisis to strike sometime this autumn, right about when our next Seminar will be going on. We will then be able to be right on top of events as they occur.

August 24, 1973

* * *

We had expected a sharp decline in the price of gold as the governments of the world threw everything they had at gold before and during the Nairobi International Monetary Fund's Conference. The fact that the price of gold has actually begun moving up again in recent weeks is probably the market's judgment that there is *no solution forthcoming*.

Precisely what they are looking for is *gold*, whose function it is to regulate and prevent the kind of imbalances which indeed broke down the old Bretton Woods system! We are quite confident that the gentlemen assembled at Nairobi, at great expense to their respective governments, will not stumble across the correct solution, namely a radical increase in the price of gold to at least $400.

Then watch the Arabs. This week's nationalization of 51% of the oil companies in Libya could cause a major panic. The Arabs, having abandoned terrorism, are now turning to oil as a weapon. They are playing with their lives, because the hand on the oil spigot had better be well-armed. This is the phase of the gold crisis we have warned about concerning the stage after tariff wars, namely shooting wars. The Arabs are not fools, and of course understand that this is a possibility. So, we now go on record as predicting they will spend a fortune on armaments in the coming period, and they will take the further step of avoiding leaving large cash balances in any potentially hostile currencies. If you were an Arab with a large account in New York City in U. S. dollars, wouldn't you be worried? You know, in wartime balances tend to be seized. Nor do they even want to worry about devaluations. Thus, we predict the Arabs will begin transferring their enormous paper holdings into gold.

If you remember our predictions in recent years, we felt the collapse of the dollar overseas would then be followed by very large U. S. trade surpluses. If this were permitted to continue, we could of course earn back all the billions we have blown overseas in recent years. For example, foreign aid, which cost us $167 billion since World War II, approximately the same amount by which the dollar is bankrupt.

Unfortunately, there is no way countries can tolerate running at such losses and we think tariff wars (already beginning), trade restrictions, currency limitations, and even shooting wars could result. This will extend to competition in interest rates, and governments are going to choose between a higher gold price or ruinously skyrocketing interest rates above 15% which could lead to a devastating world depression.

Now, you might wonder, why the hell doesn't the government just raise the price of gold and shut *The Dines Letter* up? Well, honestly, we think they would if they really understood the situation. That is why we have no hope the gold crisis will be settled without an economic smash.

September 7, 1973

*　　*　　*

[W]e are now one of the few gold bulls left, the rest having dropped out at the first sign of adversity. We think the chronic eruption of gold crises once or twice a year in the last decade will recur, probably after the Nairobi conference later this month.

Many commentators clutched at straws to indicate interest rates have crested or are already easing. We see no evidence of this, and our chart of bond yields is in a steady Uptrend. Nor do we look at the winding down of Watergate as a potential trigger of a new bull market.

Meanwhile, frantic negotiations behind the scenes in France right now are trying to get some agreement before Nairobi, but the deadlock remains complete. Jeremy Morse, of Britain, Chairman of the past conference, said it was "quite tough going." He felt it would be difficult, but not impossible to get a new monetary system fully into operation by September 1974, the original target date, at which we continue to scoff. There isn't that much time left, in our opinion.

In the view of delegates interviewed after the Paris meeting, the Nairobi ministers will probably content themselves with simply naming committees to study the monetary issues more. In other words, an agreement to agree on *nothing!*

The U. S., of course, wants the present system to continue, whereby Washington can simply run the printing presses and buy anything it damn pleases overseas. This would force countries to hold dollars instead of gold, of which we do not have enough at the present price. On the other hand, the Europeans want a monthly settlement of accounts in primary assets, such as gold or Special Drawing Rights which are convertible into gold. This must have blown Volcker's mind! Reality is always such a dreadful bore to a bankrupt.

This week the British put a guaranteed floor under its currency, pledging to increase its pound holdings if it slumped below $2.3760. This caused a temporary rally, but you would think people would have learned by now these fixed support rates simply don't work. It's going to cost the English people an awful lot of money.

The Senate approved and sent to the White House a measure authorizing Nixon to allow U. S. citizens to hold gold. The House approved last February's 10% dollar devaluation and left to Nixon the allowing of private citizens to own gold. Nixon is not expected to exercise his authority until current international monetary negotiations are completed. By that we assume until the price of gold is so high it doesn't matter whether we are allowed to buy gold or not. They simply won't let the little guy in.

September 14, 1973

*   *   *

[A] new gold crisis could materialize at any time without any more warning than there has been during any of the eruptions in the last decade. Now remains a good time to get rid of *things*, such as art you don't really want or need, co-ops, non-productive land, a business you were going to sell anyway, shaky paper such as second mortgages, and generally pay off debts and unlock your capital for the opportunities which lie somewhere ahead.

LATEST ON GOLD: THE ANTI-GOLD NUTS COUNTERATTACK

We pioneered the theory that golds move contracyclically with the market, so with the current market rally it should be no surprise golds have not been rising. Actually, they are holding up better than we had hoped for, and if this is the most a counterattack on gold can produce, then we are looking for a tremendous gold resurgence later on this year. However, those new to the ownership of gold shares are probably getting impatient and discouraged. When in doubt, just ask yourself if the entire gold crisis could possibly have been solved in the last 60 days. Absolutely not! And anyone who thinks the gold crisis has passed is operating under a self-delusion. That's how we kept our sanity in the last decade, by simply adding up the numbers again and again and again. They keep pointing to a gold crisis ahead, and those who are not prepared for it will regret it.

All we know is nothing can stop our idea, whose time has so obviously come. Nairobi will be an ineffectual failure, as we have said all along. *The Wall Street Journal* on September 20 editorialized "Nairobi as a Non-Event" and actually thought it was a good thing Nairobi would come to nothing. *The Dines Letter* has been pushing for a free float for many years, and there is still considerable resistance to it, so we are glad to have the *Journal* as an ally. They correctly point out things have calmed down in the last few months, and why end the test of a float when all the catastrophes which were predicted for it had not come true? The *Journal* also mentions excessive creation of currency in international markets, which will not serve their long-term national interests. Precisely.

Unfortunately, what few seem to understand is that the float must take place at the proper level, and for that gold will have to be many times higher than its present price. Until then, more crises, more instability, and no change in *The Dines Letter*'s predictions. The gentlemen meeting at Nairobi have been having such meetings for many years, and if they really had a solution, it is inconceivable it would not have been adopted by now. Volcker, who once said floating was an academic idea that ought to stay academic, is now pleased with the results of floating, and in fact a sizeable U. S. trade surplus is a possibility for 1974. Now Volcker is against a higher gold price because of the wild gyrations in the price of gold during the past year, which he claims makes it unsuitable as a monetary factor. Which shows how little he understands. The fluctuations are because gold is still underpriced. Set it at $500 or $1,000 an ounce, and it will stick without fluctuations. Hey, Volcker, try a higher gold price, you'll like it!

September 21, 1973

* * *

Treasury Secretary George P. Shultz said in Nairobi "I think interest rates are over the top." We strongly believe, if we may be permitted a quiet voice of dissent, that interest rates will remain quite high, despite any minor dip in the near future. Consider that any sharp drop in interest rates would once again cause dollars to flow overseas to higher-paying centers there, which would then eventually precipitate yet another gold crisis. This is what we meant when we said the dollar is trapped between a depression and a sharply higher gold price.

All told, there were enough Americans sent to the Nairobi IMF conference to fill an Air Force 707 (a second 707 was needed for Chairman Wright Patman and 21 other members of the House Banking and Currency Committee, 3 other Congressmen and 3 Senators, and a group of Congressional wives). Yet, all the king's horses and all the king's men, after the most incredible boondoggle ever thrown by a bankrupt nation, simply decided to set a new deadline of July 31, 1974, for agreement which will of course be postponed when we reach that time—unless the monetary system has collapsed by then.

September 28, 1973

*　　*　　*

The fact that nothing bullish for gold came out of Nairobi probably accounted partially for the decline in the price of gold this week, although we tend to think it is just profit-taking after the tremendous run-up from $35 to $130 earlier this year. For some time we have been looking for a dip into $80's, which would represent roughly a 50% retracement of the big rise, and pave the way for the next move over $200 towards our long-term gold targets. When all the gold shares were rising everybody claimed they had predicted it. Now, they have all turned into chrysophobes. Good news has a hundred fathers, but bad news is an orphan.

October 5, 1973

*　　*　　*

[I]t looks as if the price of oil is going much higher, and as this radiates throughout the world it is going to add to inflationary cost pressures everywhere and correspondingly put more money in the hands of Arabs, who will promptly translate this into solid gold.

The gold price broke its Downtrendline on October 8 and is now rising. Golds rapidly spurted toward their highs, illustrating why we do not like to trade the gold shares. Those who sold out at higher prices won't repurchase at lower prices, which is why we simply want you to buy golds and hold them no matter what.

Years ago we were begging people to buy U. S. double eagles for $50, and we still think they are a buy around here under $175.

October 12, 1973

*　　*　　*

We realize it's hard to resist buying in a market like this, and we know we look mistaken in our pessimism, so if necessary shut your eyes and put your fingers in your ears, but in any event resist the temptation to go back into this market which we think is a massive trap. Not only are we still working on the premise this is a Major bear market, but our pessimism is actually increasing day by day.

In our opinion, the Fed was panicked by the 10% prime rate, and a plunging stock market, so they decided to "engineer" a rally, and got it with a minimum expenditure of effort. They simply made some subtle hints and Wall Street went bananas. We have been suspicious of the Fed's inventions all along and this week our worst fears were realized when Federal Reserve Board Chairman Arthur F. Burns, in a rare press conference, offered a gloomy outlook on the Administration's anti-inflation efforts, and threw cold water on suggestions the Fed might soon loosen reins further on the nation's credit. At the press conference there were also suggestions economic controls were ready to be phased out, conservative economist Milton Friedman felt there was a good chance of a recession in 1974, and that prices next year will probably rise around 6% or 7%. William J. Casey, Undersecretary of State for Economic Affairs, said he believes the U. S. dollar "won't be further devalued and at some point is likely to be upvalued." We think he'll regret that prediction some day.

[T]he German mark looks very bad. Those of you in German stocks, bonds or currency are hereby advised by the Dines Letter to get out now. We were bullish on marks before the big upswing, but the time to take profits has arrived. Currencies are very important. If you had moved your funds from Argentina to Switzerland in 1971 the appreciation *by simply holding currency would have been 256%, and simply moving dollars to Switzerland would have given you a 40% increment not including any interest which you might have earned.* That's a lot better than some people have done in the stock market since then.

The dollar plunged against the German mark this week on news of the Mideast war showing how easy currency crises start these days. We are also getting strong rumors of German flight capital moving into Switzerland, under fears Germany will turn socialist sometime in the near future. It's a *Dines Letter* scoop—not in the newspapers yet—but that's the way smart money is moving, so we suggest you follow it.

Supposedly the main reason for the recent rally was that interest rates were on the way down. We disagree vigorously, even though it is difficult to prove our position. There might even be a tiny decline in the prime rate from the spectacularly high level of 10% but we still think levels well above 10% will be seen before the gold crisis is finished.

As the tremendous contraction in housing starts begins to cause slowdowns and radiates through the economy, we cannot see how a recession can be avoided in 1974. It might even be deeper than anyone now imagines.

Basically, however, Washington still appears to be asleep to the dangers ahead. They allowed this energy crisis to sneak up on us to begin with, and now they timidly avoid the really strong steps which are immediately necessary.

Golds have been strong the last few days, and for all we know, they have already turned up and you have missed your last buying point. As we have been saying, golds are cheap at any price near these levels, and you should take part of your position immediately.

October 19, 1973

*   *   *

Standing back from the rubble, it can be seen a mere six-week rally was enough to turn Wall Street bullish again (and then they wonder why they're going bankrupt!)

Short-term Downtrendlines are cropping up in our charts all over the place, and we predict it will become increasingly easy to lose money in the market in coming weeks and months.

We insist on standing our ground for another week, without knowing if we are correct, and with the majority of Wall Street extremely optimistic. We want you to give us a few more weeks to see if we turn out to be right or not, because you will appreciate it if we are. We still cannot envision the market holding up past Christmas, and maybe not even past November 1.

All the old problems are there, inflation, the gold crisis, oil problems, possible war, and a very shaky Wall Street.

Now that the prime rate has been cut to 9¾% the "common wisdom" is that interest rates are headed much lower. We must disagree with the consensus.

This week Chairman of the President's Council of Economic Advisors, Herbert Stein, said at a press conference that he was "looking for some rise in unemployment in 1974" and a slowdown in economic growth. Then he said, right up front, that he didn't have "any evidence that the Federal Reserve has relaxed credit reins" lately and that some observers have "greatly exaggerated" the Fed's easing of monetary policy.

Oh yes, Stein thought that inflation would continue at a rate of more than 5% well into 1974, more than double the Administration's former target. So inflation starts at 1%, then 2%, and 3%, then 5%, and eventually will get up to 7% and 10% and higher, despite everybody's good intentions to reduce inflation *mañana*. We can't see lower interest rates in this kind of climate.

October 26, 1973

* * *

Dow-Jones Indices
30 Industrials	948.83
20 Transportation	180.79
15 Utilities	98.80
Dines World Average:	295.07

As a matter of fact, we are not only still negative on the market, but we cannot remember having been this bearish since before the Crashes of 1962 and 1969. We still feel that we want you to sell everything except precious metals, and if you maneuver your portfolio 100% into gold we would not object. We want you to get out of the market and stay out. The only Wall Street advice more specific than that you're going to get is a margin call.

Keep in mind what we have been saying in recent months. *Nothing has changed with the inflation, gold, oil, food, Wall Street and interest rate crises.* Unless the Dow-Jones Transportation Average can hold above its 200-day Moving Average at 180, and the DJI hold above its at 933, there is no way this can be called a new bull market, despite the loneliness of our position.

One of the main reasons we have been negative on the market is that we are facing an oil slowdown which will permeate throughout the world's economies. The cost of energy is even more pervasive than steel. The way governments hide ruinous inflationary policies is, for example, to gut certain industries, like steel, by beating down price increases, forcing them to absorb rising costs. Unfortunately, while this can be done to domestic steel companies, it cannot to foreign oil sources which are completely beyond Washington's control. Even if we can get the overseas oil, much higher prices are ahead, and how are we going to earn enough money overseas to pay for all that oil? *We can only pay for it with gold*, and we don't have enough of that. Now we're starting to get down to brass tacks and to *real money*.

When the market plunged this week, gold shares began to skyrocket, which shows you precisely why we don't want you to trade the gold shares. When golds start running, they give no notice, and people who take profits on gold shares tend to not buy them back at lower prices. We don't know how to give you that message any more strongly.

That Dome Mines (DM - 106) could make an all-time high not only suggests another gold crisis is in the making, confirmed by this week's plunge in the Japanese yen, but that we were right in recent weeks in pointing to emerging strength in the gold group. That golds could move up despite the fact the U. S. actually reported a trade surplus of $873 million in September shows that professionals don't give a damn about what's coming out of Washington and are brushing it off as mere propaganda.

Administration officials, displaying premature elation, are now predicting U. S. trade will be in the black this year as against last year's trade deficit of $6.5 billion. The boom in U. S. agricultural products was a very big plus (we don't think that is likely to be repeated considering some of the large crops being reported around the world now) and U. S. price controls which made selling abroad more lucrative than selling domestically for some products. The impact of two dollar devaluations cannot of course be underestimated. But all this was at the expense of much higher prices domestically, which is why we do not see food prices coming down.

Farm prices continued to drop in October from their August record highs, and perhaps we look mistaken in our pessimism in the food area. But, just as we looked wrong in our prediction of high interest rates, we are going to stand our ground here also. We think runaway inflation will continue to strike at food prices in coming months and years, and the only way they can go is up—until the hangover of a depression cures the market of its inflationary binge.

This is simply a contratrend food price decline within the framework of a long and fantastic upsweep. The more paper money they print in Washington, the more pieces of paper pursuing the same egg, and you don't have to be an economist or genius to figure out there's only one way the price of that egg can go, and that is up as more and more pieces of paper are assigned to each egg. Only when they stop printing money, with nothing to back it, out of thin air, to pay for various grandiose follies, can inflation end, and the withdrawal symptoms are not going to be pleasant.

November 2, 1973

*   *   *

Between gold and oil, there must be a colossal disruption of international trade as higher fuel prices get factored into the transportation equation, which is then added on to all goods which are shipped. Just watch the disgusting scramble for oil at any price in coming months. With the world clearly on the verge of a major upheaval, anyone can see why we refused to become optimistic recently. You can't just follow charts blindly like a pack of fools and then expect to be right all the time. Sometimes common sense overrules everything. Like when we said interest rates would stay high and we were roundly ridiculed for our position. They must stay high because of the gold crisis, and only people who don't understand gold expected them to come down.

We don't want you to play bear market rallies because you are going to get into trouble. You might get disgusted with us while the market turns strongly temporarily, but those who buy and lose money will be back to us.

November 9, 1973

*   *   *

With the DJI down a sickening 123 points in the 2½ weeks since the October '73 high, it appears our lonely pessimism was justified. It was a difficult prediction, for sometimes we looked wrong but turned out right, such as was true recently. Not only did we stick to our guns, but *The Dines Letter* has never before had so many shorts outstanding at one time.

We still envision a devastating bear market ahead.

November 16, 1973

*   *   *

BULL TRAP TIGHTLY SHUT

Only 4 weeks ago Wall Street was rampantly bullish, and we disagreed vehemently. Since then there has been a decline which can only be described as *devastating*, having fallen 161 points in 3 weeks. What is interesting about this selling is that it is different from anything we have seen since the 1920's in that the selling is extremely *urgent*. Only at the beginning of the 1929 Crash was selling of this particular type seen.

That this has been no ordinary decline, and we are predicting that there is yet a far more devastating collapse ahead, is probably based on the startling existence of austerity in the United States as more and more places get cold and dark. *The gold and oil crises are converging in history in such a way that the United States can escape from neither.* The situation is really not amusing, and could well lead to World War III. Does that sound extreme to you?

This would appear to be an excellent time to postpone any major purchase, particularly automobiles, because we think prices will collapse on a broad front eventually. The natural consequence of inflation is usually a deflationary collapse, and that is somewhere ahead of us, although we are still groping for the exact dates.

Don't go through 1929 after having been warned it would happen again, because you will be kicking yourself for a long time for not following your best instincts.

We don't know how far this bear market will go, although we feel reasonably confident we'll spot Bottom when it occurs. If you don't have buying power at that time, particularly since we think it will be *the* buying point for the rest of the century, we will be most displeased.

November 23, 1973

* * *

In early 1973 our predictions the prime rate would rise above 10% were greeted by widespread scoffing. This summer when the 10% level was reached the scoffing was strangely muted, but a trivial dip in the prime rate made everyone forget our predictions and they assumed a new bull market was underway. We warned the decline in interest rates was strictly temporary, because of the gold crisis, and they would rise again soon. This was the basis of our insisting the recent rally was only a bull trap.

This week the First National Bank of Chicago announced its second straight weekly increase to 9.9%, presumably to avoid the headlines a 10% + level would generate. In our opinion, interest rates will continue to remain high until the gold crisis is solved, or until a depression removes the incentive to borrow. Then, interest rates will go down to astonishingly low levels, possibly even reminiscent of the 1940s.

Herbert Stein, chairman of the President's Council of Economic Advisers, forecasted on November 29th that unemployment would approach 6% in 1974 because of a business slowdown and fuel shortages. He did not see 6% exceeded, and assumed the oil embargo would end in 90 days or so. We disagree with Mr. Stein and, hard as it is to believe now, look for much higher unemployment as an unexpected Depression rocks everybody's forecasts.

December 7, 1973

JANUARY 1, 1974: DJI 855.32; ASA - 37⅞

* * *

[A]t our recent Seminar in November Mr. Dines warned of an imminent collapse in the English stock markets, which is now occurring.

December 14, 1973

* * *

[W]e have no choice but to stand by our pessimistic projections and we continue to urge you to use this rally to complete your liquidation and transfer into gold and silver shares.

[M]utual funds really got creamed in late 1973. The institutions and professionals have fought this decline all the way down, and at every temporary stopping place they insisted this was the time to buy, that stocks were undervalued, and so on. This is not typical of important Bottoms, which is featured more by a real aversion towards stocks by amateurs. The amount of pessimism at the real Bottom will be astronomical.

LATEST ON GOLD AND SILVER: THE INVISIBLE WINNERS. THEY ALWAYS ARE, TO LOSERS.

Gold and silver stocks have had a perfectly sensational advance since our last *Dines Letter*, and the fact that they outperformed the market by a wide margin in 1973 still does not qualify them for much Wall Street commentary. They haven't even attracted short selling yet, with Homestake the only entry on the entire short sale list. *Forbes*, in its 26th annual report on American industry, not only did not have a gold or silver category, but nowhere throughout the entire magazine, was the precious metal category even mentioned. We predict this is all going to change, particularly the way newspapers neglect to even mention gold shares were up or down for the day, or that they are in any way significant.

Actually, we are not impressed by the U. S. balance-of-payments surplus. First of all, the oil bill is not in yet, and we are going to have to start shelling out a lot more for imported oil. Second, the sharp rise in the dollar in recent weeks is going to eliminate the big trade advantage upon which our improvement was based. Third, all paper currencies are suspect, which is a long-held view of *The Dines Letter*, and we believe they will all go down the tube together.

Suppose 1973 does show a slight surplus of say $1 billion in our balance-of-payments. However, if we had paid $10 billion extra for oil imports, you can see what would happen. As the figure goes up to $20 billion, a real possibility for 1974, the collapse of the paper dollar looks virtually certain.

SILVER—THE POOR MAN'S GOLD

We have been wondering aloud in these *Letters* when silver would begin to catch up on its historic 15:1 relationship with the price of gold. With the gold price in a runaway uptrend at $121.25, silvers opened up on *breakaway gaps* this week when the May 1975 silver future deliveries sold at a record $3.538 an ounce. Silver coin futures per $1,000 face value for April delivery sold at a record $2,598 a bag.

INTERNATIONAL NEWS

The news out of England is horrendous. The government estimates 400,000 workers were being laid off because of the fuel shortage, nearly doubling the national unemployment total! As the economy grinds to a halt, additional widespread layoffs are a distinct possibility. Most firms are already on a 3-day week, which is quite a change in people's income, low enough as it was. As smaller pay envelopes begin to show up around the country it is difficult to say what the English people will do, particularly since food shortages are now expected. We cannot rule out the possibility of some form of social upheaval. The stage is set for a man on a white horse, and could be quite dangerous.

Japan is probably even worse off. A catastrophic inflation and the absence of oil could have a devastating impact on Japan's economic sitution. This is bound to affect the American and European economies, which could spread an international depression.

Just about every stock market average in the world which we follow is in a Major Downtrend. Rumors of devaluations of the English pound and the Japanese yen are abundant.

As we've been saying, paper isn't much protection, is it?

January 4, 1974

*　　*　　*

[W]e are being paid to call the shots, and we are looking for an immediate market collapse. A decline to DJI 788 for a new low would convince us we were indeed right, and we would look for a devastating plunge towards the 600 area. We know this sounds radical, and that such moves rarely happen.

We realize a reduction in margin requirements is almost always a very bullish phenomenon, but we do not think this is a prelude to easier money, the normal justification for a bull market. We believe stocks dropped so sharply there were simply too many margin accounts on the verge of being sold out, which liquidation pressure alone could cause a market collapse. Therefore, we believe the Federal authorities lowered margin requirements in *panic*, rather than in a carefully reasoned response to an extremely difficult situation.

Arabs are now talking about raising oil prices again at the beginning of the second quarter, "unless inflation is brought under control." This to us is an ominous suggestion because it indicates Arabs finally realize they are being paid in depreciating dollars by increasingly shaky governments. Their choice is either to leave the oil in the ground until fiscal sanity re-emerges, or else transfer this paper immediately into gold. This draining of purchasing power from all the oil importers will slow their real growth and accelerate inflation all over the world. The ramifications are terrifying, not only economically and politically, but militarily. Will Europe stand by and allow 3% of the Gross National Product of the entire European economy suddenly change hands from Europe to the Arabs? The increased oil cost to Japan will be $10 billion annually, which means a drastic cut in the standard of living of that country to pay for the oil. With such an ominous climate, who can look for a bull market?

The price of gold soared to $130 this week as other currencies like the yen began to collapse. We have been bearish on the yen for some time. We are surprised the dollar has been as strong as it has been, because we are quite pessimistic, frankly, on nearly all paper currencies, especially the dollar. Actually, all paper currencies are collapsing in relation to gold, every time the price of gold makes a new high. That is because more pieces of paper are required to buy one ounce of gold, the higher the price of gold goes. Eventually, the dollar will resume its plunge against srong currencies such as the South African rand, and the Swiss and French francs.

January 11, 1974

*   *   *

The majority view appears to be bullish for 1974, on the grounds most stocks have already declined to the point where they are discounting excessively the worst likely economic picture. Economists are generally looking for a decline in the first half and a rise in the second half, so we can therefore, using the Dines Theory of Positive Negativism, expect the opposite. Few seem to take the gold crisis seriously, and instead think recent market weakness was due entirely to the Arab shutoff of the oil spigots. To the contrary, we believe the oil crisis masked a deeper and more serious gold crisis, which will come into full bloom in 1974 or 1975. The oil crisis won't help, but it is not the cause of what we see coming.

The banking system is horrendously illiquid. There are substantial rumors a big eastern bank will close its doors fairly soon anyway, but along this vein it is interesting to note many individuals are already over their heads in debt and any significant decline in their income would result in massive insolvencies. This must lead to a tremendous decline in retail sales and the present worries about shortages could quickly develop into enormous price-cutting as inventories skyrocket and the resultant debt liquidation will lead to deflation. The economy is now in the final stages of inflation, the desperate runaway phase, and as we have said for nearly a decade, inflation will continue to accelerate until it arrives at its exponential peak, and then we will see deflation. There will be a desperate scramble for cash and the resultant large number of large bond flotations could result in a plunge in the bond market. Non-liquid assets like land and art will plummet and the resulting high interest rates from declining bonds will be yet another drag pulling money away from common stocks.

Wall Street is in a state of demoralized chaos. They are desperately trying to find a group popular enough to stir the public to return, but nothing seems to work. Even the oil stocks are no answer—there are no "politically safe" oils since nothing can prevent the greed and anger of losers in government from trying to grab "excess profits." We think this will happen to gold also, which is why we would like you to diversify your gold and silver portfolio *geographically.*

High interest rates have already knocked the construction industry for a loop and are quietly gnawing away at the guts of every borrower who is paying more than 10% interest. It is only a question of time before these people, and we include those overextended on margin on Wall Street, finally get cashiered.

The one event which would surprise more market analysts than anything else would be another sharp slide this year, since most of them think the market is underpriced. Hard to believe, but we think they will be surprised and that 1974 could lead to one of the worst bear markets in history. That is a strong statement, and a daring one, but that is our prediction at this time.

With higher oil prices being factored into all products throughout the economy, rising price levels are a foregone conclusion. A major distortion of trade flows is also inevitable, and as currency values lurch from one extreme to another, world trade could be disrupted as to actually come to a near-halt. Skyrocketing unemployment and declining Gross National Product will be a new experience for many countries, which could in turn lead to social upheavals not seen since the 1930's. Any event could break the back of this tense and highly unstable situation, perhaps even a simple devaluation of the English pound leading to competitive devaluations elsewhere.

Arabs, seeing their paper can be depreciated at will from the various capitals, will ultimately shift into a more secure store of value, and that must either be gold or silver. The world will lurch back towards a gold standard, willy-nilly, like it or not. The final disgrace of Keynesian economics could occur in 1974.

In the investment climate, with many companies experiencing declines in sales for the first time, with others so overextended in debt that they will not be able to survive, look for a wave of mergers as staggering companies try to hold each other up. There could be nationalization of at least one U.S. industry, perhaps oil, utilities, or railroads. Federal aid to specific companies like Lockheed, cannot be ruled out, but all we can do at this time is ask quietly, ""Who is going to bail out the U.S. government?" Anyone who is not frightened by that question does not understand the situation.

One of the worst areas to be hit will be Wall Street, which will go through a shakeout not seen since 1931.

As a rough estimate we still think the price of gold will probe the $400 area, although this is a rough estimate which we reserve the right to change.

It is evident from the letters we are receiving we have some pleased subscribers, particularly the ones with a diversified portfolio of gold and silver stocks at this time. Even if you are just breaking even you are well ahead of nearly every other investor in the world, considering the universal bear market, so the fact you are showing gains close to 1,000% on some of these gold shares should certainly be gratifying! As we look around the rubble and shambles most portfolios are in, of people who got caught by the numerous bull traps and who are now badly hurting, suffering under margin calls, while you are flush with profits, you can take the credit yourself for having faith in us. All *The Dines Letter* could possibly have done for you was to have a few dozen people and a computer spend a full-time job researching the facts and masticating it with our experience before passing it on for your own use.

Personally, we think, despite the incredible advances of gold and silver shares in recent weeks, that *you haven't seen anything yet*! We think gold and silver shares are going much higher and will pay skyrocketing dividends as their profits begin to soar. Even speculative precious metal stocks will come to life eventually, particularly when the general public climbs aboard the bandwagon. If there is any one stock *The Dines Letter* is identified with, it is ASA (ASA - 78⅜) and we are expecting radically higher dividends to be paid out in coming years.

As gold moves along on its exponential flight, it should be followed by silvers. Perhaps the move has already started. Gold has a big head start, but silver could wind up as "the poor man's gold." With the price of silver well above $3.00 an ounce, the thinly capitalized silver group could be the next major beneficiaries of the precious metal boom. The end is nowhere in sight and we think all gold and silver mining shares, including the penny speculatives Over-The-Counter, will eventually take part in this move as Johnny-come-latelies pick over the leftovers. Ultimately, gold and silver shares will become "high flyers." That will be near the Top.

*   *   *

[T]he U. S. is going to have to reduce its consumption and pay its own way in the world. We have not been paying our own way, and this staggering debt overseas of over $100 billion (which the Europeans were silly enough to accept) represents the difference between what we really produced and the amount we consumed. Now that the dollar is floating, and our credit has effectively been cut off, we are going to have to live within our means, which is certainly going to be a step down. It does not mean the end of America, but it certainly is going to take some buckling down and sacrifices.

We are asked whether or not oil shares should be bought because of the oil crisis. We don't think so. We already hear cries of "unconscionable profits" from those whose policies have already brought us to this condition. They're going to compound their mistakes and not permit the oil companies to earn enough money to find more oil. The oil industry might even degenerate towards the level of utilities and railroads, and you can see what government regulation has done there over the years.

This is the time for radical action, and *The Dines Letter* will remain fearless in warning of the perils ahead. With so many shaky margin accounts on the verge of being liquidated, a massive selling splurge cannot be ruled out. If this occurs, good stocks will go down to ridiculously cheap levels, but will nonetheless continue to make new lows because brokers will be desperately liquidating collateral to avoid being thrown into bankruptcy. What you should do therefore is to immediately place stop-losses say 25% under all your industrial stocks, even though you are confident they won't be triggered, "just in case." We think you should be 100% in precious metals stocks at this point, but if you insist on holding industrials then protect yourself with a properly-placed stop-loss order. If you do not know how to do this, read Mr. Dines' book on this subject and follow it promptly. It could mean the difference between your economic survival or not.

This could be the year when gold finally makes it. It has been a long wait for those faithful subscribers who have bought and held gold and silver shares. We have been militantly bullish on precious metals since 1961, and those who are new subscribers should switch into precious metals without delay, since we do not believe there is much time left.

How high? We have confidence we will know the Top when we get there, although we would hardly be shocked by gold well above $200 and silver about $10 sometime this year, up from present levels of $128 and $3.60. First, almost everybody said gold wouldn't rise at all; this year we think people will not buy golds because they feel they are overpriced or that it is "too late"; finally, the public will stream in to "buy before it is too late." This is a classic psychological cycle. We still predict the dollar will become convertible into gold, surely by 1975, which could end the gold play and the influence of Keynesian economics on our government.

[T]he dollar is not really rallying, because the price of gold is rising. In other words, it takes almost 4 times as many paper dollars to buy an ounce of gold now as it did at the beginning of this decade. Before it is through, we think it will take *ten* times as many paper dollars to buy an ounce of gold, and maybe even more! What this means is that *all* paper currencies are down vis-à-vis gold, which has only one price, and that is *the real one, the free one*. Each successive rise in gold prices means another massive devaluation of all paper, and a steady movement towards the day when a realistic gold price will at last be here, when profit of gold mines will be astronomical and we hope to lead you out of them with spectacular profits.

The Gold Pool established in 1961 to deliberately keep the price of gold down finally found the demand for gold so insatiable sales could not possibly continue without reducing reserves to dangerous levels. So, in 1968 the "2-tier" system was developed, keeping monetary reserves at the ridiculous price of $35 an ounce while permitting the "free price" to soar. The free market, officially ignored, has reached an all-time high at $130 and we do not think it unreasonable to conclude the recent abandonment of the "2-tier" system is an admission the free market price rather than the "official" price is the real gold price. Gold has conquered everything in its path, and soon it will conquer its wall of silence, the cold shoulder given gold by the press, governments and the investment community.

January 18, 1974

* * *

The wanton luxury of the 1960's is gone with the wind, and a new era has begun. It is one of austerity. With the losers who got us into this mess now howling for an excess-profits tax to be socked to the oil industry, and cancellation of their depletion allowance, it is clear the anti-profit Keynesians are still in charge.

The advice we offer is really radical. We have been on the outer fringe of the investment advisory spectrum for over a decade. However, now that there is the biggest financial turmoil in the world's history, as gold, like Banquo's ghost simply refuses to go away, and as more and more of our predictions turn out true, we are attracting a larger following. Sometimes loners are right.

Last week Americans were allowed to own gold coins minted up through 1959, as against the previous law of 1933. This is obviously a prelude towards allowing Americans to own gold. Reduction of the Interest Equalization Tax was the second indication that Nixon is edging towards a completely free gold market. By the time it comes, and the Wall Street Establishment, Nixon and the press all agree with us, we will be taking profits on golds and going back into the stocks we are now selling short. Everything in life is timing.

LATEST ON GOLD
WE'RE STANDING PAT AND SITTING PRETTY

Despite cooing reassurances from governments, *The Dines Letter*'s warnings of currency disasters continue to prove that it is on the correct path. Watching the price of gold skyrocket to an all-time high this week and closing close to $140 an ounce, we can only wonder why more people do not agree with our predictions. There was even some gold transacted at $160 an ounce. When the full gold panic we envision actually strikes, look for gold prices at astronomically higher levels.

The latest crisis occurred on January 19 when France announced she would float the franc "for six months" to protect French reserves, now at around $5 billion and dropping. The oil crisis alone could have brought a balance-of-payments deficit of $4 billion this year, which would have virtually wiped out reserves. The dollar, which fell to 3.8 francs last summer, is now quoted at 5.14. France, long one of the world's most adamant advocates of fixed parities, has ironically at last smashed [against] the power of gold. France no longer buys up francs as its price falls. *The Dines Letter* has long been the world's leading and most outspoken opponent of fixed parities, and instead thinks that "The Dines Plan" of allowing all currencies to float freely will eventually be adopted by everybody, albeit standing in a pool of blood. That the French rejected a West German offer of a $3 billion loan to hold up the franc and save the Common Market, shows that the French prefer to save their monetary reserves rather than engage in futile efforts to defend rigid parities. Even the *New York Times* on January 21 saw the light when it wrote "No one can know how long it may be, in this time of monetary upheaval, before France and other nations return to fixed rates. For it is now clearer than ever that currency floating under agreed rules is the only practical basis for genuine international monetary reform, with the world in so unsettled a state and the energy crisis intensifying balance-of-payments disorders. The irony of floating is that it offers the best hope of relative stability in a time of monetary turbulence."

The event shows France expects gold to be used in the ultimate currency, which is why they were absolutely unwilling to release even one ounce of gold even to buy oil. It also signals the beginning of a monetary free-for-all in Western Europe and Japan by the use of competitive devaluations. The monetary world is back to square one, before the dollar was devalued, before the Smithsonian Agreement, and before we began to predict monetary crises ahead. In the fourth quarter of 1973 the U. S. had its first quarterly trade surplus in years, but the recent rejuggling of currencies means the U. S. will slip back into deep deficit, particularly since farm exports will probably not be as high as last year because of inflation-breeding food and fiber shortages in the U. S.

On January 5, 1973, we had a major feature on silver stocks entitled "Silver is at Last a Buy Again!" This week our expectations came true with a smashing advance by silver futures into all-time high ground at $3.63. The May 1975 delivery sold at a mind-boggling record of $4.20 an ounce. Frenzied demand will drive what we have always called "the poor man's gold" into ever-higher ground.

THE CLASSIC RATIO OF GOLD TO SILVER

Brian A. Kennedy, a *Dines Letter* subscriber from Chicago, Illinois, sent us this perfectly fascinating information which we must share with you.

From about the end of the Middle Ages to the late nineteenth century, silver held its value essentially as well as gold. If you had bought gold in 1493 at 11.3 times the silver price (1493–1520 average), then in 1875 your gold would be worth the "classic" 16.0 times the silver price (1871–1875 average). Thus in 383 years, the gold-holder would achieve only a 41.59% advantage over the silver-holder—a trifling return of 0.09% compounded annually! For nearly four centuries—a reasonably long test period—silver was essentially as good a store of value as gold.

The cause, of course, was the discovery following about 1860 of the Comstock Lode and other fabulously rich silver deposits. If you trace it through, the resulting enormous silver overhang wasn't worked off for a whole century—until the U. S. government, which had heavily stockpiled the stuff, got pretty completely out. After this full-century interlude—which in historical terms is plainly an aberration—we are now stumbling (or being dragged) back into free markets in both gold and silver.

While the price of silver has been rising pretty much parallel with the rise in gold price, the gold-over-silver *multiple* is still up in the stratospheric 30–40 range attained during the glut years of the 1890's. The gold-multiple could come in for a thumping drop down towards its historic range because the century-long force (sudden silver *supply* glut) which drove and held the gold-multiple up, has been removed. So silver has both a play in the gold price and also the gold-multiple.

So, if gold "only" rises to $200, and if the gold-multiplier drops "only" to 25 (vs an historical range of 11 to 16) we'll have $8 silver. And $240 gold valued at 20 times would give us $12 silver.

January 25, 1974

* * *

Nixon has chosen the peak of this dollar rally for the latest gold development, ending an array of controls over the outflow of dollars for lending and investing abroad—some of which were more than a decade old. Superficially it looks as if we were incorrect in our warnings that such controls would increase. Nixon is premature. These controls will be removed permanently some day, but not yet—the bulk of the gold crisis is still ahead of us. Right now there is not much of a dollar outflow, because of trade balances. However, with the dollar rising and other currencies falling, one can expect a radical reversal in the trade picture sometime in March or April, which could lead to a hysterical outflow of funds, as investors trample one another to get into Switzerland before the Dollar Curtain is slammed shut. That is our position and we are sticking our necks out on it.

President Kennedy enacted the first of the control measures in 1963. The Interest Equalization Tax was designed to stem a flood of foreign bond issues in New York by making these bonds more costly to American investors. But this tax, and other controls, failed to solve the balance-of-payments problem, which did not work until the dollar was devalued (which advantage has since been nullified). Obviously, the cut in the Interest Equalization Tax a few weeks ago was simply to test the water, and when the world did not end, they went all the way.

This latest liberalization of gold rules, combined with the fact that you can now buy gold coins minted through 1959, looks as if Nixon is beginning to edge towards the door. It now appears that Nixon is beginning to understand the gold crisis, and although he has not yet kicked Volcker out of his entourage, he is nonetheless stumbling towards a free policy on gold. The U. S. might even reverse its long-standing, Keynesian-based hostility towards gold and actually encourage higher gold prices. After all, the U. S. has enormous reserves of low-grade gold ore and would be a major beneficiary of the higher gold price. More and more countries, particularly overseas, will realize that a sharply higher gold price would actually enhance their ability to pay for increased oil imports. There could well be a panic flight from paper at some point, which could translate itself into the most powerful gold fever seen since Queen Isabella threw Columbus in jail for merely discovering America instead of bringing home gold.

But it's really not amusing to see what is happening in London, including the looming prospect of an all-out British coal mine strike and continuing labor turmoil on British rails. Stock prices were knocked to a seven-year low on the London stock market this week.

Britain's finance house base rate jumped ½% to 16%. This is the highest finance house base rate ever recorded in London, and underlines the seriousness of the situation for those who think *The Dines Letter* is being alarmist. And, with paper money under enormous distrust, Britain and West Germany are now striking *barter* deals for Mideast oil! Britain is shipping Iran $242 million of textile fibers, steel, paper, chemicals and other industrial products. Doesn't the specter of people resorting to primitive barter say something to you? Doesn't it indicate that the standards of money are rotten to the core?

And Japan, licking its wounds, having suffered from a record annual balance-of-payments deficit totaling $10.07 billion in 1973, has yet to be heard from.

February 1, 1974

* * *

We think many people have been hoarding goods because they have been taken in hook, line and sinker by the so-called "shortage economy" pitch, which is why they have been suckered into buying cyclical stocks at high prices. By the time the bad news appears, and the "shortages" finally evaporate like ice on the Sahara sands, they will understand what a chartist means by a "Downtrend in the face of good news."

February 8, 1974

* * *

This week, as we saw the price of gold at a new all-time high at $146.50 it only confirmed our profound belief that gold is destined to go much higher than even now most people believe. In our considered opinion, you haven't seen anything yet, and the pyrotechnics on the upside later on this year could well become a legend well into the next century. You are watching history happening.

February 15, 1974

*   *   *

With such skyrocketing commodity prices yet to work their way through the economy towards the retail level, you can look for the most horrendous burst of inflation ever seen by this country in modern times. Ignore government projections, but just watch what commodities are doing and you know darned well that higher prices are just around the corner. This can only aggravate the gold crisis even further, and lead to the devastating Crash we envision ahead. If we are incorrect, you will be able to buy stocks back at somewhat higher prices. If we are correct, you will be grateful to us for the rest of your lives. We implore new subscribers particularly to bear with us and see.

March 1, 1974

*   *   *

A most fantastic week! With the price of gold around $178 this week, our predictions of gold exceeding $400 no longer look stupid. In fact, some day that prediction could look fuddy-duddily *conservative*. In this hectic week, with U. S. double eagles soaring towards the $400 level, some of you have already made nearly ten times your money by following us into these gold coins. Make sure you ride out the Corrections, as you have done for years, while the turkeys flee to the exits with every jiggle. The dollar has weakened steadily since our February 1 prediction that it would, and there is no change ahead. We emphasize, and re-emphasize, this is a *Major bull market* for the precious metals stocks, just as this is a *Major bear market* for most other stocks. In a Major movement, the biggest gains go to those who simply ride that Major trend all the way.

SPECIAL FEATURE: PERSPECTIVE ON ASA

The one single stock with which we are most identified is ASA Ltd (ASA - 84⅞), and while nearly all of you hold this stock at huge profits since we first recommended it at 5, some of the more recent purchasers might hold it at a loss. Since one of the functions of this letter is to prevent you from premature action, we are going to show you some perspective on ASA which we hope will persuade you to hang onto this stock no matter what. Over the years we have warned you to hold onto this stock like an English bulldog, and those who have followed this advice have profited handsomely. We still envision more stratospheric gains ahead for this one. Since it is a gold mining trust, it is obviously typical of most gold mining shares, and this feature is therefore of benefit to all gold share holders.

A Reuters Dispatch of February 28, 1974, began "Kuwaiti commercial banks have recently begun—albeit cautiously—converting part of their reserves into gold, something they have not done before, according to banking sources here. Previously, the banks had preferred to invest mainly in West European and North American currencies, or more traditional alternatives such as real estate. Now they consider it desirable to hold between 5% and 10% of their assets in gold, the sources said."

March 8, 1974

*   *   *

CURRENT MARKET ANALYSIS:
EVERYBODY OUT OF THE POOL AT ONCE!

We flashed a vigorous sell signal on January 4 at 880.23, and the market has not done very much since. In the subsequent 11 weeks since that sell signal we have repeatedly headlined this feature "The Closing Bull Trap." We have stressed that this was a *trap* that threatened to snare the unwary seeking "bargains" among depressed stocks and lead them to go against our advice to sell these stocks short. We have also been saying that there could be one more rally based on an end of the oil embargo, and with this week's information that the embargo will be ended only until June 1, the reason for the rally has now disappeared. It was a phony, superficial rally really, and one on which it was virtually impossible to make any real trading profits. In fact, we are willing to say that this week saw the end of the biggest non-rally in the market's history.

This week we headline this feature "Everybody Out of the Pool at Once." We now believe that the probable intraday DJI high for the year was reached at 904.02 on March 14 and we are willing to go on record with that evaluation.

[N]ow that golds are rallying, isn't it strongly suggestive that other stocks are about to head south precipitously? Those who were tricked out of their golds, or worse—those who sold short on gold shares and who are now developing small beadlets of perspiration on their foreheads can blame only themselves for not listening to us. Everyone must pick his mentor and follow him faithfully. If he says don't trade the golds, don't. When you have followed gold shares for as many years as we have, perhaps you will get a feel for them also—and get the message they are giving us loud and clear now—*don't sell!*

Every inflation has always ended in a severe and painful deflation. In this case we would not be surprised to see governments around the world fall as, sooner or later, the inflationary price structure topples of its own weight. Will it come when New York City grapefruits reach $1.00? Or $5.00, or $100 each? Sooner or later, you will see food riots and a complete buyers' strike by people who have run out of money. We even expect oil prices to be down by summer, regulated by supply and demand, placing tremendous strains on the new-found Arab unity. Agree or not, these are our predictions and we are putting them in print and laying our reputation on the line for them. If most of them come true, we will consider ourselves as having served you well, and indispensably so at that.

If we are correct, massive inventory liquidation is coming. Businessmen must now keep a razor-sharp eye on their inventories. All of you are advised to sell now what you eventually plan to sell anyway. We think the best prices exist now, and that selling prices will eventually come down—and hard. We are including your business itself, surplus art, land (that which is unproductive) and so on. Since consumers, as described below, are loaded up with record debt and faced with inflationary price rises and job insecurity, we have to assume they will sooner or later stop buying and step up saving. This changeover can't be good for near-term business.

With only one-third of our Technical Indicators bullish and considering the Major Downtrendline (D) on the last page of this *Letter* we have no choice but to remain militantly pessimistic on industrial stocks.

What fascinated us about this week's leap in the price of gold is that it occurred precisely at the time the U. S. announced that in 1973 it had its first yearly surplus in its "basic" balance-of-payments since it started collecting the figures in 1960. Of course, this is merely superficially good news because expensive crude oil is going to knock the daylights out of that surplus, not even counting the fact that that surplus was run up at a time when the dollar had been unfairly devalued against other currencies—an advantage which has since been cancelled. It's getting harder to fool gold speculators these days. We can remember 10 years ago when such news would knock the daylights out of gold prices. In this case, however, gold prices skyrocketed! It's high time people stopped believing the bull coming out of Washington.

Also this week French Finance Minister Valery Giscard d'Estaing announced the end of his two-tier French franc market established in

1971. Gold keeps smashing down the barriers they keep putting up, and it has not yet gotten through their thick occipital condyles that they are not going to beat gold, and it is ridiculous to keep trying.

March 22, 1974

* * *

Thursday, after Chase raised its prime rate to 9¼%, and other banks began to follow suit, the market moved down again, *raising the real possibility that the DJI high of March 14, 1974 at 904.02 will stand as the high for 1974, and in fact might not even be seen again until 1979 or so.* But we are getting ahead of the story, because later on this year we expect a Killer Wave.

Illustrating that fertilizer is scarce throughout our economy (except in Congress) there is a possibility that wage-price controls will be permitted to die an ignominious death on April 30. This will probably be greeted by another burst of inflation, which will be as useful to the economy as acupuncture to a hemophiliac. Lenders are already demanding higher interest rates to compensate for the expected burst of inflation, and this thought was greeted by utility shares promptly taking a nosedive.

[W]e remain in the painful position of expecting bad news to prove that we are right, and in a way makes us hope for bad news, which is our own special burden. All we can do is look forward to the day when we'll be leading you out of golds and silvers back into industrials in preparation for a renewed U. S. growth phase. This is a great country, and it can even survive some of the turkeys who have been running the international monetary picture.

This week the English stock market collapsed, which event begins to fulfill Mr. Dines' predictions made at the London Seminar in November 1973. It seems that this time London has lost all contact with reality, authorizing a broad range of tax increases, food subsidies, and reducing the cost of milk while raising taxes on cigarettes, ice cream, booze, gasoline, electricity and postage. This in addition to higher income tax rates, the closing of so-called tax "loopholes" and lots of talk about the people making "sacrifices." It is important to observe England here, because we think the United States is not more than a year behind. England's February deficit . . . skyrocketed to a shocking $1.01 billion, the first time England has ever recorded a billion dollar deficit in a single month. Thus the degeneration of the inflationary disease approaches its final stages, preparatory to the collapse we envision.

What really frightens us about England, and this has received virtually no publicity, is that since 1968 Britain has been guaranteeing the dollar value of the sterling holdings of 53 governments (mostly British Commonwealth members) and England has to pay the difference if the pound falls against the dollar. In 1972 England had to shell out $138 million and at the end of this month about $230 million more to governments which hold some $6.3 billion of sterling in their reserves. It has not dawned on the geniuses running the international monetary picture that if the dollar folds, there is no way the English pound can survive such guarantees.

March 29, 1974

*   *   *

The market is in an extremely precarious state, and for all we· know the DJI intraday high for the year has already been reached at 904.02 on March 14.

Not only are the Pennsylvania Railroad and Lockheed at the Federal teat, but this week TWA and Pan Am begged for a subsidy from the Federal Government in order to survive. We predict that the list of corporations lining up at the Federal trough will grow ominously long by the end of this year.

Although the DJI was off only 0.49% in the first 1974 quarter, we nonetheless think our pessimism of recent months has been correct.

French President Georges Pompidou's untimely death on April 2 somehow triggered a panic in France, which prompted the gold price in France to skyrocket to an unbelievable $197 per ounce.

On March 29 in Washington officials who were negotiating world monetary reform formally buried a return to fixed currency exchange rates for the indefinite future.

Hooray! They are finally coming to the Dines Plan—and intuitively, since they've never heard of us. It is we who have been howling for floating rates for over a decade, knowing full well that [they] could never beat gold, and they have at last thrown in the towel. Gold not only is emerging as the victor in battles like this, but has also come to mean much more. Gold is the ultimate liquidity. Gold is the ultimate freedom.

And the London Stock Exchange only prompts us to shake our heads sadly. The London Index has plummeted over 30 points in the last week, and is a long way from the 1972 high of 543.6. It is now around 263.6, its lowest level since 1962.

April 5, 1974

*   *   *

Arthur Burns, Chairman of the Federal Reserve Board, has apparently decided that it is more important to attack inflation than to worry about the damage high interest rates inflict upon the economy. After using some of the strongest language yet to describe the "ominous" nature of the world's inflation, Burns declared to a House Banking subcommittee that he was "determined to follow a course of monetary policy that will permit only moderate growth of money and credit." Burns said, indicating he is really beginning to understand the problems, "Our own and other governments have no practical choice except to put up with floating exchange rates. Faster inflation in the U. S. than abroad would tend to induce a depreciation of the dollar in exchange markets, which in turn would exacerbate our inflation problem." Burns pointed out this did not exist under the former currency system of fixed currency exchange rates, which have, as you know, been abandoned. In other words, Burns admits that when we print too much paper from now on, instead of foreigners simply accepting it in a docile fashion, as they used to, they will now throw it on the open market and, as a consequence, the dollar will decline. Furthermore, when interest rates are much lower here than abroad, money flows out, and in a floating system the effect is to push down the dollar's exchange rate and worsen the domestic inflation. Thus, the Fed must keep interest rates high enough to prevent large outflows of funds. He warned that the nation must face up squarely to the gravity of the inflation problem, and that "the pace of inflation needs to be substantially reduced, even if it cannot be halted this year." This is bad news for the economy since such high interest rates might well cause a Depression. The alternative is a gold crisis which would also cause a Depression. At least Burns, and we believe he is the only one in Washington to do so, seems to understand the dangers involved.

More big news came this week when Merrill Lynch came out with an institutional study on South African gold shares dated April 1 which began "We believe that a portfolio of leading South African gold-mining shares could provide high rates of return, in the form of high dividend income and good capital gains in the years ahead. That opinion is based on our expectation that the price of gold will continue to rise, reaching levels over the long-term that are likely to be much higher than current prices." This is the first major Wall Street broker to decisively come out in favor of golds, and is an event we have been awaiting, and predicting, for over a decade. We predict other Wall Street houses will be dragged kicking and screaming into the new monetary era. What we find fascinating is that Merrill Lynch, which is by far a distinct minority in terms of gold bullishness, is also running large ads telling how bullish they are on America and why they think the stock market is undervalued. Apparently it has not dawned on them that golds usually move in the opposite direction of industrial shares, a theory which Mr. Dines pioneered in his "Dines Rule of Gold Contracyclicality." We are glad that we will not have to hear the results of the agonizing when they discover that contradiction.

April 11, 1974

*   *   *

Our function here is to hold our readers firmly on the right path, not letting them wander off into bull traps. New subscribers have a difficult decision because they do not know us, so all we can do is ask them to give us a few months to see whether we turn out right or wrong. We are risking our reputation on the assumption that the bear market Bottom lies somewhere ahead of us, and that gold and silver shares are destined to go much higher. We don't know how to give advice more specific than that.

Remembering that not all lenders are morons, reeling under an inflation rate of nearly 15%, why on earth should they accept only 10% for their lendable money? Perhaps that is why the prime rate has been moving toward all-time high levels for this country all week, yesterday actually reaching 10¾%. At *The Dines Letter*'s last Gold Seminar a few months ago we warned of a prime rate between 10% and 15%, a prediction which no longer looks ridiculous.

What does all this mean? What is the significance of these obviously momentous changes? The implications are profound. Higher interest rates mean bonds already *on the marketplace* must decline substantially for their yields to compete with current high levels available elsewhere. That means institutions which hold bonds are taking staggering losses as the bond market collapses day after day, with no end in sight. (Now do you see why we recommended you avoid the bond market?) Furthermore, high interest rates must prove ruinous to the homebuilding industry, which will in turn harm everything which depends on home-building—from furniture to lumber. This slowdown could spread contagiously in the economy. In addition, a recent survey of some New York State savings banks shows there was an unusually heavy net deposit outflow during the first 15 days of April, and savings banks are the State's major source of home-mortgage money; we suspect people are pulling their money out of their 6% bank accounts and opting for higher interest rates readily available elsewhere. People are no longer going to stand for getting ripped off by inflation and Washington can no longer rely on investors' ignorance for its outrageous embezzlements.

April 19, 1974

* * *

The big thing we want to discuss this week is the April 22 meeting by Common Market officials who unanimously favored permitting central banks to trade gold reserves among themselves at a "market-related price" which could be far above the $42.22 "official price." At first it was unclear whether central bankers would want to buy as well as sell gold on the private market, or whether U. S. opposition might limit the plan to a regional European one. Europeans' motives are obvious. Why should they downgrade their own assets? They need to pay for the oil they are importing and therefore they must raise the gold price.

Big changes are obviously afoot. The new toughness by Burns, Simon replacing Shultz and Volcker, and now this announcement by Europeans, all leave one wondering what is going on behind the scenes. Will this lead to a new "floor" under the gold market—first the floor was $35, then $38, recently $42.22, and now to somewhere in the $150–$170 range? If so, this would be *bullish* for the gold shares, because a new floor would mean that gold mines could expand their production, confidently relying on a much higher gold price than at present.

In a particularly regressive statement, on April 24 Treasury Secretary-Designate Simon indicated he would not favor any changes in gold policies until international monetary reform negotiations have been completed. Simon said he would be glad to meet with Common Market representatives on this proposal, although he set no specific time. *However, Simon asserted that his position on gold would be consistent with the Treasury's existing stance, which is to avoid anything that might enhance the role of gold.*

The latest European move shows that gold is king! The U. S. has idiotically maintained that gold is worth $42.22, while the market value has been soaring towards the $200 level. Now the world wishes to adjust to reality and the U. S. still cannot see reason. There is a real danger Europe will walk off and leave the U. S. sulking in the corner, totally isolated. Actually, Europe has no choice since it must pay for that oil. This is a capitulation by the paper money fanatics to the ultimate realization that future money will be soundly based on gold. Some nations are realizing to their horror that they are loaded up with paper, while others are deeply involved in gold. On November 16, 1973, *The Dines Letter* wrote a major report on currencies, pointing out which ones have the most gold and which ones the most paper. We offered this article to several magazines, none of which would publish it. To our knowledge, it was published only in *The Dines Letter*. Now, it is to be the basis of the new monetary system. The Dutch, British and West Germans are against giving an important role to gold, while France and Italy are anxious to get gold into money again. Naturally, since Italy has nearly 70% of her reserves in gold, while Germany has only around 15%. We hate to tell them we told them so. . . .

On April 15, 1974, the Treasury Department imposed a ban against melting or exporting copper pennies, backed by criminal penalties. The prospect of a nationwide penny shortage is due to the fact that inflation has driven the price of copper high enough so that a penny's metal content is worth more than its face value. Unless the lunatics running those printing presses in Washington are thrown out of Washington on their ears, the prospect of an all-paper currency, without any coinage, becomes a real possibility within the next few years.

April 26, 1974

*　*　*

With auto production down by a half, and housing in a terrifying slump, why should the steel and aluminum industries be sold out? The only possibility must be a frantic inventory buildup, and these outrageously high interest rates are making it increasingly expensive for businessmen to hoard. Sooner or later, threatened with bankruptcy, they're going to dump that inventory on the open market, and with a massive economic slump underway at the time, *the inventory losses are going to be horrendous.*

The truth is, the "printomaniacs" in Washington running the printing presses, and unrestrained by gold, are causing inflation and the horrendous background for the market crash and depression which are possibly ahead. Ignore these phony rallies until there is a real sign that Washington is going back to sound money—nothing else will change our minds.

Some smug people are boasting how they sold out their gold shares at higher prices. We have heard that story before, many times in the last decade, and these people never seem to buy back at lower prices. *Remember, the person who sold gold shares at a higher price has not done the right thing until he buys back at a lower price.* Golds are notorious the way they start their moves, with blitzkrieg rapidity, and for all we know the next important upsweep has already begun. We are confident the best path is for you to ride out these declines and in fact even take advantage of weakness to complete your buying programs and average down wherever possible.

Meanwhile, as we predicted recently, the U. S. balance of trade slipped into a deficit position in March for the first time in 9 months, with the major factor being the increased price of imported oil. With no energy relief in sight for years to come, the U. S. is skidding on greased wheels towards a horrendous 1974 deficit. The soaring oil-import bill is almost unquestioningly going to keep increasing all year long. Where are we to get the money to pay for this oil? Either the U. S. cuts back industry, which will cause a Depression, and uses less oil, or we *export more* and earn enough overseas to pay for that oil. Since everybody else is scrambling to earn more to pay for oil, look for *increasing international competition*. One solution would be a markup of the price of gold to $1,000 an ounce, which would make the U. S.' pile of gold, biggest in the world, even bigger!

And England. Poor England. England could run a deficit for 1974 of anywhere between $8 to $11 billion, of which $4.75 billion comes from higher oil prices. Last year England over-stimulated demand, but because of underinvestment in the past, British suppliers could not cope with that demand, so foreign imports flooded in. Now Britain must have a recession to temper that demand back to reality.

We have been one of the world's biggest bears on the English stock market, and there is no end in sight to what we see.

Lionel D. Edie & Co. is looking "for very much higher inflation" with increases in English living costs between 18% to 22% annually. This could be the final phase of inflation and all the chaos which goes with it. England recently had a 48% leap in coal prices, a 30% boost in steel, and a 50% rise in the price of electricity. While the cost of living in Britain is only rising at an annual rate of slightly more than 10%, as the higher prices begin to work their way towards the retail level, look for horrendous demands for higher wages, withdrawal of funds from savings institutions, and, possibly, even the ultimate catastrophe— the Arabs, which traditionally hold their funds in London, might start withdrawing their funds. That would be the *coup de grâce.*

May 3, 1974

* * *

International eyebrows were raised this week when a major Swiss bank quietly decided to stop acquiring U. S. dollar Certificates of Deposit in London for fear that Britain might clamp on capital-outflow controls. The Eurodollar market, uncontrolled, and with no bank of recourse or government regulation, could be the tidal wave that sucks us into the maelstrom of economic destruction. People buzzed as to whether or not the savvy Swiss bankers began to smell signs of deterioration. Why should the Swiss suddenly insist new CD's be on the books of the U. S. bank home office rather than on the books of their London branches? Well, anything can be "explained," but still, there is a bit of an odor about this move at such a time.

Then, our new Treasury Secretary William Simon announced that he was favorably inclined to an early removal of the legal ban on ownership of gold by private U. S. citizens, and the price of gold leaped to $166.25. Simon feels any such change, however, must be "tied to broader international understandings on the role of gold in the future monetary system," which means he is going to use it as a bargaining tool! Americans have been prohibited from owning gold as private citizens since 1933. This is a blockbuster! As it stands now, Americans cannot legally avoid being embezzled by the U. S. government via inflation: all they can do is buy precious metal shares, silver futures, gold coins, or sell stocks short. Now, Americans might be able to legitimately own the real thing.

May 10, 1974

* * *

The Dines Letter would like to make what is probably the first use of the word "infression." We define an infression as an "inflationary depression," that is the state into which we think the world has already entered. It is a state similar to being in Sweden in July, where because the sun shines most of the day and night you think you should be awake, but you are actually exhausted. In an infression, people get more paper dollars, but they can buy less with them.

And, what is the proper P/E Ratio for earnings in an infression? Well, first you must deduct inflation, and we think that if that were done for most of America's leading corporations, the Dow-Jones would drop 200 points by morning! The predominant reason we are so pessimistic is the fact that nearly every chart we follow is pointing straight southeast. If you don't believe us, look at these charts yourselves. That is why we have been preaching the gospel: This is the time for capital preservation, for survival, for coming through this crisis with wherewithal—these are our paramount considerations right now. Let the greedy go on playing trivial jiggles and get caught, as they will sooner or later. We are going to hold you on the right course, away from temptations and bull traps. Our reputation is on the line on this one, and, personally, not in our 20 years of advising have we seen a more treacherous situation in which so many people, in our opinion, are going to be completely wiped out in the stock market.

After we went to press last week Franklin New York Corp, 20th largest U. S. bank, passed its common and preferred dividend and bank stocks collapsed on a broad front. At the same time First National Bank of Chicago boosted its prime rate to a shockingly high 11.40%.

[V]ery few people outside of those who read us understand that there is *a full-scale, low-key financial crisis going on right now*. To us, it is only a matter of time before a major insurance company busts, or even a savings & loan outfit.

Cost of Living Council Director John T. Dunlop predicted that the country will have a "persistent" inflation problem for the rest of this decade and into the next. We disagree. We don't think we have that much time left. We think inflation is in the terminal phase before collapse—within the next two years.

On May 13 *Barron's* editor Robert M. Bleiberg wrote a bullish editorial on gold. In it he wrote "And the latest Granville Market Letter carries a cartoon which depicts a hapless skier, tagged with a sign saying 'All the Way with ASA'—hi there, Jim Dines—falling flat on his face." Since Granville has been bearish on golds most of the way up— hi there, Bob Bleiberg—we take Granville's current pessimism towards golds as simply a confirmation that we are almost unquestionably correct in our optimism towards the gold group.

Ho hum. The U. S., West Germany and Switzerland have agreed in Basel, Switzerland, to support the dollar. This "jawboning" still apparently tricks some into thinking that the dollar's troubles are over every time a group of bankrupts get together to say that it's going to go higher. How many times have they tried to support the dollar to no avail? Nothing can stop a free market from reaching its goal ultimately, and the dollar is going to collapse down to its true value, near zero. Just watch for the rising deficits of all the oil-importing nations, and see how currency after currency collapses under the impact . . .

May 17, 1974

*   *   *

Obvious now only to insiders, there is an invisible depression going on. Sooner or later it will be visible to the masses and they will stampede toward the exits—too late. This is a Chartist's heaven and a Fundamentalist's nightmare. Shut your ears, the way the beaver does to water, to all the talk about how cheap stocks are, and how low P/E Ratios are. This is, and has been, the time to follow charts. If you haven't sold out of the market yet, get hold of a chartbook and follow your own common sense.

Some cling to the last remaining shred of hope that the DJI will hold at the important 800 level. These psychological levels are covered in Mr. Dines' book, and we are convinced that 800 will not contain this bear market.

Inflation is undermining the willingness of the American people to save and invest, and is deepening distrust of governments and paper money printed by those governments. If you would like to see a classic account of the way these things usually end (yes, inflation is not a new trick) read Andrew Dickson White's *Fiat Money Inflation in France* published in 1914. Chances are our economy could not support the existing debt burden if it were not being steadily reduced by accelerating inflation—so any significant reduction in the availability of credit will undoubtedly produce massive bankruptcies and sudden financial distress, as witness Con Edison and Franklin National Bank. Inflation has a life of its own, which nobody dares to stop. At this time the United States is extremely illiquid and the lack of faith in U. S. paper money is leading to an outbreak of international barter agreements, as more and more begin to abandon paper money as a medium of exchange and store of value.

You have a government boasting about a 12% jump in corporate profits in the first quarter, without mentioning the fact that prices jumped at an 11.5% rate. And, most of those profits were due solely to inventories. Furthermore, you've got a government which, like some thief in the night, runs two sets of books. The government boasts that its overall Gross National Product is climbing at a 4.5% annual rate. The other set of books talks about how the "real" Gross National Product (adjusted to eliminate inflation, based on 1958 dollars) actually plunged at a staggering 6.3% annual rate, even worse than the 5.8% estimated a month ago. This could be the beginning of the Depression, a situation where people get more dollars, but can get less with them.

Get out of the stock market and stay out, and don't say our advice isn't specific.

This week we finally got a short-term buy signal on the golds, but it was a weak one. We are uncertain whether or not there will be just one more shakeout but we are reasonably confident that the gold Correction will be over by July. If the current weak buy signal is correct, then gold shares should become strong no later than May 29. Don't be confused by the fluctuations in the price of gold, which have only a general mathematical correlation with gold shares. Actually, we are looking for lower gold prices possibly somewhat below $150 some time in June or July, after which a long and steady ascent should culminate in a $300 level by Christmas 1974.

The current shakeout should turn most investors pessimistic on golds again, and pave the way for the next gold bull market.

May 24, 1974

*   *   *

Clearly, people at the top are beginning to feel the alarm we have felt all along. First, Nixon has kicked out the turkeys who have been helping ruin our currency at the U. S. Treasury, and has begun replacing them with one man who talks about balancing the budget, and another who actually parts his hair in the middle! This is real progress.

And, no wonder. With the prime rate at 11¾%, and experts warning that the prime rate could even go above 12%, this is no time to dilly-dally. Considering "compensating balances" (even if you get the loan, the bank expects you to keep a portion of it in the bank, which effectively raises your interest cost) interest rates are approaching Mafia levels.

Interest rates promise to stay high because Arthur F. Burns, in some of the most ominous language he has ever employed, warned over the weekend that "the future of our country is in jeopardy" because of rampant inflation that is wreaking "havoc" on the economy. He said the potential consequences of continuing rapid inflation "threatened the very foundations of our society." Pointing to how many governments have fallen in major countries recently because they failed to cope effectively with inflation, Burns warned that "we mustn't risk the social stresses that present inflation breeds." Burns said, "I do not believe I exaggerate in saying that the ultimate consequence of inflation could well be a significant decline in economic and political freedom for the American people." He warned that unemployment would inevitably rise but this could not be helped because inflation creates "an illusory element" in business profits through its impact on the value of inventories. This element of profits, Burns said, "is not available for distribution to stockholders in view of the need to replace inventories, plants and equipment at appreciably higher prices. Worse still, the illusory part of profits is subject to the income tax, thus aggravating the deterioration in profits." Well, well, well—it seems they've got somebody up there who's at last talking *our language.*

Last week we reported a weak buy signal on the golds, suggesting gold shares should become strong no later than May 29.

On May 29 the Senate approved an amendment allowing Americans to buy and sell gold as of September 1, 1974.

Meanwhile, as more people buy gold, more vested interests develop to promulgate higher prices for it. These days, it seems, nobody benefits from a lower gold price except the politicians who have been spending themselves into reelection the last four decades. More important, governments are developing a vested interest in higher gold prices simply to pay for the oil they are going to have to have. That is why governments are not selling their gold, but will pay with anything else first—U. S. paper dollars, SDRs, IOUs on the back of a laundry slip, and so on.

There have been *no changes* in any of our positions. The gold crisis remains *unsolved*. Whether or not the U. S. will have an infression (an inflationary depression) the gold crisis means a lower stock market. We are convinced the U. S. will survive, but it's going to be a rough couple of years.

May 31, 1974

* * *

Our May 29 gold signal received more confirmation this week, which was a wild one.

Gold shares skyrocketed by Wednesday on rumors that France and West Germany were on the verge of a large, and perhaps coordinated anti-inflation and currency action. Gold pessimists must have been flabbergasted as they watched gold shares break their recent Downtrend-lines, and develop new Uptrends in many cases.

Therefore, our present position is basically unchanged. We expect the price of gold bullion itself to move lower, perhaps even towards the $140 area by mid-July after which it will begin a steady upward march which will probe the $400 area, possibly by this Christmas. Meanwhile, gold mining shares have probably completed their Correction, and are ready to march to new highs. We do not think those who sold gold shares recently will be able to buy them back at lower prices. If we are correct in this, then those who tried to trade these made a mistake adding one more to a long list in the trading annals of these shares.

June 7, 1974

* * *

WHAT THE ITALIAN COLLAPSE MEANS TO GOLD

Premier Mariano Rumor resigned this week, unable to cope with the rapidly-worsening financial and foreign trade crisis in Italy. This was Italy's 36th government in 31 years. Italy is running a foreign payments deficit of more than $1 billion a month, and has the largest foreign debt among all major industrial nations. Italy has borrowed $10.5 billion abroad in the last two years, and has exhausted its credit. However, with all else gone, Italy still has 2,500 tons of gold. While this is only worth $3.5 billion at the ridiculous U. S.-enforced price, the free market values it at more than $16 billion. Let those who scoff at gold realize that gold stood between Italy and disaster this week, and that it could happen to other countries also. Italian inflation is running at about 20%, as fewer and fewer people want to work, and the lira is floating at around 650 to the dollar.

The problem is how to evaluate that gold. There is no way gold could be sold without sharply depressing the gold market, and that would reduce the value of Italy's gold reserves! So, they finally stumbled into the solution the little *Dines Letter* has been talking about for over a decade, and are using their gold as collateral for a loan! Imagine all those high-paid officials taking so long to stumble towards such an obvious conclusion. But, if Italy is allowed to survive by using its gold, then who will be next? France? England? It is difficult to imagine a government going broke, but that is what you are watching.

The real significance of the Italian crisis is that the ten leading financial powers are at last making their monetary gold partially usable again, which to some extent is the beginning of the unfreezing of the incredible illiquidity of all the Central Banks in the world today. Apparently, only nations which are economically distressed will be allowed to expand their borrowing power by using gold as security for international loans. Thus, the U. S. is being dragged, kicking and screaming, back towards the fiscal and monetary integrity of the 19th Century. The U. S. Treasury's twisted reaction was predictable, saying that gold should be used when it is useful, but we should still try to phase it out of the world's economy.

The Dines Letter says, as it has always done, *gold will not be phased out of international monetary finance.*

June 14, 1974

* * *

ANOTHER UNDERWHELMING RALLY

We predicted the Dow-Jones Transports would not exceed 180 or that the Dow-Jones Industrials would not exceed 870, and that the recent little rally would be over by June 14. Actually, it ended on June 11, and 180 and 870 were not exceeded.

For the market to reach a lasting Bottom, the public is simply going to have to be much more pessimistic than they have been. Below, you will see that the latest Short Interest has actually declined and, studying the latest figures of stock market credit, it is clear that most small investors are simply watching their investments decline in value, month after month, with the same fascination a gopher has for an approaching rattlesnake.

Here are more solid predictions. Not only do we still think Italy (the nation itself) will go broke by the end of 1974 but, hard as it is to believe, we are looking for a major bankruptcy in either France or England. We are actually going on record as predicting that these countries will not be able to meet their payrolls for people employed by the government. This includes the military. Think about that, and what it will do to world trade.

We have long been outstanding pessimists on the London stock market, as those of you who attended our 1973 London Gold Seminar know. This week prices on the London Stock Exchange plunged to the lowest level in a dozen years amid a new wave of labor disputes, terrorist bombings, unbelievable inflation, staggeringly high interest rates, and an oil crisis which is only adding to the insoluble illiquidity of their banking system. Yesterday the London newspaper *Economist* declared Britain might be heading for an economic crisis rivaling that of 1929–32. Worse, Britain's trade deficit with the rest of the world widened in May to a record $1.15 billion. Consider this in light of the fact that for all of 1973 their deficit was $5.61 billion.

Much has been made this week of the U. S. first quarter surplus in its balance-of-payments of $2.1 billion. We must bluntly voice our suspicion that these figures are false. With an international monetary crisis going on, we can't for the life of us see how the U. S. can possibly escape unscathed. Adding credence to our doubts was a simultaneous Commerce Department announcement that last year's "surplus" of $1.2 billion has suddenly been "revised" to a *loss* of $744 million! Perhaps the figures are true, but simply unrepresentative of what the year will look like. We hope so. If not, the stench could some day be much worse than Watergate.

June 21, 1974

* * *

DJI 800 WON'T HOLD

In ancient Gaul, kings discovered the habit of assigning coins an arbitrary value in excess of their metallic worth. Ancient craftsmen then discovered they could make molds of both sides of new coins from which to make dies and could strike their own coins for a rather handsome profit. A mold made against a coin was called a *contrefet* from the Old French *contre* (against) and *faire* (to make). Some time before the thirteenth century, "counterfeit" came to stand for any fake money, no matter how it was produced.

Quite a few tricks have been learned since the thirteenth century— and the embezzlement of people's savings via inflation amounts to, in our opinion, the biggest counterfeit in the economic history of the world.

Meanwhile, our warnings about the REIT's are beginning to come true. Continental Mortgage Investors suspended dividend payments, forecasted sharply reduced profits, and outlined a rapidly deteriorating financial structure. This is the nation's second largest Real Estate Investment Trust. And, Chase Manhattan Mortgage and Realty Trust, one of the nation's largest REIT's, was suspended from trading until the auditors study the situation further. Can you hear the faint creaking sound made by a bone under pressure, just before it is about to break? That is the sickening reaction we get.

Even at this late stage, with the nation on the brink of what we believe will be considered a downside pyrotechnical catastrophe, few seem really alarmed. Nixon declared that in the fiscal year beginning this Monday the Federal government would try to limit spending to $300 billion. (*Try!* Creditors are banging on the door and the man reorders caviar!) Nixon also suggested that he would bring the budget into balance in the 1976 fiscal year, but we think these goals are too softly-stated and lackadaisical to be taken seriously.

There is no change in our expectations that the price of gold bullion will move down to test the $140 area, possibly getting as low as $127. We think gold bullion is making an important Bottom around these levels, and by early 1975 prices between $300 and $400 per ounce will be seen, hard as it is to believe now.

For gold shares themselves, we flashed a buy signal on May 29.

June 28, 1974

JULY 1, 1974: DJI 806.24; ASA - 39⅞

*   *   *

ON THE THRESHOLD OF A BUST

You are watching history being made—movies will be made and books written about this era, much as about 1929, including all the fancy dance steps. One thing we hope is explored is your editor's observation that the older he gets, the more he believes everything we do is intentional—even getting sick, especially making mistakes, and indulging in mass masochism.

As bank failures and record high interest rates prompted a smashing decline below the much-vaunted 800 level, which we warned in our last *Letter* would not hold, we can only shake our heads at our inability to get people out of the stock market before it is too late.

Who would believe *Polaroid* at 23½, down from an all-time high at 149, and *McDonald's* down 9 points in one day to 38 after *Barron's* criticized its accounting practices? Add this to the bombed-out favorites of yore, like *Levitz, Disney, Combustion Engineering, Damon, Intel,* where prices collapsed suddenly, affording no chance for investors to get out in time, and one realizes this is not an ordinary bear market. *This is a Crash.* We have been calling it a "slow-motion Crash," and now the reasons are obvious. *Polaroid* has joined that fabled elephant graveyard, that Valhalla somewhere in the sky, where rest the bones of stocks like *Winnebago, Brunswick,* and all the other stocks watered by the tears of the unsuspecting.

[W]e have felt all along that this phase of the bear market would not end until the DJI got down to the 390–626 area. That might not sound like much solace now, but when the collapse gets underway, people will be grateful for that message of hope.

GOLD HAS AT LAST BOTTOMED

We have been thinking in terms of a gold Bottom reached somewhere in our target area of $127–$140, and the plunge to $129 on July 3 probably did it.

If gold goes down too far, it will threaten Italy's recent collateralization of gold for its loans. Delivery of that gold to Italy's lenders would have to be sold on the open market, driving gold down even further. This is the remote possibility, but if it happens, then Italy will go bankrupt immediately, and the countries from whom it borrowed over $10 billion will go down the tube also. Thus, while Europe might not have figured it out yet, a decline in gold prices would signal an immediate catastrophe for Europe. When they do understand this, they will probably force the price of gold as high as possible, especially when it gets through their thick craniums that only a radically higher gold price can pay for the Arab oil bills. Italy's Council of Ministers has just announced new austerity measures, involving higher taxes, but these are only temporary, and (in our opinion) ineffectual measures.

The function of gold is to force nations to sell internationally an amount roughly equivalent to what they buy internationally. How little understood this simple concept is became clear when Britain's Chancellor of the Exchequer, Denis Healey, warned in London this week that industrial nations would be courting "disaster" if they sought to balance their increased oil-import costs with higher exports. He said it would result in "a period of cut-throat competition in world trade from which all would lose and none would gain." Such dim-witted socialist mentalities do not understand competition *helps* more people than it hurts. Those afraid to compete are usually inefficient. No wonder England is next on our bankruptcy list. (Yes, we mean the whole country.) Washington, however, seems to be getting a different message. In the fiscal year ended last month, the U. S. sold some $8.5 billion in arms, almost double the arms sales for the previous fiscal year, and almost $2 billion more than all the arms sold or given away by all nations in 1971. Most of our arms sales went to the Middle East, and the U. S. remains the world's leading arms supplier. We thought the U. S. would become the breadbasket of the world to earn our way internationally, but apparently we are becoming the gunbasket. When people have too many guns, they tend to use them. So in addition to a possible economic catastrophe, there is now possibly a military one.

July 12, 1974

*   *   *

Our predictions called for the price of gold to have a normal and perfectly healthy Technical Correction down to the $140–$150 level, later amended to the $127–$140 level to have been reached between July 15 and August 15, 1974.

In our opinion, *gold hit Bottom on July 3, 1974, at $129.* The next phase of our prediction was $400 by early 1974 and perhaps $300 by Christmas. The time is ripe for another big boom in the gold shares, considering how all the Anti-Gold Nuts have come out of the woodwork again. Having missed the previous rise, they turned bullish on gold at the Top and now this decline has evoked their smug "I told you so's." In their pathetic and desperate attempt to be right about *something*, there is widespread agreement that gold is dead. Excellent. This is the typical atmosphere which spawns big bull moves by gold bullion.

If we are correct, you will see traders and shorters scramble for cover and try to get back into the gold mining shares out of which they so smugly traded at lower prices. Everyone will suddenly become a gold mining expert, and many will advise you to take profits. Take *what* profits? Nearly all of them advised you to trade these gold shares, and now that people have been tricked out of them, more wrong advice will be offered in the form of advising taking profits. They will say "I told you so" and conveniently forget how negative they were on golds just a matter of weeks ago.

July 19, 1974

*   *   *

Charts have truly done a yeoman job of describing what we have been calling a massive redistribution of wealth among American investors. Fundamentalists continue to recommend buying. First they said there would be no recession. Then they said there would be a recession, but it would be a mild one and had already been discounted by stock prices. In the third phase they will brag about how they were pessimistic all along, and will recommend short selling right at the Bottom! Many Fundamentalists also talk about the current low Price/Earnings Ratios, the high asset value per share of many stocks, Arab oil money soon to be recycled into the U. S. market, and blah, blah, blah. To which we say, "Rubbish." Low P/E Ratios reflect earnings bloated by inflation and do not take into account the collapse in earnings probably coming in 1975. As for the high asset value per share, this is important only to raiders, for stocks are evaluated on future earnings power, as Mr. Dines pointed out so clearly in the discounting chapter of his first book.

Economists now feel that skyrocketing inventories mean slow growth for the next year or two until overstocked shelves are worked off. Once billed as a slow-growth year, 1974 has gone to no growth and now economists are talking about a slight decline in economic activity. When they begin talking depression, they will be on our wave length, and not until then!

August 9, 1974

* * *

Ford took office declaring, "Inflation is our domestic public enemy #1. It does no good to blame the public for spending too much when the government's spending too much." Well, that sounds pretty good, but let's wait to see whether or not Federal spending actually gets cut. Unfortunately Mr. Ford turned right around and declared Congress should reactivate the Cost of Living Council, which will monitor wages and prices to "expose abuses but without reimposing controls." We question this action because it is anti-capitalist. Furthermore, his immediate complaint that *General Motors* was going to raise new car prices by almost 10% has the odor of what President Kennedy in 1962 did to the steel companies' presidents when he had the IRS banging on their doors in the middle of the night (for which he, by the way, was not impeached). Kennedy's action probably precipitated that market crack, and this week's sharp decline almost certainly derives from the similar anti-profit mentality Ford displayed on his first day in office. Instead of declaring that we are in the middle of a national emergency and calling for a truly national effort for everybody to buckle down, possibly even with a "no-increase Federal budget," and a strong call back to free market principles which made this country a great capitalist nation, he resorts to the discredited "jawboning" reminiscent of Lyndon B. Johnson.

President Ford signed legislation ending the 40-year ban on gold ownership by U. S. citizens by December 31 of this year. We have been predicting this for so many years, you must forgive us if we actually include the news item itself from the *Times*:

BILL SIGNED TO ALLOW OWNING GOLD IN U.S.

WASHINGTON, Aug. 14 (Reuters)—President Ford signed a bill today allowing American citizens to buy and sell gold after Dec. 31 for the first time in 40 years.

Removal of the 40-year-old ban on gold ownership had originally been opposed by the Treasury, which believed it would encourage speculation in the gold markets and affect the international monetary situation. But the Treasury relented in its opposition this year.

The Treasury did not "relent" in its opposition. It tried every trick in the book, but simply got smashed into the ground by attempting to ignore gold.

We hope all of you have at least one gold stock by now, for all the leaders smashed into new all-time high ground this week, leaving in their wake those who scoffed at our favoring precious metal shares. *ASA* is over 100!

The House Banking Committee approved legislation authorizing the Treasury to reduce the copper content of pennies, which is currently 95%. First we lost gold, then silver, and now even lowly copper. When they run the printing presses with no sense of fiscal responsibility, then no metal can represent that paper over a long period. Would you believe paper pennies a few years from now?

Treasury Secretary William Simon predicted the high rate of inflation we now have will continue for between two to five years, declaring "It's going to be a long time before we get inflation down to an acceptable level." We don't really agree, because these results would depend on an exercise in discipline this country simply does not possess. Therefore, we think a Crash is the only way to solve the problem quickly, through massive bankruptcies wiping out excessive debt.

Inflation is a trick to cheat the workers, whose weekly pay (adjusted for rising prices and Federal taxes) is around 5% below 1973 levels. This is backfiring on the government which is forced to pay ever-higher interest rates to find any remaining suckers willing to lend money. This week the government boosted the interest-rate ceiling on home mortgages it backs to 9½% from 9%, the fourth increase in the last four months for veterans' mortgages. Of course, you know what we think of government-set interest rates in the first place: they are an abomination. The free market should determine all interest rates, all the time. Tampering with interest rates was one of the causes of the 1929 Crash.

In England nobody seems to understand just how serious the situation is, as they continue to play tic-tac-toe under an impending avalanche. Amidst constant wildcat strikes, incredible shortages, a ruinous inflation and tax structure, a collapsing stock market—all played against a Wagnerian score of IRA explosions—England is a prime candidate for a Götterdämmerung. The government, having learned nothing, moves towards additional nationalizations in a disgusting orgy of anti-capitalism and anti-profitism. The Marxists in the unions must really be happy that they are destroying England, and we sympathize with those who will suffer during the Depression they are about to undergo. Whether England will follow Chile and Russia down the totalitarian path is a moot question, but since United States history seems to be a few decades behind that of England, we find it increasingly difficult to maintain our faith in the American future. England, whose stock market is at its lowest point in 19 years, reported a July trade deficit which was just short of a record level. Poor Italy. An article in this week's *Wall Street Journal* points to the virtual complete collapse of the Italian mails. Instead of turning the mails back to individual capitalists who would run it properly, for a profit, Italian postal authorities are now handing out postage stamps as change on the toll road between Milan and Chiasso where Italians drive to Switzerland to place their mail since it is faster! Those of you who have not read Ayn Rand's book *Atlas Shrugged* should do so now.

Alan Greenspan is staying on as Chairman of the Council of Economic Advisors. Alan still favors a tight-money policy and a balanced budget to combat inflation, which he sees continuing for at least one year and possibly two. He clashes with Burns in that he is against any wage/price control policy, and he also questions the relationship of anti-trust policy to higher prices. Greenspan is upset over the high inflation rate and said that if it continues "our system will not hold together in its present form." Here is a man who knows what he's talking about. It's about time we had one such in Washington.

We have been warning for months that inventories would begin skyrocketing just as sales turned down, and sure enough June saw that happen. Businessmen have been hoarding inventory on the assumption they'll find somebody even dumber to pay an even higher price for it. As they realize they are stuck with it, they will turn around and begin to liquidate this inventory on the open market, which will lead to lower prices, layoffs, deflation, and possibly a genuine Depression. The distortion in inventories came from businessmen who were hoarding because of their experience under wage/price controls. Thus end all socialist madnesses.

August 16, 1974

*   *   *

We think the price of gold is about to have a Confirmatory Upside Breakout! According to our calculations, it should be over $200 by November and around Christmas we would hardly be surprised by $300 figures. Since silver has been selling at a 30 to 1 ratio to gold recently, this suggests a silver price around $7 as our first target.

Those of you who take our Chartbooks should carefully study how small the actual declines in gold mining shares were during their recent Corrections. After commissions, taxes, and imperfect timing, the odds were high that anyone who actually tried to trade those golds wound up getting tricked out of them. Therefore, we are glad we asked you to simply stand your ground.

Especially since there was a deep U. S. balance-of-payments deficit in the second quarter, which we predicted long ago, due to the explosion in oil payments. There were the usual excuses, but quoting them here would simply raise our blood pressure. Suffice it to say that the steady deterioration of the last 10 years is continuing in a flawless Downtrend.

Because of inflation, pressures are mounting for an increase in the prime rate from 12% to a new record of 12¼%. The rate should have been up there already, but due to behind-the-scenes government pressures, interest rates have been kept abnormally low. This is not capitalism. By holding the price down unnaturally, more loans are made than are economically advisable. When you tamper with price you tamper with the allocation mechanisms of our society, and remember there is simply not enough of everything for everybody. Meanwhile, the latest Treasury bill sale resulted in an all-time record yield of 9.564%, up from 7.836% in July.

Also note that inventory figures have been "revised" again. Whether Federal statisticians are liars or fools is beyond us, but first quarter figures showed a $5.5 billion rate of inventory accumulation, suddenly "revised" to an annual rate of $16.9 billion! To put this in perspective, an annual rise in inventory above $10 billion has only occurred three times in our history, including 1973's record $15.4 billion. When businessmen discover they can't sell this stuff they'll cut prices sharply to move it, and thus you will see the other side of inflation: deflation and depression. After all, employers with too much goods are not exactly going to keep employees working overtime, so watch unemployment start to increase soon.

August 23, 1974

*   *   *

As we see it, what will come some day to be called the "Crash of '69" squeezed all the water out of the stock market, and that was the first stage. The second stage, which is now in the process of being completed, is featured by descriptions of "stocks are underpriced." The next, and final stage, will feature panic selling at first, eventually leading into selling exhaustion and desperately low volume. That's when things will look their worst, and there will seem to be no reason whatever to ever own common stocks again. Golds will be considered "glamors" and when we lead you from golds into the discredited glamors, we are bracing ourselves for as much flak as when we recommended switching out of glamors into golds.

We feel like Noah, with an Ark of gold protecting you from the holocaust around you. Our primary job in the coming years will be to keep you on the boat, as we already have protected so many with golds in recent years.

[N]one of them are coming to grips with the concept that the government cannot cure this problem, and only a free people can. Instead, all solutions cluster around the *government doing more,* and naturally people have to be taxed to pay for it. Chances are they will seize on something "politically easy" like easing up on tight money. Thus they will probably try to force interest rates down, which will only artificially increase the demand for money, leading to malinvestment, malproduction, and more ills in the years ahead which will need to be wiped out by a devastating Depression. If interest rates want to get that high, there is a *reason* for it, and none of these otherwise intelligent economists seem to be wondering why the interest rate market has so much upside pressure. Instead, all they can think of is depressing it, or "controlling it." None of them understand how urgent these problems are. Even at this late date Harvard's Eckstein declares "Another mild recession is sort of inevitable." This *Letter* thinks that when the deterioration begins to show, it will expose itself with terrifying rapidity in the coming months.

Hotel occupancy in London and Paris is down 20% this year and luxury hotels in both cities are less than one-third full. Pan American Airways, after discovering that sharply higher prices aren't the way to get more business, is now begging for a subsidy to avert "crises." When the price of an airline ticket is up 25%, they should expect transportation, highly sensitive to prices, to respond drastically. The reason prices are so high is because of the international cartel forcing prices high. Were capitalistic competition to prevail, half the airlines would go broke, and the remainder would be profitable and solvent. But few listen to *The Dines Letter* these days, voicing such "ridiculous" economic theories.

And, the oil cartel is running into the trouble we predicted they would. Now the Arabs are trying to hold down production in an effort to keep prices at their high levels. We predict it won't help, and that the Arabs are about to get a lesson in capitalism. No cartel can work, and you don't need an international anti-trust government to enforce it. When cartel prices are too high, oversupply will force them down, sooner or later. Higher prices will eliminate the least efficient users, and people will rediscover that supply and demand really works.

Voice of the future? Economic Advisor Roy L. Ash suggested the government could not balance the budget in the next fiscal year without reducing future benefits in some social programs legislated long ago. This kind of thinking is the beginning of the wave of the future as *The Dines Letter* has warned it would be. Social Security, Medicare and Medicaid, Veterans' benefits, Welfare Retirement, all considered "untouchable" by federal budget-cutters, are going to get hit by an axe.

West Germany gave Italy a $2 billion credit to bail her out of her present financial crisis. Germany must really be worried about the loans already outstanding to grant more money to such a shaky borrower. Chancellor Helmut Schmidt said the Bank of Italy would put up its gold reserves as collateral, and it would be valued at 90% of the international free-market rate for gold during the last eight weeks. This effectively places a new floor under the price of gold and all nations will try to enforce it, because, if Italy is forced to sell its gold and the price of gold thereby drops, everybody else's reserves go down too, which could trigger a catastrophe. Germany would lose its loans, and go under with Italy. So, despite the insistence of the U. S. Treasury that gold is worth $42.22, the real floor is recognized by real people who understand markets to be really around $150, and not a dime less. This "floor" is going to rise in coming years, as the crisis we have been describing in this *Letter* unfolds to even those who are blind.

September 6, 1974

*   *   *

What we have been describing in recent years as a "slow motion Crash" is not so slow any more. We have also been calling it a "low-key panic."

Yet, isn't it amazing that as yet nobody calls this a Crash? In 1969 we wrote repeatedly that some day people would come to call that fiasco the "Crash of '69."

In coming years we are looking for a wave of reduction of utility dividends.

This is still an "infression," or inflationary depression, which will soon move into a deflationary depression.

[R]emember that *the name of the game is survival*. Almost everybody is going to be taking losses these days, because there is just about no place to hide. But, if you can preserve most of your capital intact, and get through this period, it will be the equivalent of having cash in 1932.

In our considered opinion, this is the single most oversold reading we have ever had on the precious metals. This is a God-given opportunity to all those who missed golds on the first trip around. Here is your last chance! We want our subscribers to buy the stock from Schaefer's subscribers. Perhaps we're the last of the gold bugs again, but we think the coming gold and silver rally is going to be one of the most spectacular in market history.

We know some of you have been frightened by this decline, and it is a pity Mr. Dines' book is not ready, because reading it would convert the hesitant and the unbelieving. Remember, nothing has changed. The gold crisis is still intact, there is a new "floor" under gold at $120, and gold mining share earnings are going to hold up no matter what. This decline in gold shares occurred on unbelievably high volume. For example, on Thursday, 5% of Homestake's entire capitalization was traded! The international monetary picture is increasingly shaky and England's bankruptcy could be imminent. So, hold those golds! We think the coming gold and silver rally will be led by quality, and the trigger could be a higher price for gold bullion.

We think the rally will be led first and foremost by ASA (ASA - 57⅞) and *Amgold* (OTC - 46⅞), followed hotly by *Dome Mines* (DOM - 35⅞) and *Homestake* (HM - 33⅜). In the silvers, go right after *Rosario* (ROS - 19⅛) and *Hecla* (HL - 12½) in that order. If you have more money after that, diversify into the lower-priced speculatives on our Long-Term List #5 last shown to you on September 6.

President Ford on September 6 said the nation would defeat inflation before its 200th anniversary on July 4, 1976. Ford thundered out "The tyranny of double-digit inflation is our common enemy in 1974. Inflation is the cruelest kind of taxation without representation." Hmmn. Sounds like us, doesn't it? Anyway, Ford didn't say whether inflation would be stopped by a depression, or by some other means. Inflation followed by depression has been the classic pattern of all past inflations.

Meanwhile, few among the Washington Economic Establishment have noticed that a new animal is prowling their territory. Alan Greenspan, the new Chairman of the Council of Economic Advisors, said inflation will be cut by reducing government borrowing, tightening the money supply and aiming for Federal budget surplus! Warning the short-term economic outlook was poor, suggesting rising unemployment, he said he did not consider wage increases a critical factor in current inflation—Greenspan declared with refreshing accuracy that "Inflation is essentially a financial problem." Unfortunately, Greenspan is decades too late, but what a pity we did not have someone like him up there when we needed him. Already, labor leaders are banging on President Ford's door to spur the economy and expand jobs. The housing industry, in its own private depression, is asking for relief from tight money, and so on. All the special interests want relief at the expense of the general taxpayer, and it is precisely this modus vivendi which we predict is going to be destroyed in the coming depression.

Unbelievably high unemployment will be a feature of the economic dislocations we envision ahead, and while the August jobless rate edged up to 5.4%, it is nowhere near as high as we expect its spiral to reach. Perhaps businessmen are being lulled by their higher profits into thinking they are doing better. At any rate they are not deducting inflation from their statistics, and as soon as sales turn down, they will be hit by the double whammy of both declining sales and also the inflation bite. Furthermore, while workers might be getting more paper money, they are actually getting lower "real" wages because they can buy *less* with that paper. Inflation, as Mr. Dines points out in his book, has been a way, in recent decades, of secretly and involuntarily lowering wages. Since wages are declining, we have felt freer to use labor than we ordinarily would. By supply and demand. Getting flexibility back into wages, so that they move both up *and down*, is a necessary step in assuring a healthy economy. A permanent "high" with no downs must, sooner or later, lead to catastrophic inflation. The nation must understand that their choice lies between ruinous inflation or occasionally high unemployment. Those are the facts of economic life!

Greenspan said a significant easing would produce "a short-term sense of well-being" after which "the increased money supply would put us back in the current situation, if not worse." Simon added "We must recognize that there have been years of fiscal and monetary abuse which cannot be undone overnight. Thus, fiscal and monetary restraint must be exercised patiently and consistently for a sustained period of time." It's high time somebody up there began to echo what the *Dines Letter* has been saying for many years. However, we are fearful that though it is the right solution, it is far too late.

September 13, 1974

Postscript

In October 1973 *The Dines Letter* began to predict a temporary respite from the bear market, consisting of a "series of between 2 and 5 irregularly higher market rallies which could last well into 1975 or even 1976" before the final "Killer Wave" of the bear market strikes.

After graduating from Columbia Law School and serving two years in military intelligence, James Dines joined Auerbach, Pollack & Richardson as a junior security analyst in 1958 (his predictions had been so good as a customer, they asked him to join), then switched to A. M. Kidder & Company as a senior security analyst, and then writer of their market letter. It was called *The Dines Letter* and, in 1961, Dines left Kidder to publish it on his own. It (and he) are now known world-wide. Articles by and about Mr. Dines have frequently appeared in such periodicals as *Newsweek, Time, The New York Times, Barron's, Money, Fortune, Forbes,* and *Business Week,* and he has appeared a number of times on radio and television, including CBS's *60 Minutes.* He is also the author of *How the Average Investor Can Use Technical Analysis for Stock Profits.*